Stochastic Processes

Springer
*Berlin*
*Heidelberg*
*New York*
*Hong Kong*
*London*
*Milan*
*Paris*
*Tokyo*

Kiyosi Itô playing *go* with Haruo Totoki (1934-1991) in 1969 at Itô's home at Aarhus (photo by S. A. Sørensen)

Kiyosi Itô

# Stochastic Processes

Lectures given at Aarhus University

Edited by
Ole E. Barndorff-Nielsen
Ken-iti Sato

*Kiyosi Itô*
Kyoto, Japan

*Ole E. Barndorff-Nielsen*
Department of Mathematics
University of Aarhus
Ny Munkegade
8000 Aarhus, Denmark
e-mail: oebn@imf.au.dk

*Ken-iti Sato*
Hachiman-yama 1101-5-103
Tenpaku-ku, Japan
e-mail: ken-iti.sato@nifty.ne.jp

Cataloging-in-Publication Data applied for

A catalog record for this book is available from the Library of Congress.
Bibliographic information published by Die Deutsche Bibliothek
Die Deutsche Bibliothek lists this publication in the Deutsche Nationalbibliografie; detailed bibliographic data is available in the Internet at <http://dnb.ddb.de>.

Mathematics Subject Classification (2000): 60-02, 60E07, 60G51, 60J25

ISBN 978-3-642-05805-9



Springer-Verlag is a part of Springer Science+Business Media
springeronline.com

Printed in Germany



Cover design: *design & production* GmbH, Heidelberg
Printed on acid-free paper 41/3142LK - 5 4 3 2 1 0

# Foreword

The volume *Stochastic Processes* by K. Itô was published as No. 16 of Lecture Notes Series from Mathematics Institute, Aarhus University in August, 1969, based on Lectures given at that Institute during the academic year 1968–1969. The volume was as thick as 3.5 cm., mimeographed from typewritten manuscript and has been out of print for many years. Since its appearance, it has served, for those able to obtain one of the relatively few copies available, as a highly readable introduction to basic parts of the theories of additive processes (processes with independent increments) and of Markov processes. It contains, in particular, a clear and detailed exposition of the Lévy–Itô decomposition of additive processes.

Encouraged by Professor Itô we have edited the volume in the present book form, amending the text in a number of places and attaching many footnotes. We have also prepared an index.

Chapter 0 is for preliminaries. Here centralized sums of independent random variables are treated using the dispersion as a main tool. Lévy's form of characteristic functions of infinitely divisible distributions and basic properties of martingales are given.

Chapter 1 is analysis of additive processes. A fundamental structure theorem describes the decomposition of sample functions of additive processes, known today as the Lévy-Itô decomposition. This is thoroughly treated, assuming no continuity property in time, in a form close to the original 1942 paper of Itô, which gave rigorous expression to Lévy's intuitive understanding of path behavior.

Chapter 2 is an introduction to the theory of Markov processes. The setup is a system of probability measures on path space, and the strong Markov property, generators, Kac semigroups, and connections of Brownian motion with potential theory are discussed. Here the works of Kakutani, Doob, Hunt, Dynkin, and many others including Itô himself are presented as a unity.

Following these chapters, there is a collection of exercises containing about 70 problems. An Appendix gives their complete solutions. This material should be helpful to students.

In the 34 years since these lectures were given by Itô, the theory of stochastic processes has made great advances and many books have been published in the field. Below we mention some of them, continuing the list of the four books in the Preface to the Original:

5. F. Spitzer: Principles of Random Walk, Van Nostrand, Princeton, 1964 (Second edition, Springer, New York, 1976).

6. P. Billingsley: Convergence of Probability Measures, Wiley, New York, 1968.

7. H. P. McKean: Stochastic Integrals, Academic Press, New York, 1969.

8. I. Karatzas and S. E. Shreve: Brownian Motion and Stochastic Calculus, Springer, New York, 1988.

9. N. Ikeda and S. Watanabe: Stochastic Differential Equations and Diffusion Processes, North-Holland/Kodansha, Amsterdam/Tokyo, 1989.

10. J.-D. Deuschel and D. W. Stroock: Large Deviations, Academic Press, Boston, 1989.

11. H. Kunita: Stochastic Flows and Stochastic Differential Equations, Cambridge Univ. Press, Cambridge, 1990.

12. D. Revuz and M. Yor: Continuous Martingales and Brownian Motion, Springer, Berlin, 1991.

13. M. Fukushima, Y. Oshima, and M. Takeda: Dirichlet Forms and Symmetric Markov Processes, Walter de Gruyter, Berlin, 1994.

14. J. Bertoin: Lévy Processes, Cambridge Univ. Press, Cambridge, 1996.

15. O. Kallenberg: Foundations of Modern Probability, Springer, New York, 1997.

16. K. Sato: Lévy Processes and Infinitely Divisible Distributions, Cambridge Univ. Press, Cambridge, 1999.

17. D. W. Stroock: Markov Processes from K. Itô's Perspective, Princeton Univ. Press, Princeton, 2003.

Itô defined stochastic integrals and discovered the so-called Itô formula, which is the analogue for stochasic analysis of Taylor's formula in classical analysis. This celebrated work is not touched upon in these lectures. However the reader can study it in many books such as 7, 8, 9, 11, 12, 15, and 17.

We are grateful to the Mathematics Institute, University of Aarhus, as well as to Anders Grimvall, who assisted Professor Itô in the preparation of the Lecture Notes, for readily giving their consent to this book project, and to Catriona Byrne, Springer Verlag, for her support and helpfulness.

It has been a great pleasure for us to revisit Kiyosi Itô's Aarhus Lecture Notes and especially to carry out the editing in close contact with him.

August 31, 2003

O. E. Barndorff-Nielsen (Aarhus)
K. Sato (Nagoya)

# Preface

For taking on the arduous task of editing and preparing my *Aarhus Lecture Notes on Stochastic Processes* for publication from Springer Verlag, I wish to extend my deepest gratitude to Ole Barndorff-Nielsen and Ken-iti Sato. I am truly grateful and delighted to see the manuscript reappearing thirty-four years later in this neat professional edition.

I spent three happy years at Aarhus University, from 1966 to 1969. Professor Bundgaard, who was the Director of the Mathematics Institute at the time, made sure that I could pursue my research from the very beginning of my stay within the active local mathematics community, and I profited immensely from my interactions with faculty, visitors, and students. Many mathematicians passed through Aarhus during those years, including Professors Cramér, Getoor, Kesten, Totoki, and Yosida. I feel grateful to this day that I was able to be reacquainted with Professor Cramér, whom I had met at the 1962 Clermont-Ferrand symposium in Southern France commemorating the $300^{\text{th}}$ anniversary of the death of Blaise Pascal.

My lasting impression of Denmark is that of a highly civilized country. This is felt by every visitor walking the beautiful city streets, and becomes even more evident to the foreign resident from more subtle aspects of daily life. For example, postal workers, librarians, police officers, and others working in public service are not only very efficient, but also extremely kind and go out of their way to help people. To cite one memorable incident among many: When Professor Yosida was visiting Aarhus, he asked a policeman for direction to our house. My wife and I were very surprised when he arrived in a patrol car accompanied by a police officer, who was just making sure that our dinner guest safely reached his destination. Professor Yosida remarked that Denmark is like the legendary Yao-Shun 堯舜 period (Gyou-Shun in Japanese), the golden age of Chinese antiquity (ca. $24^{\text{th}}$ century BC) noted for inspiring virtue, righteousness, and unselfish devotion.

My stay in Aarhus was productive as well as enjoyable. My wife, who came from the snow country in Northern Japan, knew how to enjoy the Scandinavian climate, and we often took walks in the park nearby with a beautiful forest. Working on this book after thirty-four years has brought back joyous memories of those days, and I wish to dedicate this book to my wife, Shizue Ito, who passed away three years ago. She would have been delighted to see the Aarhus lecture notes published in book form by Springer Verlag. If this book helps future students of mathematics become interested in stochastic processes and probability theory, this will be a source of immense pleasure to me.

Kyoto, August 11, 2003

K. Itô

# Preface to the Original

These lecture notes are based on my lectures on stochastic processes given to the third year students in the Mathematical Institute at Aarhus University. The reader is assumed to know basic facts on measure-theoretic probability theory such as Kolmogorov's extension theorem, characteristic functions, law of large numbers, random walks, Poisson processes, Brownian motion and Markov chains.

Chapter 0 is to remind the reader of some basic notions which might not be familiar.

Chapter 1 is devoted to processes with independent increments with particular emphasis on the properties of the sample functions.

Chapter 2 is devoted to the modern set-up of the theory of Markov processes. The main emphasis is put on generators and strong Markov property. The relation of the Markov processes to potentials and harmonic functions is explained. I hope that the reader who has read this chapter will have no trouble in reading the following books that treat the current problems in this field:

1. E. B. Dynkin: Markov Processes, Moscow, 1963. English translation: Springer, Berlin, 1965.
2. K. Itô and H. P. McKean: Diffusion Processes and their Sample Paths, Springer, Berlin, 1965.
3. P. Meyer: Probability and Potentials, Ginn (Blaisdell), Boston, 1966.
4. R. M. Blumenthal and R. Getoor: Markov Processes and Potential Theory, Academic Press, New York, 1968.

Because of limitation of time the theory of stationary processes has not been covered in these notes in spite of its importance.

At the end of the notes I have listed 68 problems that are proposed to the students as exercises. As appendices I have put a note on compactness and conditional compactness[1] and the solutions to the exercises. These appendices are written by Mr. A. Grimvall.

I would like to express my hearty thanks to Mr. A. Grimvall, who corrected many mistakes in the original version of the notes.

Aarhus, June 4, 1969 K. Itô

[1] The note has been omitted, as the reader can refer to textbooks such as Billingsley's book mentioned in the Foreword. (Editors)

# Table of Contents

# 0 Preliminaries

## 0.1 Independence

Let $(\Omega, \mathcal{B}, P)$ be the basic probability space. The family $\mathfrak{B}$ of all sub-$\sigma$-algebras of $\mathcal{B}$ is a *complete lattice* with respect to the usual set-theoretic inclusion relation.[1] For $\{\mathcal{B}_\lambda : \lambda \in \Lambda\} \subset \mathfrak{B}$ the least upper bound is denoted by $\bigvee_\lambda \mathcal{B}_\lambda$ and the greatest lower bound by $\bigwedge_\lambda \mathcal{B}_\lambda$. The latter, $\bigwedge_\lambda \mathcal{B}_\lambda$, is the same as the set-theoretic intersection $\bigcap_\lambda \mathcal{B}_\lambda$ but the former, $\bigvee_\lambda \mathcal{B}_\lambda$, is not the set-theoretic union $\bigcup_\lambda \mathcal{B}_\lambda$ but the $\sigma$-algebra generated by the union.

For a class $\mathcal{C}$ of subsets of $\Omega$ we denote by $\mathcal{B}[\mathcal{C}]$ the $\sigma$-algebra generated by $\mathcal{C}$. For example

$$\bigvee_\lambda \mathcal{B}_\lambda = \mathcal{B}\left[\bigcup_\lambda \mathcal{B}_\lambda\right].$$

For a class $\mathcal{M}$ of random variables on $\Omega$ we denote by $\mathcal{B}[\mathcal{M}]$ the $\sigma$-algebra generated by all sets of the form $B = \{\omega : X(\omega) < c\}$, $X \in \mathcal{M}$, $-\infty < c < \infty$.

A finite family $\{\mathcal{B}_1, \mathcal{B}_2, \ldots, \mathcal{B}_n\} \subset \mathfrak{B}$ is called *independent* if we have

$$P\left(\bigcap_i B_i\right) = \prod_i P(B_i), \qquad B_i \in \mathcal{B}_i .$$

An infinite family $\mathfrak{C} \subset \mathfrak{B}$ is called *independent* if every finite subfamily of $\mathfrak{C}$ is independent.

Let $\mathcal{X}_\lambda$ be a family of random variables on $\Omega$ or measurable subsets of $\Omega$ for each $\lambda \in \Lambda$. Then $\{\mathcal{X}_\lambda\}_{\lambda \in \Lambda}$ is called *independent* if $\{\mathcal{B}[\mathcal{X}_\lambda]\}_\lambda$ is.[2]

The following theorem follows at once from the definition.

**Theorem 1.** *(i) Every subfamily of an independent family is independent.*

*(ii) If $\{\mathcal{B}_\lambda\}_\lambda$ is independent and if $\mathcal{B}'_\lambda \subset \mathcal{B}_\lambda$ for each $\lambda$, then $\{\mathcal{B}'_\lambda\}_\lambda$ is independent.*

---

(All footnotes are written by the Editors.)

[1] See Problem 0.45 in Exercises on complete lattices. But any knowledge of lattices is not required for further reading.

[2] The notation $\{\mathcal{B}[\mathcal{X}_\lambda]\}_\lambda$ means $\{\mathcal{B}[\mathcal{X}_\lambda]\}_{\lambda \in \Lambda}$. Such abbreviation will be used frequently.

A class $\mathcal{C}$ of subsets of $\Omega$ is called *multiplicative* if $\mathcal{C}$ contains $\Omega$ and is closed under finite intersections. A class $\mathcal{D}$ of subsets of $\Omega$ is called a *Dynkin class* if it is closed under proper differences and countable disjoint unions. For a given class $\mathcal{C}$ of subsets of $\Omega$, the smallest Dynkin class including $\mathcal{C}$, $\mathcal{D}[\mathcal{C}]$ in notation, is said to be *generated* by $\mathcal{C}$. By the same idea in proving the monotone class theorem we get[3]

**Lemma 1 (Dynkin).** *If $\mathcal{C}$ is multiplicative, then we have $\mathcal{D}[\mathcal{C}] = \mathcal{B}[\mathcal{C}]$.*

Since the class of all sets for which two given bounded measures have the same value is a Dynkin class, Lemma 1 implies

**Lemma 2.** *Let $\mu$ and $\nu$ be bounded measures on $\Omega$. If $\mu = \nu$ on a multiplicative class $\mathcal{C}$, then $\mu = \nu$ on $\mathcal{B}[\mathcal{C}]$.*

Using this we will prove

**Theorem 2.** *Let $\mathcal{C}_i, i = 1, 2, \ldots, n$, be all multiplicative. If we have*

$$P\left(\bigcap_i C_i\right) = \prod_i P(C_i) \qquad \text{for } C_i \in \mathcal{C}_i,\ i = 1, 2, \ldots, n\,,$$

*then $\{\mathcal{B}[\mathcal{C}_i]\}_i$ is independent.*

*Proof.* First we want to prove that

$$P\left(\bigcap_{i=1}^{n-1} C_i \cap B_n\right) = \left(\prod_{i=1}^{n-1} P(C_i)\right) P(B_n)$$

for $C_i \in \mathcal{C}_i$, $i \leq n-1$, and $B_n \in \mathcal{B}[\mathcal{C}_n]$. Since both sides are bounded measures in $B_n$ and since they are equal for $B_n \in \mathcal{C}_n$ by the assumption, they are equal for $B_n \in \mathcal{B}[\mathcal{C}_n]$ by Lemma 2.

Using this fact and making the same argument we have

$$P\left(\bigcap_{i=1}^{n-2} C_i \cap B_{n-1} \cap B_n\right) = \left(\prod_{i=1}^{n-2} P(C_i)\right) P(B_{n-1})P(B_n)$$

for $C_i \in \mathcal{C}_i$, $i \leq n-2$, and $B_j \in \mathcal{B}[\mathcal{C}_j]$, $j = n-1, n$.

Repeating this, we arrive at the identity

$$P\left(\bigcap_{i=1}^{n} B_i\right) = \prod_{i=1}^{n} P(B_i), \qquad B_i \in \mathcal{B}[\mathcal{C}_i],\ i = 1, 2, \ldots, n\,,$$

which completes the proof.

---

[3] See Problem 0.3 in Exercises. This lemma will be referred to as Dynkin's theorem.

A family $\{\mathcal{B}_\lambda\}_{\lambda\in\Lambda} \subset \mathfrak{B}$ is called *directed up* if for every $\lambda, \mu \in \Lambda$ we have $\nu \in \Lambda$ such that $\mathcal{B}_\nu \supset \mathcal{B}_\lambda, \mathcal{B}_\mu$.

**Theorem 3.** *Let $\{\mathcal{B}_{i\lambda}\}_{\lambda\in\Lambda_i}$ be directed up for each $i = 1, 2, \ldots, n$ and let $\mathcal{B}_i = \bigvee_{\lambda\in\Lambda_i} \mathcal{B}_{i\lambda}$. If $\{\mathcal{B}_{i\lambda_i}\}_i$ is independent for every choice of $\lambda_i \in \Lambda_i$, then $\{\mathcal{B}_i\}_i$ is also independent.*

*Proof.* Since $\{\mathcal{B}_{i\lambda}\}_{\lambda\in\Lambda_i}$ is directed up, the set-theoretic union $\mathcal{C}_i = \bigcup_{\lambda\in\Lambda_i} \mathcal{B}_{i\lambda}$ is a multiplicative class for each $i$, and we have

$$P\left(\bigcap_i C_i\right) = \prod_i P(C_i), \qquad C_i \in \mathcal{C}_i,\ i = 1, 2, \ldots, n\,.$$

Since $\mathcal{B}_i = \mathcal{B}[\mathcal{C}_i]$, $\{\mathcal{B}_i\}_i$ is independent by Theorem 2.

**Theorem 4.** *Let $\Lambda = \bigcup_{\mu\in M} \Lambda_\mu$ (disjoint union) and suppose that we are given $\mathcal{B}_\lambda \in \mathfrak{B}$ for each $\lambda \in \Lambda$.*

*(i) If $\{\mathcal{B}_\lambda\}_{\lambda\in\Lambda}$ is independent, then $\{\bigvee_{\lambda\in\Lambda_\mu} \mathcal{B}_\lambda\}_{\mu\in M}$ is independent.*

*(ii) If $\{\mathcal{B}_\lambda\}_{\lambda\in\Lambda_\mu}$ is independent for each $\mu \in M$ and if $\{\bigvee_{\lambda\in\Lambda_\mu} \mathcal{B}_\lambda\}_{\mu\in M}$ is independent, then $\{\mathcal{B}_\lambda\}_{\lambda\in\Lambda}$ is also independent.*

*Proof of (i).* We can assume that $M$ is a finite set, say $M = \{1, 2, \ldots, n\}$. Fix $i \in M$ for the moment, take an arbitrary finite set $\{\lambda_1, \lambda_2, \ldots, \lambda_m\} \subset \Lambda_i$ and $B_j \in \mathcal{B}_{\lambda_j}$ for each $j = 1, 2, \ldots, m$ and then form a set

$$C = \bigcap_{j=1}^{m} B_j\,.$$

Let $\mathcal{C}_i$ denote all such sets $C$'s. Then $\mathcal{C}_i$ is a multiplicative class generating $\bigvee_{\lambda\in\Lambda_i} \mathcal{B}_\lambda$. Since $\{\mathcal{B}_\lambda\}_{\lambda\in\Lambda}$ is independent, it is easy to see that

$$P\left(\bigcap_{i=1}^{n} C_i\right) = \prod_{i=1}^{n} P(C_i), \qquad C_i \in \mathcal{C}_i,\ i = 1, 2, \ldots, n\,.$$

This implies the independence of $\{\mathcal{B}[\mathcal{C}_i]\}_i$ by Theorem 2, completing the proof.

The proof of (ii) is easy and so omitted.

The *tail $\sigma$-algebra* of a sequence $\{\mathcal{B}_n\}_n \subset \mathfrak{B}$ is defined to be $\limsup \mathcal{B}_n$, that is, $\bigwedge_k \bigvee_{n\geq k} \mathcal{B}_n$. A $\sigma$-algebra $\mathcal{C} \in \mathfrak{B}$ is called *trivial* if $P(C) = 1$ or $0$ for every $C \in \mathcal{C}$.

**Theorem 5 (Kolmogorov's 0-1 law).** *If $\{\mathcal{B}_n\}_n$ is independent then its tail $\sigma$-algebra is trivial.*

*Proof.* Write $\mathcal{T}$ for $\limsup \mathcal{B}_n$ and $\mathcal{T}_k$ for $\bigvee_{n>k} \mathcal{B}_n$. Since $\{\mathcal{B}_n\}_n$ is independent, $\{\mathcal{B}_1, \mathcal{B}_2, \ldots, \mathcal{B}_k, \mathcal{T}_k\}$ is independent by Theorem 4 (i). Since $\mathcal{T} \subset \mathcal{T}_k$, $\{\mathcal{B}_1, \mathcal{B}_2, \ldots, \mathcal{B}_k, \mathcal{T}\}$ is independent by Theorem 1. Hence it follows from definition of independence that $\{\mathcal{B}_1, \mathcal{B}_2, \ldots, \mathcal{T}\}$ is independent. Therefore $\{\bigvee_n \mathcal{B}_n, \mathcal{T}\}$ is independent by Theorem 4 (i). Since $\mathcal{T} \subset \bigvee_n \mathcal{B}_n$, $\{\mathcal{T}, \mathcal{T}\}$ is independent. Suppose $C \in \mathcal{T}$. Then

$$P(C) = P(C \cap C) = P(C)P(C)$$

and so $P(C) = 1$ or $0$.

**Theorem 6 (Borel).** *If $\{A_n\}$ is an independent sequence of sets (= events), then*[4]

$$P(\limsup_{n\to\infty} A_n) = 1 \qquad \text{in case} \sum P(A_n) = \infty\,,$$

$$P(\liminf_{n\to\infty} A_n^{\mathrm{C}}) = 1 \qquad \text{in case} \sum P(A_n) < \infty\,.$$

*The second assertion holds without the assumption of independence* (**Borel–Cantelli**).

*Proof.* Suppose that $\sum P(A_n) < \infty$. Without using the assumption of independence we have

$$P(\limsup A_n) \le P\left(\bigcup_{n\ge k} A_n\right) \le \sum_{n\ge k} P(A_n) \to 0, \qquad k \to \infty\,,$$

and so

$$P(\limsup A_n) = 0 \quad \text{i.e.} \quad P(\liminf A_n^{\mathrm{C}}) = 1\,.$$

Suppose that $\{A_n\}$ is independent and that $\sum_n P(A_n) = \infty$. Then

$$P(\limsup A_n^{\mathrm{C}}) = P\left(\bigcup_k \bigcap_{n\ge k} A_n^{\mathrm{C}}\right) \le \sum_k P\left(\bigcap_{n\ge k} A_n^{\mathrm{C}}\right)\,.$$

But

$$P\left(\bigcap_{n\ge k} A_n^{\mathrm{C}}\right) \le P\left(\bigcap_{n=k}^{k+m} A_n^{\mathrm{C}}\right) = \prod_{n=k}^{k+m} P(A_n^{\mathrm{C}})$$

$$= \prod_{n=k}^{k+m} (1 - P(A_n)) \to 0 \qquad \text{as } m \to \infty$$

by virtue of $\sum_{n=k}^{\infty} P(A_n) = \infty$. Therefore

$$P(\liminf A_n^{\mathrm{C}}) = 0 \qquad \text{i.e.} \qquad P(\limsup A_n) = 1\,.$$

[4] Definitions: $\limsup_{n\to\infty} A_n = \bigcap_k \bigcup_{n\ge k} A_n$, $\liminf_{n\to\infty} A_n^{\mathrm{C}} = \bigcup_k \bigcap_{n\ge k} A_n^{\mathrm{C}}$ where $A^{\mathrm{C}}$ denotes the complement of $A$ in $\Omega$.

## 0.2 Central Values and Dispersions

To indicate the center and the scattering degree of a one-dimensional probability measure $\mu$ with the second order moment finite, we have the *mean value*[5]

$$m = m(\mu) = \int_{R^1} x\mu(\mathrm{d}x)$$

and the *standard deviation*footnoteThe variance of $\mu$ is denoted by $V(\mu)$.

$$\sigma = \sigma(\mu) = \sqrt{V(\mu)} = \left( \int_{R^1} (x-m)^2 \mu(\mathrm{d}x) \right)^{1/2} .$$

By virtue of the celebrated inequality of Chebyshev

$$\mu[m - \alpha\sigma,\ m + \alpha\sigma] \geq 1 - \frac{1}{\alpha^2} ,$$

we can derive several important properties of 1-dimensional probability measures. However, $\sigma$ is useless in case the probability measure in consideration has no finite second order moment. Here we will introduce the central value and the dispersion that will play roles similar to $m$ and $\sigma$ for general 1-dimensional probability measures.

The *central value* $\gamma = \gamma(\mu)$ is defined by J. L. Doob to be the real number $\gamma$ such that

$$\int_{R^1} \arctan(x - \gamma)\mu(\mathrm{d}x) = 0 .$$

The existence and the uniqueness of $\gamma$ follow from the fact that for $x$ fixed, $\arctan(x - \gamma)$ decreases strictly and continuously from $\pi/2$ to $-\pi/2$ as $\gamma$ moves from $-\infty$ to $+\infty$.

The *dispersion* $\delta = \delta(\mu)$ is defined by

$$\delta(\mu) = -\log \iint_{R^2} \mathrm{e}^{-|x-y|} \mu(\mathrm{d}x)\mu(\mathrm{d}y) .$$

For a 1-dimensional probability measure $\mu$ we introduce the following operations[6]

$$\begin{aligned} (\alpha\mu + \beta)(E) &= \mu\{x\colon \alpha x + \beta \in E\} && \text{(linear transformations)},\\ \check{\mu} &= (-1)\mu && \text{(reflection)},\\ \widetilde{\mu} &= \mu * \check{\mu} && \text{(symmetrization)}. \end{aligned}$$

[5] The set of real numbers is denoted by $R^1$.

[6] The convolution $\mu * \nu$ of probability measures $\mu$ and $\nu$ is defined by

$$(\mu * \nu)(B) = \int_{R^1} \mu(B - x)\,\nu(dx) = \int_{R^1} \nu(B - x)\,\mu(dx)$$

for Borel sets $B$, where $B - x = \{y - x\colon y \in B\}$.

If the probability law of a random variable $X$ is $\mu$, then $\alpha\mu+\beta$ is the law of $\alpha X+\beta$ and $\check{\mu}$ is that of $-X$. If $X$ and $Y$ are independent and have the same probability law $\mu$, then $X-Y$ has the law $\widetilde{\mu}$.

It is easy to see that

$$\delta(\mu) = -\log \int_{R^1} \mathrm{e}^{-|x|}\widetilde{\mu}(\mathrm{d}x)\,.$$

Recalling the identity

$$\mathrm{e}^{-|x|} = \frac{1}{\pi}\int_{R^1} \mathrm{e}^{\mathrm{i}\,\xi x}\frac{\mathrm{d}\xi}{1+\xi^2}\,,$$

we have

$$\begin{aligned}\delta(\mu) &= -\log\left(\frac{1}{\pi}\int_{R^1}\varphi(\xi;\widetilde{\mu})\frac{\mathrm{d}\xi}{1+\xi^2}\right)\\ &= -\log\left(\frac{1}{\pi}\int_{R^1}|\varphi(\xi;\mu)|^2\frac{\mathrm{d}\xi}{1+\xi^2}\right)\end{aligned}$$

where $\varphi(\xi;\mu)$ denotes the characteristic function of $\mu$, that is,

$$\varphi(\xi;\mu) = \int_{R^1} \mathrm{e}^{\mathrm{i}\,\xi x}\mu(\mathrm{d}x)\,.$$

As in the case of the mean value, the *central value* $\gamma(X)$ of a random variable $X$ is defined to be that of the probability law $P_X$ of $X$. Similarly for the *dispersion* $\delta(X)$. The central value $\gamma(X)$ is the solution of

$$E(\arctan(X-\gamma)) = 0\,.$$

Since $\varphi(\xi;P_X) = E(\mathrm{e}^{\mathrm{i}\xi X})$, $\delta(X)$ is expressible as

$$\delta(X) = -\log\left(\frac{1}{\pi}\int_{R^1}|E(\mathrm{e}^{\mathrm{i}\,\xi X})|^2\frac{\mathrm{d}\xi}{1+\xi^2}\right)\,.$$

Corresponding to the facts

$$m(\alpha\mu+\beta) = \alpha m(\mu)+\beta, \qquad \sigma(\alpha\mu+\beta) = |\alpha|\sigma(\mu)\,,$$

we have

**Theorem 1.** $\gamma(\pm\mu+\beta) = \pm\gamma(\mu)+\beta, \qquad \delta(\pm\mu+\beta) = \delta(\mu).$

*Proof.* Obvious by the definitions.

Corresponding to the fact that $\mu = \delta_c$ (= the *$\delta$-distribution concentrated at $c$*) if and only if $m(\mu) = c$ and $\sigma(\mu) = 0$, we have

**Theorem 2.** *$\mu = \delta_c$ if and only if $\gamma(\mu) = c$ and $\delta(\mu) = 0$.*

*Proof.* The "only if" part is trivial. Suppose that $\gamma(\mu) = c$ and $\delta(\mu) = 0$. From $\delta(\mu) = 0$ we get

$$\iint \mathrm{e}^{-|x-y|}\mu(\mathrm{d}x)\mu(\mathrm{d}y) = 1 .$$

Since $\mu$ is a probability measure, there exists a $c'$ with

$$\int \mathrm{e}^{-|x-c'|}\mu(\mathrm{d}x) \geq 1 \quad (= 1 \text{ in fact}).$$

Since $\mathrm{e}^{-|x-c'|} < 1$ for $x \neq c'$, $\mu$ is concentrated at $c'$. Then $c' = \gamma(\mu) = c$.

The weak* topology in the set $\mathfrak{P}$ of all 1-dimensional probability measures is metrizable by the *Lévy metric* $\rho$, with respect to which $\mathfrak{P}$ is a complete metric space[7]. Now we want to prove the continuity of $\gamma(\mu)$ and $\delta(\mu)$ in $\mu$.

**Theorem 3.** *Both $\gamma(\mu)$ and $\delta(\mu)$ are continuous in $\mu \in \mathfrak{P}$ in the metric $\rho$.*

*Proof.* Suppose that $\mu_n \to \mu$. Then $\varphi(\xi; \mu_n) \to \varphi(\xi; \mu)$ and so $\delta(\mu_n) \to \delta(\mu)$ by the bounded convergence theorem. By $\mu_n \to \mu$ we have a constant $K$ such that

$$\mu_n(-K, K) > 2/3, \qquad n = 1, 2, \ldots .$$

If $\{\gamma(\mu_n)\}$ is unbounded below, we have a subsequence $\{\nu_n\}$ of $\{\mu_n\}$ such that[8] $\gamma_n \equiv \gamma(\nu_n) \to -\infty$. By the definition of $\gamma$ we have

$$\int_{(-K,K)^{\mathrm{C}}} \arctan(x - \gamma_n)\nu_n(\mathrm{d}x) + \int_{(-K,K)} \arctan(x - \gamma_n)\nu_n(\mathrm{d}x) = 0 .$$

Since the first integral $I_1(n)$ is bounded in modulus by $\pi/6$, the second integral $I_2(n)$ must be bounded in modulus by $\pi/6$. But

$$I_2(n) \geq \arctan(-K - \gamma_n)\nu_n(-K, K)$$

and so

$$\limsup_{n\to\infty} I_2(n) \geq (\pi/2)(2/3) = \pi/3 ,$$

in contradiction with $|I_2(n)| < \pi/6$. Therefore $\{\gamma(\mu_n)\}$ is bounded below, i. e.

$$\underline{\gamma} = \liminf_{n\to\infty} \gamma(\mu_n) > -\infty .$$

---

[7] See Problems 0.7 and 0.43 in Exercises for definitions of the weak* topology and the Lévy metric. A sequence $\mu_n$ in $\mathfrak{P}$ converges to $\mu \in \mathfrak{P}$ in the Lévy metric (written $\mu_n \to \mu$) if and only if $\int f(x)\mu_n(\mathrm{d}x) \to \int f(x)\mu(\mathrm{d}x)$ for all bounded continuous functions $f(x)$ on $R^1$.

[8] The equality $\gamma_n \equiv \gamma(\nu_n)$ means that $\gamma_n$ is defined to be equal to $\gamma(\nu_n)$. The same symbol $\equiv$ is sometimes used in the meaning "constantly equal to".

Similarly $\overline{\gamma} = \limsup_{n\to\infty} \gamma(\mu_n) < \infty$. Thus we have

$$-\infty < \underline{\gamma} \le \overline{\gamma} < \infty .$$

Taking a subsequence $\{\nu_n\}$ of $\{\mu_n\}$ such that $\gamma_n \equiv \gamma(\nu_n) \to \overline{\gamma}$ and observing the inequality

$$|\arctan x - \arctan y| \le |x-y| ,$$

we obtain

$$\left| \int_{R^1} \arctan(x-\gamma_n)\nu_n(\mathrm{d}x) - \int_{R^1} \arctan(x-\overline{\gamma})\nu_n(\mathrm{d}x) \right| \\ \le |\gamma_n - \overline{\gamma}| \to 0 \qquad (n\to\infty) .$$

Since $\nu_n \to \mu$, we have

$$\int_{R^1} \arctan(x-\overline{\gamma})\nu_n(\mathrm{d}x) \to \int_{R^1} \arctan(x-\overline{\gamma})\mu(\mathrm{d}x) .$$

Therefore

$$\int_{R^1} \arctan(x-\overline{\gamma})\mu(\mathrm{d}x) = \lim_n \int_{R^1} \arctan(x-\gamma_n)\nu_n(\mathrm{d}x) = 0 ,$$

since $\gamma_n = \gamma(\nu_n)$. This implies $\overline{\gamma} = \gamma(\mu)$. Similarly $\underline{\gamma} = \gamma(\mu)$. This proves $\gamma(\mu_n) \to \gamma(\mu)$.

**Theorem 4.** $\delta(\mu_n) \to 0 \iff \mu_n - \gamma(\mu_n) \to \delta_0$ .

*Proof.* Suppose that $\mu_n - \gamma(\mu_n) \to \delta_0$. Then

$$\delta(\mu_n) = \delta(\mu_n - \gamma(\mu_n)) \to \delta(\delta_0) = 0$$

by Theorems 1 and 3.

For the proof of the converse, suppose that $\delta(\mu_n) \to 0$. Then we have

$$\alpha_n \equiv \iint \mathrm{e}^{-|x-y|}\mu_n(\mathrm{d}x)\mu_n(\mathrm{d}y) \to 1 .$$

Since $\mu_n(R^1) = 1$, we have $c_n \in R^1$ with

$$\int \mathrm{e}^{-|x-c_n|}\mu_n(\mathrm{d}x) \ge \alpha_n .$$

But the left-hand side is not more than

$$\mu_n(c_n - \varepsilon, c_n + \varepsilon) + \mathrm{e}^{-\varepsilon}(1 - \mu_n(c_n - \varepsilon, c_n + \varepsilon)) .$$

Therefore we get

$$\mu_n(c_n - \varepsilon, c_n + \varepsilon) \geq \frac{\alpha_n - \mathrm{e}^{-\varepsilon}}{1 - \mathrm{e}^{-\varepsilon}} \to 1 \qquad (n \to \infty)$$

for every $\varepsilon > 0$. This proves that $\mu_n - c_n \to \delta_0$. Hence it follows by Theorems 1 and 3 that

$$\gamma(\mu_n) - c_n = \gamma(\mu_n - c_n) \to \gamma(\delta_0) = 0 \, .$$

Therefore $\mu_n - \gamma(\mu_n) \to \delta_0$.

Let $\mathcal{R}$ be the set of all random variables where we identify two random variables equal to each other a. s.[9] Set

$$\|X\| = E\left(\frac{|X|}{1+|X|}\right) .$$

Then $\|X\| \geq 0$; $\|X\| = 0$ implies $X = 0$ a. s.; $\|-X\| = \|X\|$ and $\|X+Y\| \leq \|X\| + \|Y\|$. Therefore

$$d(X,Y) = \|X - Y\|$$

defines a metric in $\mathcal{R}$. With this metric $\mathcal{R}$ is a complete metric space. The convergence "$d(X_n, X) \to 0$" is equivalent to "$X_n \to X$ i. p."[10]

Since $d(X,Y) \to 0$ implies $\rho(P_X, P_Y) \to 0$ and since $\gamma(X) = \gamma(P_X)$ and $\delta(X) = \delta(P_X)$, we have

**Theorem 5.** *(i) Both $\gamma(X)$ and $\delta(X)$ are continuous in $X \in \mathcal{R}$.*
*(ii)* $\delta(X_n) \to 0 \iff X_n - \gamma(X_n) \to 0$ *i. p.*

The assertion (i) is by Theorem 3 and (ii) by Theorem 4.

For square integrable random variables we have

$X_n \to X$ in square mean

$$\iff m(X_n) \to m(X), \quad \sigma(X_n - X) \to 0 \, ,$$

$X_n$ is convergent in square mean

$$\iff |m(X_m) - m(X_n)| + \sigma(X_m - X_n) \to 0 \quad (m, n \to \infty) \, .$$

We will generalize this fact to $\mathcal{R}$ in terms of $\gamma$ and $\delta$.

**Theorem 6.** *$X_n \to X$ i. p. if and only if $\gamma(X_n) \to \gamma(X)$ and $\delta(X_n - X) \to 0$.*

---

[9] a. s. = almost surely = with probability one.

[10] See Problem 0.12 in Exercises for the equivalence. A sequence $\{X_n\}$ is said to converge to $X$ in probability (i. p.) if, for every $\varepsilon > 0$, $P(|X_n - X| > \varepsilon) \to 0$ as $n \to \infty$. Completeness of the space $\mathcal{R}$ with the metric $d$ is proved by an argument similar to the proof of Theorem 7.

*Proof.* If $X_n \to X$ i. p., then $\|X_n - X\| \to 0$. By Theorem 5 (i) we have $\gamma(X_n) \to \gamma(X)$ and $\delta(X_n - X) \to 0$.

Suppose conversely that $\gamma(X_n) \to \gamma(X)$ and $\delta(X_n - X) \to 0$. By the second condition and Theorem 5 (ii), we have

$$X_n - X - c_n \to 0 \quad \text{i. p.,} \quad c_n = \gamma(X_n - X) .$$

Therefore $X_n - c_n \to X$ i. p. By Theorem 5 (i), this implies $\gamma(X_n) - c_n \to \gamma(X)$. Since $\gamma(X_n) \to \gamma(X)$, we get

$$c_n \to 0 .$$

Thus we have $X_n \to X$ i. p.

**Theorem 7.** *A sequence $\{X_n\}$ is convergent i. p. if and only if*

$$\gamma(X_m) - \gamma(X_n) \to 0, \quad \delta(X_m - X_n) \to 0 \qquad (m, n \to \infty) .$$

*Proof.* Suppose that $\{X_n\}$ is convergent i. p. and write $X$ for the limit. Then $\gamma(X_n) \to \gamma(X)$ and so

$$\gamma(X_m) - \gamma(X_n) \to 0 \qquad (m, n \to \infty) .$$

Since $\|X_m - X_n\| \to 0$ $(m, n \to \infty)$ by our assumption, we have $\|X_m - X_n - \gamma(X_m - X_n)\| \to 0$ and $\delta(X_m - X_n) \to 0$ by Theorem 5 (ii).

Suppose that $\gamma(X_m) - \gamma(X_n) \to 0$, $\delta(X_m - X_n) \to 0$ $(m, n \to \infty)$. By Theorem 5 (ii) we find by $\delta(X_m - X_n) \to 0$ a sequence $p(1) < p(2) < \cdots$ such that

$$P(|X_{p(n+1)} - X_{p(n)} - d_n| \geq 1/2^n) < 1/2^n, \qquad d_n = \gamma(X_{p(n+1)} - X_{p(n)}) .$$

Using the Borel–Cantelli lemma we see that

$$X_{p(1)} + \sum_n Y_n \qquad (Y_n = X_{p(n+1)} - X_{p(n)} - d_n)$$

is convergent a. s. In other words

$$X_{p(n)} - e_n = X_{p(1)} + \sum_{i=1}^{n-1} Y_i \qquad \left( e_n = \sum_{i=1}^{n-1} d_i \right), \qquad n = 1, 2, \ldots$$

is convergent a. s. Write $X$ for the limit. Then we have

$$\delta(X_{p(m)} - e_m - X_n) = \delta(X_{p(m)} - X_n) \to 0 \qquad (m, n \to \infty) .$$

Since $X_{p(m)} - e_m \to X$ a. s., we get

$$\delta(X - X_n) \to 0 \qquad (n \to \infty)$$

by Theorem 5 (i). Using Theorem 5 (ii) we get

$$X_n - X - \gamma_n \to 0 \quad \text{i. p.} \quad (\gamma_n = \gamma(X_n - X))$$

and so

$$\gamma(X_n) - \gamma_n \to \gamma(X)$$

by Theorem 5 (i). But $\gamma(X_n)$ is convergent since $\lim_{m,n\to\infty} |\gamma(X_m) - \gamma(X_n)| = 0$. Therefore $\gamma_n$ is also convergent. Thus we see that $X_n$ is convergent i. p.

Let $\mathfrak{M}$ be a subset of the set $\mathfrak{P}$ of all 1-dimensional probability measures. If $\{\sigma(\mu)\}_{\mu\in\mathfrak{M}}$ is bounded (say $< c$), then

$$(\mu - m(\mu))(-K, K) \geq 1 - (c^2/K^2)$$

by the Chebyshev inequality, so that $\{\mu - m(\mu)\}_{\mu\in\mathfrak{P}}$ is conditionally compact[11]. Corresponding to this fact we have

**Theorem 8.** *Let $\mathfrak{M}$ be a subset of $\mathfrak{P}$. Then $\{\mu - \gamma(\mu)\}_{\mu\in\mathfrak{M}}$ is conditionally compact if and only if*

$$\lim_{\alpha\downarrow 0} \sup_{\mu\in\mathfrak{M}} \delta(\alpha\mu) = 0.$$

*Proof of necessity.* Suppose that

$$a = \limsup_{\alpha\downarrow 0} \sup_{\mu\in\mathfrak{M}} \delta(\alpha\mu) > 0\,.$$

Then we have $\{\mu_n\} \subset \mathfrak{M}$ and $\alpha_n \downarrow 0$ such that

$$\delta(\alpha_n\mu_n) \geq a/2$$

and so

$$\delta(\alpha_n(\mu_n - \gamma(\mu_n))) \geq a/2\,.$$

If $\{\mu - \gamma(\mu)\}_{\mu\in\mathfrak{M}}$ is conditionally compact, then $\{\mu_n - \gamma(\mu_n)\}_n$ has a convergent subsequence $\{\nu_n \equiv \mu_{p(n)} - \gamma(\mu_{p(n)})\}_n$. Since $\{\nu_n\}$ is convergent, we have

$$\alpha_{p(n)}\nu_n \to \delta_0\,,$$

so that

$$\delta(\alpha_{p(n)}\nu_n) \to 0$$

by Theorem 3. This is a contradiction.

---

[11] A subset $\mathfrak{M}$ of $\mathfrak{P}$ is called conditionally compact if the closure of $\mathfrak{M}$ is compact. Since $\mathfrak{P}$ is a metric space, $\mathfrak{M}$ is conditionally compact if and only if any sequence $\{\mu_n\}$ in $\mathfrak{M}$ has a subsequence convergent in $\mathfrak{P}$.

*Proof of sufficiency.* Suppose that $f(\alpha) \equiv \sup_{\mu\in\mathfrak{M}} \delta(\alpha\mu) \to 0$ as $\alpha \downarrow 0$. For $\mu \in \mathfrak{M}$ we have

$$\begin{aligned} e^{-f(\alpha)} &\le e^{-\delta(\alpha\mu)} \\ &= \iint e^{-\alpha|x-y|}\mu(dx)\mu(dy) \\ &\le \int e^{-\alpha|x-c(\mu)|}\mu(dx) \qquad \text{(for some } c(\mu)) \\ &= \int e^{-\alpha|x|}(\mu - c(\mu))(dx) \\ &\le (\mu - c(\mu))(-K,K) + e^{-\alpha K}(1-(\mu-c(\mu))(-K,K)) . \end{aligned}$$

Hence it follows that

$$(\mu - c(\mu))(-K,K) \ge \frac{e^{-f(\alpha)} - e^{-\alpha K}}{1-e^{-\alpha K}} ,$$

so that

$$\lim_{K\to\infty} \inf_{\mu\in\mathfrak{M}} (\mu - c(\mu))(-K,K) \ge e^{-f(\alpha)} \to 1 \qquad (\alpha \downarrow 0) .$$

This implies that $\{\mu - c(\mu)\}_{\mu\in\mathfrak{M}}$ is conditionally compact.[12] Let $\{\mu_n\} \subset \mathfrak{M}$. Then we can find a subsequence $\{\nu_n\}$ such that $\{\nu_n - c_n\}_n$ ($c_n = c(\nu_n)$) is convergent. Then $\{\gamma(\nu_n) - c_n\}_n$ is convergent by Theorems 1 and 3. Therefore $\{\nu_n - \gamma(\nu_n)\}_n$ is also convergent. This completes the proof.

**Theorem 9.** *A subset $\mathfrak{M}$ of $\mathfrak{P}$ is conditionally compact if and only if $\{\gamma(\mu)\}_{\mu\in\mathfrak{M}}$ is bounded and*

$$\lim_{\alpha\downarrow 0} \sup_{\mu\in\mathfrak{M}} \delta(\alpha\mu) = 0.$$

*Proof.* Since $\gamma(\mu)$ is continuous in $\mu$ by Theorem 3, $\gamma(\mu)$ is bounded on a conditionally compact set $\mathfrak{M}$. The rest of the proof follows easily from Theorem 8.

## 0.3 Centralized Sum of Independent Random Variables

In the theory of sums of independent square integrable random variables the Kolmogorov inequality plays an important role. For example we can use it to prove

[12] See Problem 0.44 in Exercises.

**Proposition 1.** *Let $\{X_n\}$ be an independent sequence of square integrable random variables and let $S_n = \sum_1^n X_i$. If $\lim_n \sigma(S_n) < \infty$, then $S_n - E(S_n)$ is convergent a. s.*

To extend this fact to the case that the random variables are not necessarily square integrable, P. Lévy used his *concentration functions.* Here we will use the dispersion $\delta$ for the same purpose.

If $X, Y$ are independent and square integrable, then $\sigma(X+Y) \geq \sigma(X)$ and the equality holds if and only if $Y = \text{const}$ a. s. Corresponding to this fact we have

**Theorem 1.** *If $X$ and $Y$ are independent, then*

$$\delta(X+Y) \geq \delta(X)$$

*and the equality holds if and only if $Y = \text{const}$ a. s.*

*Remark.* This fact is stated in terms of probability measures as follows:

$$\delta(\mu * \nu) \geq \delta(\mu)$$

where the equality holds if and only if $\nu$ is concentrated at a single point.

*Proof of Theorem 1.* Notice that

$$\begin{aligned} \mathrm{e}^{-\delta(X+Y)} &= \frac{1}{\pi} \int_{R^1} |\varphi(\xi; X+Y)|^2 \frac{\mathrm{d}\xi}{1+\xi^2} \\ &= \frac{1}{\pi} \int_{R^1} |\varphi(\xi; X)|^2 |\varphi(\xi; Y)|^2 \frac{\mathrm{d}\xi}{1+\xi^2} \\ &\leq \frac{1}{\pi} \int_{R^1} |\varphi(\xi; X)|^2 \frac{\mathrm{d}\xi}{1+\xi^2} = \mathrm{e}^{-\delta(X)} \end{aligned}$$

and so $\delta(X+Y) \geq \delta(X)$. If $Y = \text{const}$ a. s., then $\delta(X+Y) = \delta(X)$. If $\delta(X+Y) = \delta(X)$, then $|\varphi(\xi; Y)| = 1$ a. e. near $\xi = 0$, because $|\varphi(\xi; X)| > 0$ near $\xi = 0$. Since $\varphi(\xi; Y)$ is continuous in $\xi$, $|\varphi(\xi; Y)| = 1$ near $\xi = 0$. Using the inequality for characteristic functions[13]

$$1 - |\varphi(2\xi)|^2 \leq 4(1 - |\varphi(\xi)|^2),$$

we see that $|\varphi(\xi; Y)| = 1$. Therefore $\delta(Y) = 0$ i. e. $Y = \text{const}$ a. s.

Let $\{X_n\}_n$ be an independent sequence and let $S_n = \sum_1^n X_i$. Then $\delta(S_n)$ increases[14] with $n$ as Theorem 1 says. We have two cases:

convergent case $\quad \lim \delta(S_n) < \infty$,
divergent case $\quad \lim \delta(S_n) = \infty$.

[13] See Problem 0.1 in Exercises.

[14] The words "increase" and "decrease" are always used in the meaning allowing flatness.

**Theorem 2.** *(i) In the convergent case, $\{S_n - \gamma(S_n)\}$ is convergent a. s.*

*(ii) In the divergent case, $\{S_n - c_n\}$ is divergent (= not convergent) a. s. for every choice of $\{c_n\}$.*

*Remark.* Since the event that $\{S_n - c_n\}$ is convergent belongs to the tail $\sigma$-algebra of the independent sequence $\{\mathcal{B}(X_n)\}_n$, its probability is 0 or 1.

*Proof of (ii).* Suppose that $S_n - c_n \to S$ a. s. Then

$$\lim_n \delta(S_n) = \lim_n \delta(S_n - c_n) = \delta(S) < \infty .$$

By the remark mentioned above, this proves (ii).

For the proof of (i) we will use

**Proposition 2.** *If $B$ is a bounded subset of $R^1$ with the Lebesgue measure $0 < |B| < \infty$, then we have a positive constant $K = K(B)$ such that*

$$\int_B (1 - \cos \xi x) \mathrm{d}\xi \geq K \frac{x^2}{1 + x^2} \qquad \textit{for every } x .$$

*Proof.* Consider the function

$$f(x) = \frac{1 + x^2}{x^2} \int_B (1 - \cos \xi x) \mathrm{d}\xi, \qquad x \neq 0 .$$

This is continuous and positive. We have

$$\lim_{x \to 0} f(x) = \int_B \frac{\xi^2}{2} \mathrm{d}\xi > 0$$

and

$$\lim_{|x| \to \infty} f(x) = |B| > 0$$

by the Riemann–Lebesgue theorem. Therefore $f(x)$ has a positive infimum, say $K > 0$. This proves our proposition.

*Proof of (i).* By the assumption $\lim \delta(S_n) < \infty$ we have

$$\int \prod_1^\infty |\varphi(\xi; X_n)|^2 \frac{\mathrm{d}\xi}{1 + \xi^2} > 0$$

and so

$$\prod_1^\infty |\varphi(\xi; X_n)|^2 > 0$$

i. e.

$$\sum_1^\infty (1 - |\varphi(\xi; X_n)|^2) < \infty$$

on a $\xi$-set $A$ with $|A| > 0$. Then we have a bounded subset $B$ of $A$ with $|B| > 0$ and a positive constant $K_1$ such that

$$\sum_1^\infty (1 - |\varphi(\xi; X_n)|^2) < K_1 \qquad \text{for every } \xi \in B .$$

This can be written as

$$\sum_1^\infty \int_{R^1} (1 - \cos \xi x)\widetilde{\mu}_n(\mathrm{d}x) < K_1 \qquad \text{for every } \xi \in B ,$$

where $\widetilde{\mu}_n$ is the symmetrization of the probability law $\mu_n$ of $X_n$. Integrating both sides over $B$ we have

$$\sum_1^\infty \int_{R^1} \int_B (1 - \cos \xi x)\mathrm{d}\xi \, \widetilde{\mu}_n(\mathrm{d}x) < K_1 |B| ,$$

so that

$$\sum_1^\infty \int_{R^1} \frac{x^2}{1+x^2} \widetilde{\mu}_n(\mathrm{d}x) < \frac{K_1|B|}{K} \equiv K_2$$

by Proposition 2. Observing

$$\begin{aligned} \int_{R^1} \frac{x^2}{1+x^2} \widetilde{\mu}_n(\mathrm{d}x) &= \int_{R^1} \int_{R^1} \frac{(x-y)^2}{1+(x-y)^2} \mu_n(\mathrm{d}x)\, \mu_n(\mathrm{d}y) \\ &\geq \int_{R^1} \frac{(x-c_n)^2}{1+(x-c_n)^2} \mu_n(\mathrm{d}x) \qquad \text{for some } c_n , \end{aligned}$$

we get

$$\sum_1^\infty \int_{R^1} \frac{(x-c_n)^2}{1+(x-c_n)^2} \mu_n(\mathrm{d}x) < \infty$$

i. e.

$$\sum_1^\infty E\left[\frac{(X_n - c_n)^2}{1 + (X_n - c_n)^2}\right] < \infty .$$

Therefore[15]

$$\sum_1^\infty E[(X_n - c_n)^2, \, |X_n - c_n| \leq 1] < \infty, \qquad \sum_n P(|X_n - c_n| > 1) < \infty .$$

---

[15] For a random variable $X$ and an event $B$, $E(X, B)$ means $\int_B X(\omega) P(\mathrm{d}\omega)$.

Setting

$$Y_n = \begin{cases} X_n - c_n & \text{if } |X_n - c_n| \le 1 \\ 0 & \text{otherwise} \end{cases}$$

we have

$$\sum_n E(Y_n^2) < \infty, \qquad \sum_n P(Y_n \ne X_n - c_n) < \infty. \tag{1}$$

By the first condition of (1), we get

$$\lim_{n\to\infty} \sigma^2\left(\sum_1^n Y_i\right) = \lim_{n\to\infty} \sum_1^n \sigma^2(Y_i) \le \lim_{n\to\infty} \sum_1^n E(Y_i^2).$$

As $\{Y_n\}$ is independent by the independence of $\{X_n\}$, $\{\sum_1^n (Y_i - d_i)\}_n$ ($d_i = E(Y_i)$) is convergent a. s. by Proposition 1. By the second condition of (1) we can use the Borel–Cantelli lemma to get

$$P(Y_n = X_n - c_n \text{ for } n \text{ big enough}) = 1.$$

Therefore $\{\sum_1^n (X_i - c_i - d_i)\}_n$ is also convergent a. s. Writing $\sum_1^n (c_i + d_i)$ as $e_n$, we see that $\{S_n - e_n\}_n$ converges to a random variables $S$ a. s. Then

$$\gamma(S_n) - e_n \to \gamma(S).$$

This implies that $S_n - \gamma(S_n) \to S - \gamma(S)$ a. s, completing the proof.

Let $\{X_\alpha\}_{\alpha\in A}$ be an independent family where $A$ is a *countable* set. For a finite set $F \subset A$ we set

$$S_F = \sum_{\alpha\in F} X_\alpha, \qquad S_F^\circ = S_F - \gamma(S_F).$$

We call $S_F$ the *partial sum* over $F$ and $S_F^\circ$ the *centralized partial sum* over $F$. We often write $S_F^\circ$ as $\sum_{\alpha\in F}^\circ X_\alpha$. Let $\delta(F)$ denote $\delta(S_F)$. For an infinite subset $B$ of $A$ we define $\delta(B)$ to be $\sup_F \delta(F)$, $F$ ranging over all finite subsets of $B$.

**Theorem 3.** *Suppose that $\delta(B) < \infty$. For every sequence of finite sets $\{F_n\}$ such that $F_1 \subset F_2 \subset \cdots \to B$, $S_{F_n}^\circ$ is convergent a. s. The limit $S_B^\circ$ is independent of the choice of $\{F_n\}$. We have*

$$\gamma(S_B^\circ) = 0 \qquad \text{and} \qquad \delta(S_B^\circ) = \delta(B).$$

**Definition 1.** The limit $S_B^\circ$ is called the *centralized sum* of $X_\alpha$ over $\alpha \in B$ and is often denoted by $\sum_{\alpha\in B}^\circ X_\alpha$.

*Proof of Theorem 3.* Let $Y_1 = S_{F_1}$, $Y_{n+1} = S_{F_{n+1}} - S_{F_n}$, $n = 1, 2, \ldots$. Then $\{Y_n\}$ is independent. Applying Theorem 2 (i) to $\{Y_n\}$ we get the a. s. convergence of $\{S^\circ_{F_n}\}$. Write the limit as $S^\circ_B$. Then $\gamma(S^\circ_B) = \lim_n \gamma(S^\circ_{F_n}) = 0$ and $\delta(S^\circ_B) = \lim_n \delta(S^\circ_{F_n})$.

Let $\{G_n\}$ be another sequence of finite sets such that $G_1 \subset G_2 \subset \cdots \to B$. If $F_k \supset G_m \supset F_n$, then we have

$$\delta(S^\circ_{F_k} - S^\circ_{G_m}) \leq \delta(S^\circ_{F_k} - S^\circ_{F_n}) .$$

Letting $k \to \infty$, we have

$$\delta(S^\circ_B - S^\circ_{G_m}) \leq \delta(S^\circ_B - S^\circ_{F_n}) .$$

Letting $m \to \infty$, we have

$$\delta(S^\circ_B - S^\circ) \leq \delta(S^\circ_B - S^\circ_{F_n}), \qquad S^\circ = \lim_n S^\circ_{G_n} .$$

Letting $n \to \infty$, we get

$$\delta(S^\circ_B - S^\circ) = 0 \quad \text{i.e.} \quad S^\circ_B - S^\circ = \text{const} .$$

But $\gamma(S^\circ_B) = \gamma(S^\circ) = 0$ and so the constant must be 0.

By the same argument we have[16]

**Theorem 4.** *(i) If $B = \bigcup B_n$ (disjoint) and $\delta(B) < \infty$, then $S^\circ_B = \sum^\circ_n S^\circ_{B_n}$.*
*(ii) If $B_n \uparrow B$ and $\delta(B) < \infty$, then $S^\circ_{B_n} \to S^\circ_B$ a. s.*
*(iii) If $B_n \downarrow B$ and $\delta(B_1) < \infty$, then $S^\circ_{B_n} \to S^\circ_B$ a. s.*

If $\mu_1 * \mu_2 = \mu$, then $\mu_1$ is called a factor of $\mu$. If $\mu_1$ is a factor of $\mu$, we have $\delta(\mu_1) \leq \delta(\mu)$ by Theorem 1 and the Remark to it. Using Theorem 0.2.8[17], we have

**Theorem 5.** *Let $\mathfrak{M}$ be a set of 1-dimensional probability measures and $\mathfrak{M}'$ the set of all factors of probability measures in $\mathfrak{M}$. If $\mathfrak{M}$ is conditionally compact, then $\{\mu' - \gamma(\mu')\}_{\mu' \in \mathfrak{M}'}$ is also conditionally compact.*

*Proof.* For every $\mu' \in \mathfrak{M}'$ we have $\mu \in \mathfrak{M}$ such that $\mu'$ is a factor of $\mu$. Then $\alpha\mu'$ is also a factor of $\alpha\mu$, so that $\delta(\alpha\mu') \leq \delta(\alpha\mu)$. Therefore

$$\lim_{\alpha \downarrow 0} \sup_{\mu' \in \mathfrak{M}'} \delta(\alpha(\mu' - \gamma(\mu'))) \leq \lim_{\alpha \downarrow 0} \sup_{\mu \in \mathfrak{M}} \delta(\alpha\mu) = 0 .$$

Apply Theorem 0.2.8 to complete the proof.

---

[16] See Problem 0.9 in Exercises.

[17] Theorem 0.2.8 indicates Theorem 8 of Section 0.2. We will use the same convention throughout.

## 0.4 Infinitely Divisible Distributions

The Gauss distribution $N(m, v)$ and the Poisson distribution $p(\lambda)$ have the following property[18]:

$$N(m, v) = N\left(\frac{m}{n}, \frac{v}{n}\right)^{n*}, \qquad p(\lambda) = p\left(\frac{\lambda}{n}\right)^{n*}.$$

In other words both have the $n$-th convolution root for every $n = 1, 2, \ldots$. Observing this fact P. Lévy introduced the notion of *infinitely divisible distributions.*

**Definition 1.** A one-dimensional distribution $\mu$ is called an infinitely divisible distribution if for every $n = 1, 2, \ldots$ we can find a distribution $\mu_n$ such that

$$\mu = {\mu_n}^{n*}.$$

It is easy to see

**Theorem 1.** *The family of all infinitely divisible distributions is closed under (a) linear transformation, (b) convolutions and (c) convergence.*[19]

A characteristic function (= the Fourier transform of a distribution in $R^1$) is called *infinitely divisible* if it corresponds to an infinitely divisible distribution. A characteristic function $\varphi(z)$ is infinitely divisible if and only if for every $n = 1, 2, \ldots$ we can find a characteristic function $\varphi_n(z)$ such that

$$\varphi(z) = \varphi_n(z)^n.$$

Theorem 1 can be stated in terms of characteristic functions as follows.

**Theorem 1′.** *The family of all infinitely divisible characteristic functions is closed under (a) the operation* $\varphi \longrightarrow \mathrm{e}^{\mathrm{i}mz}\varphi(az)$*, (b) multiplications and (c) limits of uniform convergence on compacts.*

Since both Gauss distributions and Poisson distributions are infinitely divisible, the following functions are infinitely divisible characteristic functions:

$$\exp\left(\mathrm{i}mz - \frac{v}{2}z^2\right), \qquad m \in R^1, \quad v \geq 0,$$
$$\exp\left(\lambda(\mathrm{e}^{\mathrm{i}z} - 1)\right), \qquad \lambda \geq 0.$$

Applying Theorem 1′ to these functions, we can get a large class of infinitely divisible characteristic functions; for example,

$$\exp\left\{\mathrm{i}mz - \frac{v}{2}z^2 + \sum_{k=1}^{K}(\mathrm{e}^{\mathrm{i}zu_k} - 1)\lambda_k\right\}, \qquad 0 < \lambda_k < \infty, \quad u_k \in R^1,$$

[18] For $\mu \in \mathfrak{P}$, $\mu^{n*}$ denotes the $n$-fold convolution of $\mu$.

[19] See Problem 0.42 in Exercises for a proof of closedness under (c).

$$\exp\left\{\mathrm{i}mz - \frac{v}{2}z^2 + \int_{-\infty}^{\infty}(\mathrm{e}^{\mathrm{i}zu} - 1)\,n(\mathrm{d}u)\right\}, \quad (n(\mathrm{d}u) = \text{bounded measure}).$$

Observing that

$$|\mathrm{e}^{\mathrm{i}zu} - 1 - \mathrm{i}zu| \le z^2u^2/2$$

and recalling the closedness under (c) in Theorem 1′, we can easily see that the function

$$\exp\left\{\mathrm{i}mz - \frac{v}{2}z^2 + \int_{|u|>1}(\mathrm{e}^{\mathrm{i}zu} - 1)\,n(\mathrm{d}u) + \int_{|u|\le 1}\left(\mathrm{e}^{\mathrm{i}zu} - 1 - \frac{\mathrm{i}zu}{1+u^2}\right)n(\mathrm{d}u)\right\}$$

is also an infinitely divisible characteristic function as long as the measure $n(\mathrm{d}u)$ satisfies

$$\int_{|u|>1} n(\mathrm{d}u) < \infty, \qquad \int_{|u|\le 1} u^2 n(\mathrm{d}u) < \infty\,.$$

Writing the two integrals in one integral with $m$ modified we get

$$\exp\left\{\mathrm{i}mz - \frac{v}{2}z^2 + \int_{-\infty}^{\infty}\left(\mathrm{e}^{\mathrm{i}zu} - 1 - \frac{\mathrm{i}zu}{1+u^2}\right)n(\mathrm{d}u)\right\}, \tag{1}$$

where the measure $n$ satisfies

$$\int_{-\infty}^{\infty}\frac{u^2}{1+u^2}n(\mathrm{d}u) < \infty\,;$$

since the integrand vanishes at $u = 0$, the jump of $n$ at $u = 0$ has nothing to do with the whole expression in (1) and so can be assumed to be 0. Thus we have the following

**Theorem 2.** *Let $m$ be real, $v \ge 0$ and $n$ a measure on $(-\infty, \infty)$ satisfying*

$$n(\{0\}) = 0, \qquad \int_{-\infty}^{\infty}\frac{u^2}{1+u^2}n(\mathrm{d}u) < \infty\,.$$

*Then (1) gives an infinitely divisible characteristic function and so corresponds to an infinitely divisible distribution.*

The important fact is the converse of Theorem 2, which is due to P. Lévy and reads as follows.

**Theorem 3 (P. Lévy's formula).** *Every infinitely divisible characteristic function can be written in the form (1) with $m, v$ and $n$ satisfying the conditions in Theorem 2. Here $m$, $v$ and $n$ are uniquely determined by the infinitely divisible characteristic function (or distribution) and are called its* three components. *The measure $n$ is called its* Lévy measure.[20]

[20] The formula (1) is sometimes called Lévy's canonical form of an infinitely divisible characteristic function. This was derived by P. Lévy in his 1934 paper cited in the footnote to Theorem 1.7.1.

Before proving P. Lévy's formula let us remind the reader of some preliminary facts.

1. To indicate that $\varphi_n(z)$ converges to $\varphi(z)$ uniformly on $|z| \le a$, we write

$$\varphi_n(z) \rightrightarrows \varphi(z) \quad \text{on } |z| \le a \, .$$

To indicate that $\varphi_n(z)$ converges to $\varphi(z)$ uniformly on compacts, we write

$$\varphi_n(z) \underset{\mathrm{c}}{\rightrightarrows} \varphi(z) \, .$$

2. If $\varphi(z)$ is a characteristic function, then[21]

$$|\varphi(z+h) - \varphi(z)| \le \sqrt{2|\varphi(h) - 1|} \, .$$

Hence we have

$$|\varphi(z) - 1| \le k\sqrt{2|\varphi(z/k) - 1|}, \qquad k = 1, 2, \ldots \, .$$

3. We define the indeterminate values of a function $f(x)$ at $x = 0$ by the limit value if it exists. For example

$$(1 - \cos x)\frac{1+x^2}{x^2} = \frac{1}{2} \qquad \text{at } x = 0,$$
$$\left(1 - \frac{\sin x}{x}\right)\frac{1+x^2}{x^2} = \frac{1}{6} \qquad \text{at } x = 0.$$

4. There exists $c$ such that

$$\left(1 - \frac{\sin x}{x}\right)\frac{1+x^2}{x^2} > c > 0$$

for every $x$; see 3 for the value at $x = 0$.

5. P. Lévy's formula is equivalent to *A. Khinchin's formula*:

$$\exp\left\{\mathrm{i}mz + \int_{-\infty}^{\infty}\left(\mathrm{e}^{\mathrm{i}zu} - 1 - \frac{\mathrm{i}zu}{1+u^2}\right)\frac{1+u^2}{u^2}\,G(\mathrm{d}u)\right\}, \tag{1'}$$

where $m \in R^1$ and $G$ is a bounded measure on $(-\infty, \infty)$. The connection of $G$ with $v$ and $n$ is as follows:

$$G(\mathrm{d}u) = \frac{u^2}{1+u^2}n(\mathrm{d}u), \qquad u \ne 0,$$
$$G(\{0\}) = v.$$

*Proof of Theorem 3.* Let us start with the general idea of deriving (1′). Let $\varphi(z)$ be an infinitely divisible characteristic function. Then we have

[21] See the solution of Problem 0.16 in Exercises.

$$\varphi(z) = \varphi_n(z)^n, \qquad n = 1, 2, \dots ,$$

where $\{\varphi_n(z)\}$ are characteristic functions. Writing $\mu_n$ for the distribution corresponding to $\varphi_n$, we have

$$\begin{aligned}
\varphi(z) &= \left(\int_{-\infty}^{\infty} \mathrm{e}^{\mathrm{i}zu} \mu_n(\mathrm{d}u)\right)^n \\
&= \left(1 + \int_{-\infty}^{\infty} (\mathrm{e}^{\mathrm{i}zu} - 1)\mu_n(\mathrm{d}u)\right)^n \\
&\sim \exp\left\{n \int_{-\infty}^{\infty} (\mathrm{e}^{\mathrm{i}zu} - 1)\mu_n(\mathrm{d}u)\right\} \\
&= \exp\left\{\mathrm{i}\, m_n z + \int_{-\infty}^{\infty} \left(\mathrm{e}^{\mathrm{i}zu} - 1 - \frac{\mathrm{i}zu}{1+u^2}\right) n\, \mu_n(\mathrm{d}u)\right\} \\
&= \exp\left\{\mathrm{i}\, m_n z + \int_{-\infty}^{\infty} \left(\mathrm{e}^{\mathrm{i}zu} - 1 - \frac{\mathrm{i}zu}{1+u^2}\right) \frac{1+u^2}{u^2}\, G_n(\mathrm{d}u)\right\} ,
\end{aligned}$$

where

$$m_n = \int_{-\infty}^{\infty} \frac{u}{1+u^2}\, n\, \mu_n(\mathrm{d}u) ,$$

$$G_n(\mathrm{d}u) = \frac{u^2}{1+u^2}\, n\, \mu_n(\mathrm{d}u) .$$

By letting $n \to \infty$, we shall obtain (1′). We will elaborate this idea.

First we will prove that $\varphi(z)$ never vanishes. By $\varphi = \varphi_n{}^n$ we have $|\varphi| = |\varphi_n|^n$. Since $\varphi$ is a characteristic function, we have $a > 0$ such that

$$|\varphi(z)| \neq 0 \qquad \text{for } |z| < a .$$

Then

$$|\varphi_n(z)| = |\varphi(z)|^{1/n} \to 1 \quad (n \to \infty) \qquad \text{for } |z| < a .$$

Let $z$ be an arbitrary real number. Then we can find an integer $k > 0$ such that

$$|z/k| < a .$$

Since $|\varphi_n(z)|^2$ is also a characteristic function, the above remark (b) ensures that

$$\big||\varphi_n(z)|^2 - 1\big| \leq k\sqrt{2\,\big||\varphi_n(z/k)|^2 - 1\big|} \to 0 \qquad (n \to \infty)$$

by $|z/k| < a$. Thus $\varphi_n(z) \neq 0$ for some big $n$. Therefore $\varphi(z) = \varphi_n(z)^n \neq 0$.

Since $\varphi(z)$ does not vanish, we can consider $\log \varphi(z)$. The function $\varphi(z)$ being complex-valued in general, $\log \varphi(z)$ is determined up to multiples of $\mathrm{i}\,2\pi$. Observing that $\varphi(z)$ is continuous and takes the value 1 at $z = 0$, we can use the analytic continuation theorem to get a unique branch of $\log \varphi(z)$ starting with $\log \varphi(0) = 0$ and changing continuously with $z$. From now on $\log \varphi(z)$ will indicate this branch. Then we have

$$(\mathrm{e}^{(1/n)\log\varphi(z)})^n = \mathrm{e}^{\log\varphi(z)} = \varphi(z) = \varphi_n(z)^n$$

namely

$$(\varphi_n(z)\mathrm{e}^{-(1/n)\log\varphi(z)})^n = 1$$

and so

$$\varphi_n(z)\mathrm{e}^{-(1/n)\log\varphi(z)} = 1, \zeta, \zeta^2, \ldots, \text{ or } \zeta^n, \qquad \zeta = \mathrm{e}^{\mathrm{i}\,2\pi/n}\,.$$

As the left side is continuous in $z$ and equals 1 at $z = 0$, it must be identically equal to 1. This proves

$$\varphi_n(z) = \mathrm{e}^{-(1/n)\log\varphi(z)}$$

and so

$$n(\varphi_n(z) - 1) = \frac{\mathrm{e}^{(1/n)\log\varphi(z)} - 1}{1/n} \underset{\mathrm{c}}{\rightrightarrows} \log\varphi(z)\,; \tag{2}$$

the convergence " $\underset{\mathrm{c}}{\rightrightarrows}$ " follows from the fact that $\log\varphi(z)$ is continuous in $z$ and so bounded on compacts.

Taking the real part of (2), we can get

$$n\int_{-\infty}^{\infty}(1-\cos zu)\mu_n(\mathrm{d}u) \underset{\mathrm{c}}{\rightrightarrows} -\log|\varphi(z)|\,. \tag{3}$$

Introducing the measures

$$G_n(\mathrm{d}u) = \frac{nu^2}{1+u^2}\,\mu_n(\mathrm{d}u)$$

we can write (3) as

$$\int_{-\infty}^{\infty}(1-\cos zu)\frac{1+u^2}{u^2}\,G_n(\mathrm{d}u) \underset{\mathrm{c}}{\rightrightarrows} -\log|\varphi(z)|\,. \tag{3'}$$

Now we will verify the following properties for $\{G_n\}$:

$$\sup_n \int_{-\infty}^{\infty} G_n(\mathrm{d}u) < \infty\,, \tag{4}$$

$$\lim_{A\to\infty}\sup_n \int_{|u|>A} G_n(\mathrm{d}u) = 0\,. \tag{5}$$

For this purpose we operate $(1/\delta)\int_0^\delta \mathrm{d}z$ to both sides of (3′). Then we have

$$\int_{-\infty}^{\infty}\left(1-\frac{\sin\delta u}{\delta u}\right)\frac{1+u^2}{u^2}\,G_n(\mathrm{d}u) \to -\frac{1}{\delta}\int_0^\delta \log|\varphi(z)|\,\mathrm{d}z\,.$$

Putting $\delta = 1$, we have the boundedness of

$$\int_{-\infty}^{\infty} \left(1 - \frac{\sin u}{u}\right) \frac{1+u^2}{u^2} G_n(\mathrm{d}u), \qquad n = 1, 2, 3, \ldots$$

and so the boundedness of $\int_{-\infty}^{\infty} G_n(\mathrm{d}u)$ by the above remark (d). This proves (4). Given $\varepsilon > 0$, taking $\delta_0 = \delta_0(\varepsilon)$ small enough, we have

$$-\frac{1}{\delta_0} \int_0^{\delta_0} \log |\varphi(z)| \mathrm{d}z < \varepsilon .$$

Therefore taking $n_0 = n_0(\varepsilon)$ large enough, we have

$$\int_{-\infty}^{\infty} \left(1 - \frac{\sin \delta_0 u}{\delta_0 u}\right) \frac{1+u^2}{u^2} G_n(\mathrm{d}u) < \varepsilon \qquad \text{for } n \geq n_0(\varepsilon) .$$

Since the integrand is $> 1/2$ for $|u| > 2/\delta_0(\varepsilon) \equiv A_0(\varepsilon)$ we get

$$\int_{|u| > A_0(\varepsilon)} G_n(\mathrm{d}u) < 2\varepsilon \qquad \text{for } n \geq n_0(\varepsilon) .$$

Since each $G_n$ is a bounded measure, we can find, for each $n$, $A_n(\varepsilon)$ such that

$$\int_{|u| > A_n(\varepsilon)} G_n(\mathrm{d}u) < 2\varepsilon .$$

Writing $A = A(\varepsilon)$ for $\max\{A_0(\varepsilon), A_1(\varepsilon), \ldots, A_{n_0(\varepsilon)}(\varepsilon)\}$ we get

$$\int_{|u| > A(\varepsilon)} G_n(\mathrm{d}u) < 2\varepsilon \qquad \text{for every } n ,$$

which proves (5).

By (4) we can apply Helly's theorem to find a subsequence $\{G_{p(n)}\}$ and a bounded measure $G$ such that

$$\int f(u) G_{p(n)}(\mathrm{d}u) \to \int f(u) G(\mathrm{d}u) \tag{6}$$

for every continuous function $f$ with compact support. By (5) we have (6) also for every bounded continuous function $f$.

By (2) we have

$$\begin{aligned}
\log \varphi(z) &= \lim_n p(n)(\varphi_{p(n)}(z) - 1) \\
&= \lim_n p(n) \int_{-\infty}^{\infty} (\mathrm{e}^{\mathrm{i}zu} - 1)\, \mu_{p(n)}(\mathrm{d}u) \\
&= \lim_n \left\{ \mathrm{i}\, m_{p(n)} z + p(n) \int_{-\infty}^{\infty} \left( \mathrm{e}^{\mathrm{i}zu} - 1 - \frac{\mathrm{i}zu}{1+u^2} \right) \mu_{p(n)}(\mathrm{d}u) \right\} \\
&= \lim_n \left\{ \mathrm{i}\, m_{p(n)} z + \int_{-\infty}^{\infty} \left( \mathrm{e}^{\mathrm{i}zu} - 1 - \frac{\mathrm{i}zu}{1+u^2} \right) \frac{1+u^2}{u^2} G_{p(n)}(\mathrm{d}u) \right\} .
\end{aligned}$$

Since the integrand is bounded and continuous in $u$,

$$\lim_n \int_{-\infty}^{\infty} \left( e^{izu} - 1 - \frac{izu}{1+u^2} \right) \frac{1+u^2}{u^2} G_{p(n)}(du)$$
$$= \int_{-\infty}^{\infty} \left( e^{izu} - 1 - \frac{izu}{1+u^2} \right) \frac{1+u^2}{u^2} G(du) .$$

Therefore $\lim_n (im_{p(n)} z)$ exists for every $z$ and so $\lim_{n\to\infty} m_{p(n)}$ exists. Writing this limit as $m$, we have

$$\log \varphi(z) = i mz + \int_{-\infty}^{\infty} \left( e^{izu} - 1 - \frac{izu}{1+u^2} \right) \frac{1+u^2}{u^2} G(du) , \tag{7}$$

which proves (1).

Now we will prove the uniqueness of the expression. Suppose that we have Lévy's formula

$$\log \varphi(z) = i mz - \frac{v}{2} z^2 + \int_{-\infty}^{\infty} \left( e^{izu} - 1 - \frac{izu}{1+u^2} \right) n(du) .$$

Introducing $G$ by $G(du) = \dfrac{u^2}{1+u^2} n(du)$ for $u \neq 0$ and $G\{0\} = v$, we have

$$\log \varphi(z) = i mz + \int_{-\infty}^{\infty} \left( e^{izu} - 1 - \frac{izu}{1+u^2} \right) \frac{1+u^2}{u^2} G(du) .$$

Setting

$$\psi(z) = 2 \log \varphi(z) - \int_{z-1}^{z+1} \log \varphi(t) dt,$$

we get

$$\psi(z) = 2 \int_{-\infty}^{\infty} e^{izu} \left( 1 - \frac{\sin u}{u} \right) \frac{1+u^2}{u^2} G(du) .$$

By the inversion formula for the Fourier transform of bounded measures, the measure

$$H(du) = 2 \left( 1 - \frac{\sin u}{u} \right) \frac{1+u^2}{u^2} G(du)$$

is determined by $\psi(z)$ and so by $\varphi(z)$. Since $G$ is determined by $H$, $\varphi(z)$ determines $G$ and so $v$, $n$, and finally $m$. This completes the proof of the uniqueness part.

We will later discuss infinitely divisible distributions in connection with Lévy processes.

## 0.5 Continuity and Discontinuity of Infinitely Divisible Distributions

Let $\mu$ be infinitely divisible with characteristic function[22]

$$\varphi(z;\mu) = \exp\left\{\mathrm{i}\, mz - \frac{v}{2}z^2 + \int_{-\infty}^{\infty}\left(\mathrm{e}^{\mathrm{i}zu} - 1 - \frac{\mathrm{i}zu}{1+u^2}\right)\lambda(\mathrm{d}u)\right\}. \tag{1}$$

Then we have[23]

**Theorem.**[24] *(i) If $v = 0$ and if $\lambda$ is a purely discontinuous bounded measure, then $\mu$ is purely discontinuous.*

*(ii) If $v > 0$ or if $\lambda$ is unbounded, then $\mu$ is continuous.*

*(iii) If $v = 0$ and if $\lambda$ is a bounded measure which is not purely discontinuous, then $\mu$ is neither continuous nor purely discontinuous.*

Before going into the proof, we will make some preliminary observations. If $\mu = \nu * \theta$ namely $\mu(E) = \int \nu(E - x)\theta(\mathrm{d}x)$, $\nu$ is called a *factor* of $\mu$. The following lemma is obvious.

**Lemma.** *Suppose that $\nu$ is a factor of $\mu$. Then*

$$\sup_x \mu(\{x\}) \le \sup_x \nu(\{x\}) .$$

*In particular, if $\nu$ is continuous, then $\mu$ is also continuous.*

Let $\mu$ be defined by (1). Then the Gauss distribution $N(0, v)$ is a factor of $\mu$. If $\sigma$ is a bounded measure $\le \lambda$, then the compound Poisson distribution

$$\nu \equiv \sum_{k=0}^{\infty} \mathrm{e}^{-\sigma(R^1)} \frac{\sigma(R^1)^k}{k!} \widetilde{\sigma}^{k*}, \qquad \widetilde{\sigma}(\,\cdot\,) = \frac{\sigma(\,\cdot\,)}{\sigma(R^1)} \tag{2}$$

is a factor of $\mu$, because the characteristic function of $\nu$ is

$$\varphi(z;\nu) = \exp\left(\int_{-\infty}^{\infty} (\mathrm{e}^{\mathrm{i}zu} - 1)\,\sigma(\mathrm{d}u)\right)$$

and

$$\varphi(z;\mu) = \varphi(z;\nu)\exp\left\{\mathrm{i}\, m'z - \frac{v}{2}z^2 + \int_{-\infty}^{\infty}\left(\mathrm{e}^{\mathrm{i}zu} - 1 - \frac{\mathrm{i}zu}{1+u^2}\right)(\lambda - \sigma)(\mathrm{d}u)\right\},$$

where $m'$ is a constant.

---

[22] We assume that $m \in R^1$ and $v \ge 0$ and that $\lambda$ is a measure on $(-\infty, \infty)$ satisfying $\int u^2/(1+u^2)\lambda(\mathrm{d}u) < \infty$ and $\lambda(\{0\}) = 0$. By Theorems 0.4.2 and 0.4.3, (1) is the general form of an infinitely divisible characteristic function.

[23] A $\sigma$-finite measure $\mu$ on $(-\infty, \infty)$ is called continuous if $\mu(\{x\}) = 0$ for all $x$; purely discontinuous if, for all Borel sets $E$ with $\mu(E) < \infty$, $\mu(E) = \sum \mu(E \cap \{x\})$ where the sum ranges over all (necessarily countable) $x$ with $\mu(\{x\}) > 0$.

[24] The original theorem has been corrected by the Editors.

*Proof of Theorem (i).* Under the assumption, the compound Poisson distribution $\nu$ defined for $\sigma = \lambda$ by (2) is a shifted measure of $\mu$. Since $\lambda$ is purely discontinuous, $\lambda$ is situated on a countable set $S$ and so is $\widetilde{\lambda}$. Then $\widetilde{\lambda}^{n*}$ is situated on the $n$-fold algebraic sum of $S$, say $S_n$. Then $\nu$ is situated on the union $\bigcup_n S_n$, which is countable. Therefore $\nu$ is purely discontinuous and so is $\mu$.

*Proof of Theorem (ii).* If $v > 0$, then $\mu$ has a factor $N(0, v)$ which is continuous and so $\mu$ is also continuous by the Lemma.

If $\lambda$ is purely discontinuous with $\lambda(R^1) = \infty$, then $\mu$ is continuous. To prove this, let us denote the discontinuity points of $\lambda$ by $\{u_n\}$ and set $\lambda_n = \lambda(\{u_n\})$ and $\lambda'_n = \min\{\lambda_n, 1\}$. Then $\sum_n \lambda'_n = \infty$. Now consider the following bounded measures:

$$\sigma_n = \sum_1^n \lambda'_i \delta_{u_i}, \qquad n = 1, 2, \dots .$$

Since $\sigma_n \leq \lambda$, the compound Poisson distribution $\nu_n$ defined for $\sigma = \sigma_n$ by (2) is a factor of $\mu$. Therefore

$$\sup_x \mu(\{x\}) \leq \sup_x \nu_n(\{x\}), \qquad n = 1, 2, \dots . \tag{3}$$

But

$$\nu_n(\{x\}) = \sum_k \mathrm{e}^{-(\lambda'_1 + \cdots + \lambda'_n)} \frac{(\lambda'_1 + \cdots + \lambda'_n)^k}{k!} \widetilde{\sigma}_n^{k*}(\{x\}) .$$

Notice that

$$\sup_x \widetilde{\sigma}_n(\{x\}) \leq \max_{i=1,\dots,n} \frac{\lambda'_i}{\lambda'_1 + \cdots + \lambda'_n} \leq \frac{1}{\lambda'_1 + \cdots + \lambda'_n}$$

and so

$$\sup_x (\widetilde{\sigma}_n)^{k*}(\{x\}) \leq \sup_x \widetilde{\sigma}_n(\{x\}) \leq \frac{1}{\lambda'_1 + \cdots + \lambda'_n}, \qquad k = 1, 2, \dots$$

by the Lemma. Observing (2), we can easily derive from this that

$$\sup_x \nu_n(\{x\}) \leq \mathrm{e}^{-(\lambda'_1 + \cdots + \lambda'_n)} + \frac{1}{\lambda'_1 + \cdots + \lambda'_n} \to 0, \qquad n \to \infty .$$

This, combined with (3), implies $\sup_x \mu(\{x\}) = 0$, namely, the continuity of $\mu$.

Next, let us show that if $\lambda$ is continuous with $\lambda(R^1) = \infty$, then $\mu$ is continuous. In this case, let $\sigma_n$ be the restriction of $\lambda$ to $|x| > 1/n$ and let $\nu_n$ be the compound Poisson distribution defined by (2) with $\sigma = \sigma_n$. Then $(\widetilde{\sigma}_n)^{k*}$ is continuous for $k = 1, 2, \dots$. Then we have (3) again and

$$\sup_x \nu_n(\{x\}) = \nu_n(\{0\}) = \mathrm{e}^{-\lambda\{|x|>1/n\}} \to 0, \quad n \to \infty.$$

Hence $\mu$ is continuous.

Finally, let us show that if $\lambda(R^1) = \infty$, then $\mu$ is continuous. In this case, $\lambda = \lambda_1 + \lambda_2$ with a purely discontinuous $\lambda_1$ and a continuous $\lambda_2$. Let $\mu_1$ and $\mu_2$ be the infinitely divisible distributions defined by (1) with $\lambda = \lambda_1$ and $\lambda_2$, respectively. Then $\mu_1$ and $\mu_2$ are factors of $\mu$. Since $\lambda_1$ or $\lambda_2$ is unbounded, $\mu_1$ or $\mu_2$ is continuous. In any case, $\mu$ is continuous.

*Proof of Theorem (iii).* We assume that $v = 0$ and that $\lambda$ is a bounded measure, not purely discontinuous. Let $\nu$ be the compound Poisson distribution (2) with $\sigma = \lambda$. Then $\mu$ is a shifted measure of $\nu$. Since $\widetilde{\sigma}^{0*} = \delta_0$, we have $\nu(\{0\}) > 0$. Thus $\mu$ is not continuous. Since $\widetilde{\sigma}^{1*}$ is not purely discontinuous, the formula (2) shows that $\nu$ is not purely discontinuous.

Thus our proof of the Theorem is completed.

## 0.6 Conditional Probability and Expectation

Let $(\Omega, \mathcal{B}, P)$ stand for the basic probability space. Let $\mathcal{C}$ be a sub-$\sigma$-algebra of $\mathcal{B}$ and $L^p$ be the space of all $p$-th order integrable real functions on $(\Omega, \mathcal{B}, P)$[25]. Let $L^p(\mathcal{C})$ denote the closed linear subspace of $L^p$ that consists of all functions $\in L^p$ measurable $(\mathcal{C})$.[26] The space $L^2$ is a real Hilbert space with the inner product

$$(X, Y) = E(XY) .$$

Let $\Pi_{\mathcal{C}}$ stand for the projection operator $L^2 \longrightarrow L^2(\mathcal{C})$.

We will define $E(X \mid \mathcal{C})$, the *conditional expectation* of $X$ under $\mathcal{C}$ for $X \in L^1$. In elementary probability theory $E(X \mid C)$, $C \in \mathcal{B}$, was defined by

$$E(X \mid C) = E(X,\, C)/P(C) .$$

Suppose that $X \in L^2$ and that $\mathcal{C}$ is finite. Then we have a finite disjoint subfamily $\{C_i\}_{i=1}^n$ of $\mathcal{C}$ such that $P(C_i) > 0$ and that every $C \in \mathcal{C}$ is equal a. s. to the union of a subfamily of $\{C_i\}$. In particular

$$\Omega = \bigcup_{i=1}^n C_i \quad \text{a. s.}$$

Each $C_i$ is called an *atom* in $\mathcal{C}$. In this case the conditional expectation $E(X \mid \mathcal{C})$ is defined to be a random variable which takes value $E(X \mid C_i)$ in each $C_i$ namely[27]

[25] A function $X(\omega)$ on $(\Omega, \mathcal{B}, P)$ is said to be $p$-th order integrable if $E(|X|^p) < \infty$.

[26] A function $X(\omega)$ is called measurable $(\mathcal{C})$ or $\mathcal{C}$-measurable if $\{\omega \colon X(\omega) \in B\} \in \mathcal{C}$ for all Borel sets $B$.

[27] The function $e(\omega)$ is called the indicator of a set $C$ if $e(\omega) = 1$ on $C$ and $e(\omega) = 0$ on the complement of $C$. The indicator of $C$ is sometimes denoted by $e_C$.

$$
\begin{aligned}
E(X \mid \mathcal{C}) &= \sum_i E(X \mid C_i)\, e_i(\omega) \qquad (e_i = \text{ indicator of } C_i) \\
&= \sum_i \frac{E(X, C_i)}{P(C_i)}\, e_i
\end{aligned}
$$

i. e.

$$
E(X \mid \mathcal{C}) = \sum_i \frac{(X, e_i)}{\|e_i\|^2}\, e_i, \qquad (\|\cdot\| = \text{ norm in } L^2)\,. \tag{1}
$$

Since $\{C_i\}$ are disjoint, the $e_i$ are mutually orthogonal. Since every $\mathcal{C}$-measurable function is equal a. s. to a function constant on each $C_i$, i. e. to a linear combination of $\{e_i\}_i$, $\{e_i\}_i$ constitute a complete orthogonal system in $L^2(\mathcal{C})$. Therefore (1) shows that

$$
E(X \mid \mathcal{C}) = \Pi_{\mathcal{C}} X\,.
$$

Now we want to define $E(X \mid \mathcal{C})$ for $\mathcal{C}$ general and $X \in L^2$ as the limit of $E(X \mid \mathcal{F})$, $\mathcal{F}$ finite, as $\mathcal{F} \uparrow \mathcal{C}$. In fact this limit is $\Pi_{\mathcal{C}} X$. The precise meaning is as follows.

**Theorem 1.** *Let $\mathcal{C}$ be an arbitrary sub-$\sigma$-algebra of $\mathcal{B}$, and $X \in L^2$. Then for every $\varepsilon > 0$ we have a finite $\sigma$-algebra $\mathcal{F}_\varepsilon$ such that*

$$
\|E(X \mid \mathcal{F}) - \Pi_{\mathcal{C}} X\| < \varepsilon \qquad \textit{for } \mathcal{F}_\varepsilon \subset \mathcal{F} \subset \mathcal{C}\,.
$$

*Proof.* Let $Y = \Pi_{\mathcal{C}} X$. Then

$$
\begin{aligned}
E(X \mid \mathcal{F}) &= \Pi_{\mathcal{F}} X = \Pi_{\mathcal{F}} \Pi_{\mathcal{C}} X \qquad (\text{by } \mathcal{F} \subset \mathcal{C}) \\
&= \Pi_{\mathcal{F}} Y\,.
\end{aligned}
$$

Let $\{r_n\}$ be a numbering of all rationals and set

$$
\begin{aligned}
\mathcal{D}_n &= \sigma\text{-algebra generated by the set } \{\omega\colon Y \le r_n\}\,, \\
\mathcal{C}_n &= \mathcal{D}_1 \vee \mathcal{D}_2 \vee \cdots \vee \mathcal{D}_n\,, \\
\mathcal{C}_\infty &= \bigvee_n \mathcal{D}_n\,.
\end{aligned}
$$

Then $Y$ is measurable $(\mathcal{C}_\infty)$ and so

$$
\Pi_{\mathcal{C}_\infty} Y = Y\,.
$$

Since every $C \in \mathcal{C}_\infty$ can be approximated in measure by a sequence $C_n \in \mathcal{C}_n$, the spaces $L^2(\mathcal{C}_n)$, $n = 1, 2, \ldots$, span the space $L^2(\mathcal{C}_\infty)$. Clearly $L^2(\mathcal{C}_n)$ increases with $n$. Therefore

$$\|\Pi_{\mathcal{C}_n} Y - \Pi_{\mathcal{C}_\infty} Y\| \to 0$$

i. e.

$$\|\Pi_{\mathcal{C}_n} Y - Y\| \to 0 \, .$$

Hence we can find $\mathcal{C}_{n(\varepsilon)} = \mathcal{C}_{n(\varepsilon)}(X)$ such that

$$\|\Pi_{\mathcal{C}_{n(\varepsilon)}} Y - Y\| < \varepsilon \, .$$

Set $\mathcal{F}_\varepsilon = \mathcal{F}_\varepsilon(X) = \mathcal{C}_{n(\varepsilon)}$. If $\mathcal{F}_\varepsilon \subset \mathcal{F} \subset \mathcal{C}$, then $L^2(\mathcal{F}_\varepsilon) \subset L^2(\mathcal{F}) \subset L^2(\mathcal{C})$ and

$$\|\Pi_{\mathcal{F}} X - \Pi_{\mathcal{C}} X\| \le \|\Pi_{\mathcal{F}_\varepsilon} X - \Pi_{\mathcal{C}} X\| = \|\Pi_{\mathcal{F}_\varepsilon} Y - Y\| < \varepsilon \, .$$

Here the first inequality is valid because $Y = \Pi_{\mathcal{C}} X$ and $\Pi_{\mathcal{F}_\varepsilon} Y = \Pi_{\mathcal{F}_\varepsilon} \Pi_{\mathcal{C}} X = \Pi_{\mathcal{F}_\varepsilon} X$ by $\mathcal{F} \subset \mathcal{C}$. This completes the proof.

Now we will consider the case $X \in L^1$. Let $\mathcal{F}$ be finite. Then $E(X \mid \mathcal{F})$ is meaningful and it is easy to see that

$$E(X \mid \mathcal{F}) \in L^1 \, .$$

Write $X_n$ for the truncation $((-n) \vee X) \wedge n$. Then

$$\|X_n - X\|_1 < \varepsilon \qquad (\|\cdot\|_1 = L^1\text{-norm})$$

for $n$ big enough. Let $\{F_i\}$ be the atoms of $\mathcal{F}$ and let $e_i =$ indicator of $F_i$. Then it follows that

$$\begin{aligned}
&\|E(X_n \mid \mathcal{F}) - E(X \mid \mathcal{F})\|_1 \\
&\quad = \int |E(X_n - X \mid \mathcal{F})| P(\mathrm{d}\omega) \\
&\quad = \int \sum_i \frac{1}{P(F_i)} \left| \int_{F_i} (X_n - X) P(\mathrm{d}\omega') \right| e_i(\omega) P(\mathrm{d}\omega) \\
&\quad \le \sum_i \int_{F_i} |X_n - X| P(\mathrm{d}\omega) = \int |X_n - X| P(\mathrm{d}\omega) \\
&\quad = \|X_n - X\|_1 < \varepsilon \, .
\end{aligned}$$

If $\mathcal{F}_\varepsilon \subset \mathcal{F}_1 \subset \mathcal{C}$ and $\mathcal{F}_\varepsilon \subset \mathcal{F}_2 \subset \mathcal{C}$ ($\mathcal{F}_\varepsilon = \mathcal{F}_\varepsilon(X_n)$ as above), we have

$$\begin{aligned}
&\|E(X \mid \mathcal{F}_1) - E(X \mid \mathcal{F}_2)\|_1 \\
&\quad \le \|E(X_n \mid \mathcal{F}_1) - \Pi_{\mathcal{C}} X_n\|_1 + \|E(X_n \mid \mathcal{F}_2) - \Pi_{\mathcal{C}} X_n\|_1 \\
&\qquad + \|E(X_n \mid \mathcal{F}_1) - E(X \mid \mathcal{F}_1)\|_1 + \|E(X_n \mid \mathcal{F}_2) - E(X \mid \mathcal{F}_2)\|_1 \\
&\quad < 4\varepsilon \, .
\end{aligned}$$

This observation justifies the following definition.

**Definition 1.** For $X \in L^1$, $E(X \mid \mathcal{C})$ is defined to be

$$L^1\text{-}\lim_{\substack{\mathcal{F} \uparrow \mathcal{C} \\ \mathcal{F} \text{ finite}}} E(X \mid \mathcal{F}) \,.$$

**Theorem 2.** *$Y = E(X \mid \mathcal{C})$ is characterized by the following two conditions:*

(a) $\qquad Y$ *is measurable* $(\mathcal{C})$,

(b) $\qquad E(Y, C) = E(X, C) \qquad$ *for every* $C \in \mathcal{C}$.

*Proof.* It is obvious that $Y = E(X \mid \mathcal{C})$ satisfies these conditions. If $Z$ satisfies these, then

$$N = \{\omega \colon Y(\omega) < Z(\omega)\} \in \mathcal{C} \,.$$

If $P(N) > 0$, then

$$E(X, N) = E(Y, N) < E(Z, N) = E(X, N) \,,$$

which is a contradiction. Therefore

$$Y \geq Z \qquad \text{a. s.}$$

Similarly

$$Z \geq Y \qquad \text{a. s.}$$

*Remark.* The usual definition of $E(X \mid \mathcal{C})$ is given by the conditions (a) and (b). The existence is proved by the Radon–Nikodym theorem.

The conditional probability of an event $A$ under $\mathcal{C}$ is defined in terms of conditional expectation.

**Definition 2.** The conditional probability $P(A \mid \mathcal{C})$ of $A$ under $\mathcal{C}$ is defined to be $E(e_A \mid \mathcal{C})$.

We will list basic properties of conditional expectation. The properties of conditional probability are derived as special cases. We assume the $X$ in $E(X \mid \mathcal{C})$ is always in $L^1$ and omit the phrase a. s.

**Theorem 3.** *(i)* $E(1 \mid \mathcal{C}) = 1$.

*(ii)* $E(X \mid \mathbf{2}) = E(X)$, *where* $\mathbf{2} =$ *trivial* $\sigma$*-algebra.*

*(iii)* $E(E(X \mid \mathcal{C}_2) \mid \mathcal{C}_1) = E(X \mid \mathcal{C}_1)$ *if* $\mathcal{C}_1 \subset \mathcal{C}_2$.

*(iv)* $E(X \mid \mathcal{C}) = E(X)$ *if* $X$ *is independent of* $\mathcal{C}$.

*(v)* $E(X \mid \mathcal{C}) = X$ *if* $X$ *is measurable* $(\mathcal{C})$.

*(vi)* $E(XY \mid \mathcal{C}) = XE(Y \mid \mathcal{C})$ *if* $X$ *is measurable* $(\mathcal{C})$.

*(vii)* $E(E(X \mid \mathcal{C})) = E(X)$.

**Theorem 4.** *(i) If $X \leq Y$, then $E(X \mid \mathcal{C}) \leq E(Y \mid \mathcal{C})$.*

*(ii) $E(\alpha X + \beta Y \mid \mathcal{C}) = \alpha E(X \mid \mathcal{C}) + \beta E(X \mid \mathcal{C})$.*

*(iii) (Jensen's inequality) If $\varphi$ is convex in $R^1$, and if $\varphi(X) \in L^1$, then*

$$\varphi(E(X \mid \mathcal{C})) \leq E(\varphi(X) \mid \mathcal{C}) .$$

*In particular*

$$|E(X \mid \mathcal{C})| \leq E(|X| \mid \mathcal{C}) .$$

*(iv) If $X_n \to X$ in $L^1$, then $E(X_n \mid \mathcal{C}) \to E(X \mid \mathcal{C})$ in $L^1$.*

*(v) If $0 \leq X_n \uparrow X \in L^1$, then $E(X_n \mid \mathcal{C}) \uparrow E(X \mid \mathcal{C})$.*

*(vi) If $|X_n| \leq Y \in L^1$ and if $X_n \to X$ a. s., then*

$$E(X_n \mid \mathcal{C}) \to E(X \mid \mathcal{C}) .$$

*Proof.* (i), (ii) Trivial.

(iii) Since $\varphi$ is convex, we have an increasing right-continuous function $\alpha(m)$ such that

$$\alpha(m)(x - m) + \varphi(m) \leq \varphi(x) \qquad \text{for every } (x, m) ;$$

in fact $\alpha(m)$ is the right derivative of $\varphi(x)$ at $x = m$. Using this inequality, we have

$$\alpha(m)[E(X \mid \mathcal{C}) - m] + \varphi(m) = E[\alpha(m)(X - m) + \varphi(m) \mid \mathcal{C}] \leq E(\varphi(X) \mid \mathcal{C}) .$$

When $X$ is bounded, the same reasoning can be made for $m = E(X \mid \mathcal{C})$, which gives $\varphi(m) \leq E(\varphi(X) \mid \mathcal{C})$. For general $X$, use approximation.

(iv) Observe that

$$|E(X_n \mid \mathcal{C}) - E(X \mid \mathcal{C})| = |E(X_n - X \mid \mathcal{C})| \leq E(|X_n - X| \mid \mathcal{C}) ,$$
$$E(|E(X_n \mid \mathcal{C}) - E(X \mid \mathcal{C})|) \leq E(|X_n - X|) \to 0 .$$

(v) Observe that $E(X_n \mid \mathcal{C})$ increases as $n \uparrow \infty$. Let $Y$ denote the limit. It follows from

$$E(E(X_n \mid \mathcal{C}), C) = E(X_n, C), \qquad C \in \mathcal{C} ,$$

that

$$E(Y, C) = E(X, C), \qquad C \in \mathcal{C} ,$$

which shows $Y = E(X \mid \mathcal{C})$.

(vi) Set $Z_n = \sup_{k \geq n} |X_k - X| \leq 2Y$. Then $0 \leq 2Y - Z_n \uparrow 2Y$. By (v), $E(2Y - Z_n \mid \mathcal{C}) \uparrow E(2Y \mid \mathcal{C})$, and hence $E(Z_n \mid \mathcal{C}) \downarrow 0$. Thus

$$|E(X_n \mid \mathcal{C}) - E(X \mid \mathcal{C})| = |E(X_n - X \mid \mathcal{C})|$$
$$\leq E(|X_n - X| \mid \mathcal{C}) \leq E(Z_n \mid \mathcal{C}) \to 0 .$$

**Definition 3.** Denote

$$E(X \mid Y) = E(X \mid \mathcal{B}[Y]),$$
$$E(X \mid Y_\lambda, \lambda \in \Lambda) = E(X \mid \mathcal{B}[Y_\lambda : \lambda \in \Lambda]).$$

Similarly for conditional probability.

**Theorem 5.** *Suppose that $X$ and $Y$ are independent and that $\varphi(x,y)$ is a Borel measurable function on $R^2$ such that $E(|\varphi(X,Y)|) < \infty$. Then*[28]

$$E(\varphi(X,Y) \mid Y) = E(\varphi(X,y))_{y=Y}.$$

*Proof.* For $C \in \mathcal{B}[Y]$ we have $C = \{Y \in \Gamma\}$ with some $\Gamma$. Denote $\mu(\cdot) = P(X \in \cdot)$ and $\nu(\cdot) = P(Y \in \cdot)$. Then

$$\begin{aligned} E(\varphi(X,Y),\, C) &= \iint_{x \in R^1,\, y \in \Gamma} \varphi(x,y)\mu(\mathrm{d}x)\nu(\mathrm{d}y) \\ &= \int_{y \in \Gamma} \left[ \int_{x \in R^1} \varphi(x,y)\mu(\mathrm{d}x) \right] \nu(\mathrm{d}y) \\ &= \int_{y \in \Gamma} E(\varphi(X,y))\nu(\mathrm{d}y) = \int_{y \in \Gamma} \psi(y)\nu(\mathrm{d}y) \\ &= E(\psi(Y),\, C) \end{aligned}$$

with $\psi(y) = E(\varphi(X,y))$. Since $\psi(Y)$ is measurable ($\mathcal{B}[Y]$), we have $\psi(Y) = E(\varphi(X,Y) \mid Y)$.

Note that $E(\varphi(X,y))_{y=Y} \neq E(\varphi(X,Y))$ in general.

## 0.7 Martingales

Let $(\Omega, \mathcal{B}, P)$ stand for the basic probability space and $\{\mathcal{B}_n\}$ for a finite or infinite increasing sequence of sub-$\sigma$-algebras of $\mathcal{B}$. We write $X \in \mathcal{B}_n$, meaning that $X$ is measurable ($\mathcal{B}_n$). A sequence $\{X_n\}$ is said to be *adapted* to $\{\mathcal{B}_n\}$ if $X_n \in \mathcal{B}_n$ for every $n$. We will omit the phrase a. s. if there is no possibility of confusion.

### a. Definition and elementary properties.

**Definition 1.** Let $\{X_n\}$ be a sequence with $E|X_n| < \infty$ for every $n$.

(i) $\{X_n\}$ is called a *submartingale* if

$$E(X_{n+1} \mid \mathcal{B}[X_1, \ldots, X_n]) \geq X_n, \qquad n = 1, 2, \ldots.$$

[28] Meaning: $E(\varphi(X,y))_{y=Y} = \psi(Y)$, where $\psi(y) = E(\varphi(X,y))$.

(ii) $\{X_n\}$ is called a *submartingale* (relative to) $\{\mathcal{B}_n\}$ if $\{X_n\}$ is adapted to $\{\mathcal{B}_n\}$ and if

$$E(X_{n+1} \mid \mathcal{B}_n) \geq X_n, \qquad n = 1, 2, \ldots .$$

By replacing $\geq$ with $\leq$ or $=$, we define a *supermartingale* or a *martingale*, respectively.

The case (i) is a special case of (ii) with $\mathcal{B}_n = \mathcal{B}[X_1, X_2, \ldots, X_n]$. If $\{X_n\}$ is a submartingale (supermartingale, martingale) relative to $\{\mathcal{B}_n\}$ in the sense of (ii), $\{X_n\}$ is so in the sense of (i).

It follows at once from the definition that

(i) $\{X_n\}$ is a supermartingale $\iff$ $\{-X_n\}$ is a submartingale,

(ii) $\{X_n\}$ is a martingale

$\iff$ $\{X_n\}$ and $\{-X_n\}$ are submartingales

$\iff$ $\{X_n\}$ is a submartingale as well as a supermartingale.

Because of this fact every property of supermartingales or martingales follows at once from its corresponding property of submartingales. We will therefore state mainly the properties of submartingales.

If $\{X_n\}_{n=1}^m$ is a submartingale $\{\mathcal{B}_n\}_{n=1}^m$ $(m < \infty)$, then $\{X_{n \wedge m}\}_{n=1,2,\ldots}$ is a submartingale $\{\mathcal{B}_{n\wedge m}\}_{n=1,2,\ldots}$. It is therefore enough to discuss only the infinite sequence case.

**Example 1.** Let $\{Y_n\}_n$ be independent and $EY_n = 0$. Set $X_n = \sum_1^n Y_i$. Then (i) $\{X_n\}$ is a martingale, (ii) $\{X_n^2\}$ is a submartingale.

**Example 2.** If $E|X| < \infty$, then $\{X_n \equiv E(X \mid \mathcal{B}_n)\}_n$ is a martingale $\{\mathcal{B}_n\}$.

**Example 3.** $X_n \equiv$ const is a martingale $\{\mathcal{B}_n\}$.

**Theorem 1.** *(i) $\{X_n\}$ is a submartingale $\{\mathcal{B}_n\}$ if and only if*

$$E(X_{n+1}, B) \geq E(X_n, B) \qquad \textit{for all } B \in \mathcal{B}_n \textit{ and } n = 1, 2, \ldots .$$

*(ii) If $\{X_n\}$ is a submartingale $\{\mathcal{B}_n\}$, then*

$$E(X_{n+k} \mid \mathcal{B}_n) \geq X_n \qquad (k = 0, 1, 2, \ldots, \ n = 1, 2, \ldots)$$

*and*

$$E(X_1) \leq E(X_2) \leq \cdots .$$

*(iii) If $\{X_n\}$, $\{Y_n\}$ are submartingales $\{\mathcal{B}_n\}$, then $\{aX_n + bY_n + c\}$ is a submartingale $\{\mathcal{B}_n\}$ provided $a, b \geq 0$ and $c$ is real.*

*If $\{X_n\}$, $\{Y_n\}$ are martingales $\{\mathcal{B}_n\}$, then $\{aX_n + bY_n + c\}$ is a martingale $\{\mathcal{B}_n\}$ provided $a, b, c$ are real.*

*(iv) If $\{X_n\}$ is a submartingale $\{\mathcal{B}_n\}$, if $\varphi$ is convex and increasing and if $E|\varphi(X_n)| < \infty$, then $\{\varphi(X_n)\}_n$ is a submartingale. For example: $\varphi(\xi) = \xi^+$, $\xi \vee a$, $e^\xi$.*

*If $\{X_n\}$ is a martingale $\{\mathcal{B}_n\}$, if $\varphi$ is convex and $E|\varphi(X_n)| < \infty$, then $\{\varphi(X_n)\}_n$ is a submartingale. Note that $\varphi$ is not assumed increasing here. For example: $\varphi(\xi) = \xi^+$, $\xi \vee a$, $e^\xi$, $|\xi|^p$ $(p \geq 1)$.*

*Proof.* (i) Obvious by the definition of conditional expectation.

(ii) Observe that

$$\begin{aligned} X_n &\leq E(X_{n+1} \mid \mathcal{B}_n) \\ &\leq E(E(X_{n+2}|\mathcal{B}_{n+1})|\mathcal{B}_n) = E(X_{n+2} \mid \mathcal{B}_n) \\ &\leq \cdots \\ &\leq E(X_{n+k} \mid \mathcal{B}_n) \end{aligned}$$

and

$$E(X_{n+k}) = E(E(X_{n+k} \mid \mathcal{B}_n)) \geq E(X_n)\,.$$

(iii) Obvious by the linearity of conditional expectation.

(iv) Use Jensen's inequality for conditional expectation.

**b. Time change.**

**Definition 2.** A random variable $T$ with values in $\{1, 2, 3, \ldots, \infty\}$ is called a *stopping time* (relative to) $\{\mathcal{B}_n\}$ if

$$\{T \leq n\} \in \mathcal{B}_n \text{ for every } n\,.$$

Note that

$$\begin{aligned} \{T \leq n\} \in \mathcal{B}_n \text{ for every } n \quad &\iff \quad \{T = n\} \in \mathcal{B}_n \text{ for every } n \\ &\iff \quad \{T > n\} \in \mathcal{B}_n \text{ for every } n\,. \end{aligned}$$

**Example 4.** $T \equiv n$ (*deterministic time*) is a stopping time $\{\mathcal{B}_n\}$.

**Example 5.** Let $\{X_n\}$ be adapted to $\{\mathcal{B}_n\}$ and $E$ be a Borel set in $R^1$. Set

$$T_E = \begin{cases} \min\, \{k \colon X_k \in E\} & \text{if there is such } k \\ \infty & \text{if otherwise}\,. \end{cases}$$

Then $T_E$ is a stopping time $\{\mathcal{B}_n\}$ called the *entrance time* to $E$.

*Proof.* $\{T_E \leq n\} = \bigcup_{k=1}^n \{X_k \in E\} \in \mathcal{B}_n$.

**Theorem 2.** *If $T$ and $S$ are stopping times $\{\mathcal{B}_n\}$, then $T \vee S$, $T \wedge S$ and $T + S$ are stopping times $\{\mathcal{B}_n\}$.*[29]

**Definition 3.** Define

$$\begin{aligned} \mathcal{B}_T &= \{B \in \mathcal{B} \colon B \cap \{T \leq n\} \in \mathcal{B}_n\} \\ &= \{B \in \mathcal{B} \colon B \cap \{T = n\} \in \mathcal{B}_n\}\,. \end{aligned}$$

This notation is consistent with $\mathcal{B}_n$, because $\mathcal{B}_T = \mathcal{B}_n$ for $T \equiv n$ (deterministic time).

[29] Proof is as in the solution of Problem 2.6.

**Theorem 3.** *Let $T$ and $S$ be stopping times $\{\mathcal{B}_n\}$.*
*(i) The class $\mathcal{B}_T$ is a sub-$\sigma$-algebra of $\mathcal{B}$, and $T \in \mathcal{B}_T$.*
*(ii) $\mathcal{B}_S \subset \mathcal{B}_T$ for $S \le T$.*
*(iii) If $\{X_n\}$ is adapted to $\{\mathcal{B}_n\}$, then $X_T \in \mathcal{B}_T$.* [30]
*(iv) $\{T < S\}, \{T \le S\}, \{T = S\} \in \mathcal{B}_T \wedge \mathcal{B}_S$.*

*Proof.* (i) To see that $T \in \mathcal{B}_T$, observe

$$\{T = k\} \cap \{T \le n\} = \begin{cases} \{T = k\} \in \mathcal{B}_n & \text{for } k \le n \\ \emptyset \in \mathcal{B}_n & \text{for } k > n \,. \end{cases}$$

(ii) Obvious from definition.
(iii) $\{X_T \le a\} \cap \{T = n\} = \{X_n \le a\} \cap \{T = n\} \in \mathcal{B}_n$.
(iv) Since

$$\{T < S\} \cap \{T = n\} = \{S > n\} \cap \{T = n\} \in \mathcal{B}_n$$

and

$$\begin{aligned} \{T < S\} \cap \{S = n\} &= \{T < n\} \cap \{S = n\} \\ &= \{T \le n-1\} \cap \{S = n\} \in \mathcal{B}_n \,, \end{aligned}$$

we see that $\{T < S\} \in \mathcal{B}_S \wedge \mathcal{B}_T$. It follows that $\{T \le S\} = \{S < T\}^{\mathrm{C}} \in \mathcal{B}_S \wedge \mathcal{B}_T$ and hence $\{T = S\} = \{T \le S\} - \{T < S\} \in \mathcal{B}_S \wedge \mathcal{B}_T$.

**Definition 4.** Let $\{T_k\}$ be an increasing sequence of finite stopping times $\{\mathcal{B}_n\}$. The operation

$$\{X_n\} \longrightarrow \{X_{T_n}\}$$

is called *time change*, where $\{X_n\}$ is adapted to $\{\mathcal{B}_n\}$. (Note that $\{X_{T_n}\}$ is adapted to $\{\mathcal{B}_{T_n}\}$ by Theorem 3 (iii).)

**Theorem 4.** *Submartingale property (martingale property, supermartingale property as well) are preserved by time change, provided $T_n(\omega)$ is uniformly bounded in $(n, \omega)$.*

*Proof.* Let $\{X_n\}$ be a submartingale $\{\mathcal{B}_n\}$, and $\{T_n\}$ be an increasing sequence of stopping times $\{\mathcal{B}_n\}$ such that $T_n(\omega) \le m$, $m$ being a finite constant. First observe that

$$E|X_{T_n}| \le \sum_1^m E|X_k| < \infty \,.$$

[30] Since $X_\infty$ is undefined, $X_T$ is defined only on $\{T < \infty\}$. The assertion $X_T \in \mathcal{B}_T$ is understood as $\{X_T \le a,\, T < \infty\} \in \mathcal{B}_T$ for every $a$.

It is sufficient to prove that

$$E(X_{T_{n+1}}, B) \geq E(X_{T_n}, B) \qquad \text{for } B \in \mathcal{B}_{T_n} \ .$$

Write $S$ for $T_n$ and $T$ for $T_{n+1}$. Then

$$m \geq T \geq S \geq 1 \ .$$

Take an arbitrary $B \in \mathcal{B}_S$ and write $B_k$ for $B \cap \{S = k\}$. If $k \leq l \leq m$, then $B_k \cap \{T \leq l\} \in \mathcal{B}_k \vee \mathcal{B}_l \subset \mathcal{B}_l$, $B_k \cap \{T > l\} \in \mathcal{B}_k \vee \mathcal{B}_l \subset \mathcal{B}_l$ , and

$$\begin{aligned} E(X_{T\wedge(l+1)}, B_k) &= E(X_T, B_k \cap \{T \leq l\}) + E(X_{l+1}, B_k \cap \{T > l\}) \\ &\geq E(X_T, B_k \cap \{T \leq l\}) + E(X_l, B_k \cap \{T > l\}) \\ &\quad \text{(because } \{X_n\} \text{ is a submartingale } \{\mathcal{B}_n\} \text{ and } B_k \cap \{T > l\} \in \mathcal{B}_l) \\ &= E(X_{T\wedge l}, B_k) \ . \end{aligned}$$

Thus $E(X_{T\wedge l}, B_k)$ is increasing with $l = k, k+1, \ldots, m$, so that

$$E(X_{T\wedge m}, B_k) \geq E(X_{T\wedge k}, B_k) \ .$$

Since $T \wedge m = T$ by $T \leq m$ and $T \wedge k = k = S$ on $B_k$ by $T \geq S$, we see that

$$E(X_T, B_k) \geq E(X_S, B_k) \ .$$

Adding both sides over $k$, we obtain

$$E(X_T, B) \geq E(X_S, B), \qquad B \in \mathcal{B}_S \ .$$

That is,

$$E(X_{T_{n+1}}, B) \geq E(X_{T_n}, B), \qquad B \in \mathcal{B}_{T_n} \ .$$

This proves that $\{X_{T_n}\}_n$ is a submartingale $\{\mathcal{B}_{T_n}\}$.

**c. Fundamental inequalities.**

**Theorem 5.** *Let $\{X_n\}_n$ be a submartingale and $c > 0$. Then*

$$c\,P\left(\sup_n X_n > c\right) \leq \sup_n E[X_n^+]$$

*and*

$$c\,P\left(\inf_n X_n \leq -c\right) \leq \sup_n E[X_n^+] - E[X_1] \ .$$

*Proof.* Write $\mathcal{B}_n$ for $\mathcal{B}[X_1, \ldots, X_n]$. Let

$$\begin{aligned} A_1 &= \{X_1 > c\} \in \mathcal{B}_1 \ , \\ A_2 &= \{X_1 \leq c,\ X_2 > c\} \in \mathcal{B}_2 \ , \\ &\cdots \\ A_n &= \{X_1, X_2, \ldots, X_{n-1} \leq c,\ X_n > c\} \in \mathcal{B}_n \ . \end{aligned}$$

Then

$$
\begin{aligned}
E(X_n^+) &\geq \sum_{k=1}^{n} E(X_n^+,\, A_k) && (\{A_k\} \text{ disjoint}) \\
&\geq \sum_{k=1}^{n} E(X_k^+,\, A_k) && (\{X_n^+\} \text{ submartingale}) \\
&\geq c \sum_{k=1}^{n} P(A_k) \\
&= c\, P\left(\max_{1\leq k\leq n} X_k > c\right) .
\end{aligned}
$$

Let $n \uparrow \infty$ to get the first inequality. Similarly for the second inequality.

Let $f(n)$ be a function of $n$. The *upcrossing number* of $[a, b]$ for $f$ is defined to be the supremum of $m$ for which there exists $n_1 < n_2 < \cdots < n_{2m}$ such that $f(n_{2i-1}) < a$ and $f(n_{2i}) > b$ for $i = 1, 2, \ldots m$. This number is denoted by $U_{ab}(f)$. The upcrossing number of $[a, b]$ for $f(n)$, $n \leq N$, is denoted by $U_{ab}^N(f)$.

**Theorem 6.** *Let $\{X_n\}$ be a submartingale. Then*

$$
(b-a)E[U_{ab}(X)] \leq \sup_n E[(X_n - a)^+] .
$$

*Proof.* We will define an increasing sequence of stopping times $\{\mathcal{B}_n \equiv \mathcal{B}[X_1, \ldots, X_n]\}$:

$$
T_1 = \begin{cases} \min\{n \leq 2m\colon X_n < a\} \\ 2m & \text{if there is no such } n, \end{cases}
$$

$$
T_2 = \begin{cases} \min\{n \leq 2m\colon n > T_1,\, X_n > b\} \\ 2m & \text{if there is no such } n, \end{cases}
$$

$$
T_{2k-1} = \begin{cases} \min\{n \leq 2m\colon n > T_{2k-2},\, X_n < a\} \\ 2m & \text{if there is no such } n, \end{cases}
$$

$$
T_{2k} = \begin{cases} \min\{n \leq 2m\colon n > T_{2k-1},\, X_n > b\} \\ 2m & \text{if there is no such } n\ , \end{cases}
$$

where $m$ is any fixed number $< \infty$. Then $T_1 \leq T_2 \leq \cdots \leq 2m < \infty$. Now let $Y(n) = X_n \vee a$ and observe

$$
\begin{aligned}
(b-a)\,U_{ab}^{2m}(X) &\le \sum_{k=1}^{m}(Y(T_{2k})-Y(T_{2k-1})) \\
&= Y(T_{2m})-Y(T_1)+\sum_{k=1}^{m-1}(Y(T_{2k})-Y(T_{2k+1})) \\
&\le Y(m)-a+\sum_{k=1}^{m-1}(Y(T_{2k})-Y(T_{2k+1}))\,.
\end{aligned}
$$

Since $\{X_n\}$ is a submartingale, so is $\{Y(n)\}$. Therefore

$$E(Y(T_{2k})-Y(T_{2k+1}))\le 0\,.$$

Using $Y(m)-a=X_m\vee a-a=(X_m-a)^+$, we get

$$(b-a)E[U_{ab}^{2m}(X)]\le E[(X_m-a)^+]\le \sup_n E[(X_n-a)^+]\,.$$

Finally, let $m\uparrow\infty$.

**d. Case of continuous time.**

Martingales, submartingales and supermartingales with continuous time parameter can be defined exactly in the same way as in the discrete time case: $E(X_t \mid \mathcal{B}_s)\ge X_s$ $(t>s)$ in place of $E(X_{n+1}\mid\mathcal{B}_n)\ge X_n$.

Let $X_t$ be a martingale with continuous time.[31] Then for $t_1<t_2<\ldots<t_n$, $\{Y_k=X(t_k)\}_k$ is a martingale with discrete time. Similarly for sub- or supermartingales.

The following theorem will be most useful later.

**Theorem 7 (Doob).** *Let $\{X_t\}_{t\ge 0}$ be a submartingale (or supermartingale or martingale) and suppose that it is continuous in probability. Then there exists a stochastic process $\{Y_t(\omega)\}_{t\ge 0}$ with the following properties:*

*(a) almost every sample path of $\{Y_t\}$ is right continuous and has left limit at every $t$;*

*(b) $P(X_t=Y_t)=1$ for each $t$.*

*This $\{Y_t\}$ is called the* Doob modification *of $\{X_t\}$.*

*Proof.* Use Theorems 5 and 6 to prove this theorem in the same way as in the proof of a similar fact for additive processes (Section 1.3).

N. B. See Problems 0.34–0.39 in Exercises for other important properties of martingales and submartingales.

[31] $X_t$ is sometimes written as $X(t)$.

# 1 Additive Processes (Processes with Independent Increments)

## 1.1 Definitions

Let $(\Omega, \mathcal{B}, P)$ stand for the basic probability space as before. We assume that $\mathcal{B}$ is *complete* with respect to $P$, namely every subset of $N \in \mathcal{B}$ with $P(N) = 0$ belongs to $\mathcal{B}$ and so automatically has $P$-measure (= probability) 0. Let $\omega$ stand for a generic element of $\Omega$.

The *closure* of a sub-$\sigma$-algebra $\mathcal{F}$ of $\mathcal{B}$ is denoted by $\overline{\mathcal{F}}$, which consists of all sets in $\mathcal{B}$ that differ from sets in $\mathcal{F}$ by sets in $\mathcal{B}$ of probability 0. The closure $\overline{\mathcal{F}}$ should be distinguished from the completion of $\mathcal{F}$ with respect to the restriction of $P$ to $\mathcal{F}$; the latter is smaller than the former $\overline{\mathcal{F}}$ in general. We write "$\mathcal{F}_1 = \mathcal{F}_2$ a. s." to indicate $\overline{\mathcal{F}}_1 = \overline{\mathcal{F}}_2$. It is obvious from the definition that if $\{\mathcal{B}_\lambda\}$ is independent, so is $\{\overline{\mathcal{B}}_\lambda\}$.

A stochastic process $\{X_t(\omega)\}_{t\in T}$ is a family of real random variables indexed with a time parameter $t$. Throughout this chapter, $T$ is assumed to be $[0, \infty)$ unless stated otherwise. Sometimes we write $X_t(\omega)$ as $X_t$, $X(t)$, or $X(t, \omega)$.

A stochastic process $\{X_t\}_{t\in T}$ is called an *additive process* or a *process with independent increments* if $X_0 = 0$ a. s. and if for every $n$ and every $\{t_0 < t_1 < \cdots < t_n\} \subset T$ the family $\{X(t_i) - X(t_{i-1})\}_i$ is independent. The concept of additive processes is a *continuous time version* of that of sums of independent variables.

Let $\{\mathcal{B}_{st} : s < t,\ s, t \in T\}$ be a family of $\sigma$-algebras. It is called *additive* if the following two conditions are satisfied:

(a) $\quad \mathcal{B}_{st} \vee \mathcal{B}_{tu} = \mathcal{B}_{su}$ a. s. for $s < t < u$,

(b) $\quad \{\mathcal{B}_{t_0t_1}, \mathcal{B}_{t_1t_2}, \ldots, \mathcal{B}_{t_{n-1}t_n}\}$ is independent for every $n$ and every $\{t_0 < t_1 < \cdots < t_n\} \subset T$.

If $\{\mathcal{B}_{st}\}_{s,t}$ is additive, so is $\{\overline{\mathcal{B}}_{st}\}_{s,t}$ .

For a stochastic process $\{X_t\}_{t\in T}$ we will introduce the *differential $\sigma$-algebras*

$$\mathcal{B}_{st}[\mathrm{d}X] = \mathcal{B}[X_v - X_u : s \le u < v \le t], \qquad 0 \le s < t < \infty .$$

A stochastic process $\{X_t\}_{t\in T}$ is called *adapted* to an additive family $\{\mathcal{B}_{st}\}_{s,t}$ if $X_t - X_s$ is measurable $(\mathcal{B}_{st})$ for $s < t$. In this situation $\{X_t\}$

is obviously an additive process and

$$\mathcal{B}_{st}[\mathrm{d}X] \subset \mathcal{B}_{st} ,$$

provided that $X_0 = 0$ a. s.

**Theorem 1.** *If $\{X_t\}_t$ is an additive process, then $\{\mathcal{B}_{st}[\mathrm{d}X]\}_{s,t}$ is additive and $\{X_t\}_t$ is adapted to $\{\mathcal{B}_{st}[\mathrm{d}X]\}_{s,t}$ .*

*Proof.* Write $\mathcal{B}_{st}$ for $\mathcal{B}_{st}[\mathrm{d}X]$. We will verify (b) for $\mathcal{B}_{st}$; the rest will be trivial. Fix $i$ for the moment and consider for every finite subset $F$ of $[t_{i-1}, t_i]$ the following $\sigma$-algebra

$$\mathcal{B}(F) = \mathcal{B}[X_v - X_u \colon u, v \in F] .$$

Then $\{\mathcal{B}(F)\}_F$ is directed up because $F \subset F'$ implies $\mathcal{B}(F) \subset \mathcal{B}(F')$. It is obvious that

$$\bigvee_F \mathcal{B}(F) = \mathcal{B}_{t_{i-1}t_i}[\mathrm{d}X] .$$

By Theorem 0.1.3 it remains only to prove that for every choice of a finite subset $F_i$ of $[t_{i-1}, t_i]$, $i = 1, 2, \ldots, n$, $\{\mathcal{B}(F_i)\}_i$ is independent. Let

$$F_i = \{t_{i0} < t_{i1} < \cdots < t_{i,m(i)}\}$$

and set

$$X_{ij} = X(t_{ij}) - X(t_{i,j-1}) .$$

Then $\{\mathcal{B}[X_{ij}]\}_{i,j}$ is independent by the additivity of the process $\{X_t\}$ and we have

$$\mathcal{B}(F_i) = \bigvee_j \mathcal{B}[X_{ij}], \qquad i = 1, 2, \ldots, n .$$

Therefore $\{\mathcal{B}(F_i)\}_i$ is also independent by Theorem 0.1.4 (i).

**Theorem 2.** *Suppose that $\{\mathcal{B}^i_{st}\}_{s,t}$ , $i = 1, 2, \ldots, n$, are all additive and that $\{\bigvee_{s,t} \mathcal{B}^i_{st}\}_i$ is independent. Then the family*

$$\mathcal{B}_{st} = \bigvee_{i=1}^{n} \mathcal{B}^i_{st} , \qquad s < t ,$$

*is also additive.*

*Proof.* Use Theorems 0.1.1 and 0.1.4.

**Theorem 3.** *Every linear combination of independent additive processes is additive.*

*Proof.* Let $\{X_t^i\}_t$, $i = 1, 2, \ldots, n$, be an independent family of additive processes and let

$$X_t = \sum_{i=1}^n c_i X_t^i \qquad (c_i = \text{constant}) .$$

By Theorem 2 the family

$$\mathcal{B}_{st} = \bigvee_{i=1}^n \mathcal{B}_{st}[\mathrm{d}X^i], \qquad s < t ,$$

is additive and $\{X_t\}$ is obviously adapted to this family. Therefore $\{X_t\}$ is additive.

If $\{X_n, Y_n, \ldots, W_n\}$ is independent for each $n$ and if $X_n \to X$, $Y_n \to Y$, ..., $W_n \to W$ i. p., then $\{X, Y, \ldots, W\}$ is also independent. Therefore we have

**Theorem 4.** *The limit in probability (for each t) of a sequence of additive processes is also additive.*

**Example 1.** A trivial example of additive processes is a *deterministic process* $X_t(\omega) \equiv f(t)$.

**Example 2.** The Brownian motion is additive.[1]

**Example 3.** The Poisson process is additive.

## 1.2 Decomposition of Additive Processes

In this section a stochastic process $\{X_t\}_{t \in T}$ is considered as a funtion

$$t \in T \longrightarrow X_t \in \mathcal{R}$$

where $\mathcal{R}$ is the set of all random variables modulo *trivial random variables* (= random variables equal to 0 a. s.). The continuity of $X_t$ in $t$ is the continuity in probability (= *stochastic continuity*), namely the continuity with respect to the metric

$$d(X, Y) = \|X - Y\| = E\left[\frac{|X - Y|}{1 + |X - Y|}\right] .$$

This is different from the continuity of the sample functions (= sample paths = sample processes). For example the Poisson process $\{X_t\}$ has discontinuous (in fact step-wise) sample paths but is continuous i. p. by virtue of

[1] Definitions of the Brownian motion and the Poisson process will be given at the end of Section 1.4.

$$P(|X_t - X_s| < \varepsilon) = P(X_t - X_s = 0) = \mathrm{e}^{-(t-s)} \quad (0 < \varepsilon < 1,\ s < t)\,.$$

Properties of the sample processes will be discussed in the subsequent sections. The main purpose of this section is to prove that every additive process is the sum of three independent parts, i. e. the *deterministic part*, the *discontinuous part* and the *continuous part.*

An additive process $\{X_t\}$ is called *centralized* if $\gamma(X_t) \equiv 0$. Every additive process is expressed as its centralized part $\{X_t - \gamma(X_t)\}_t$ plus a deterministic process $\{\gamma(X_t)\}_t$. Thus the study of additive processes is reduced to that of centralized ones.

**Theorem 1.** *If $\{X_t\}$ is a centralized additive process, then the right limit $X_{t+} = \operatorname*{l.i.p.}_{s \to t+} X_s$ $(t \geq 0)$ and the left limit $X_{t-} = \operatorname*{l.i.p.}_{s \to t-} X_s$ $(t > 0)$ both exist.*[2]

*Proof.* If suffices to show that for $s_1 < s_2 < \cdots \to t$ (or $s_1 > s_2 > \cdots \to t$) $\{X_{s_n}\}$ is convergent i. p.

Consider the case $s_1 < s_2 < \cdots \to t$. Then $X_{s_n}$ is the $n$-th partial sum of an independent sequence

$$X_{s_1},\, X_{s_2} - X_{s_1},\, X_{s_3} - X_{s_2},\, \ldots\,.$$

Since $\delta(X_{s_n}) \leq \delta(X_t) < \infty$ and $\gamma(X_{s_n}) = 0$ for every $n$, $\{X_{s_n}\}$ is convergent a. s. by Theorem 0.3.2 (i), a fortiori convergent i. p.

To discuss the case $s_1 > s_2 > \cdots \to t$, we will consider the independent sequence

$$X_{s_1} - X_{s_2},\, X_{s_2} - X_{s_3},\, \ldots\,.$$

Its $n$-th partial sum is $X_{s_1} - X_{s_n}$ and

$$\delta(X_{s_1} - X_{s_n}) \leq \delta(X_{s_1}) < \infty, \qquad n = 1, 2, \ldots\,.$$

Therefore $\{X_{s_1} - X_{s_n} - c_n\}_n$ $(c_n = \gamma(X_{s_1} - X_{s_n}))$ converges a. s. to a random variable $Y$. Therefore $\{X_{s_n} + c_n\}_n$ converges a. s. to $X_{s_1} - Y$. This implies that

$$\gamma(X_{s_n}) + c_n \to \gamma(X_{s_1} - Y) = c\,.$$

Since $\gamma(X_{s_n}) = 0$, we have $c_n \to c$. Therefore $X_{s_n} \to X_{s_1} - Y - c$ a. s. and so i. p. This completes the proof.

*Remark.* The proof above shows that $X_{s_n} \to X_{t-}$ a. s. for $s_n \uparrow t$, but the exceptional $\omega$-set may vary with $\{s_n\}$. The proof does not claim that a. s. $\lim_{s \uparrow t} X_s$ exists. In fact the $\omega$-set for which $X_s(\omega)$ is convergent as $s \uparrow t$ may not be even measurable, unless we restrict the process as we will do in the subsequent sections. Similarly for $X_{t+}$.

---

[2] l.i.p. = limit in probability.

**Theorem 2.** *Suppose that $\{X_t\}$ is a centralized additive process. For $t_0 < t_1 < \cdots < t_n$ the following family is independent:*

$$X_{t_0},\, X_{t_0+} - X_{t_0},\, X_{t_1-} - X_{t_0+},\, X_{t_1} - X_{t_1-},\, X_{t_1+} - X_{t_1},$$
$$X_{t_2-} - X_{t_1+},\, X_{t_2} - X_{t_2-},\, X_{t_2+} - X_{t_2},\, \ldots,\, X_{t_n+} - X_{t_n}\,.$$

*Proof.* It suffices to recall the fact that if $\{X_n, Y_n, \ldots, W_n\}$ is independent for every $n$ and if $X_n \to X$, $Y_n \to Y, \ldots, W_n \to W$ i. p., then $\{X, Y, \ldots, W\}$ is also independent.

**Theorem 3.** $\gamma(X_{t\pm}) = 0$ *and* $\delta(X_{t\pm}) = \lim_{s\to t\pm} \delta(X_s)$.

*Proof.* Obvious from the continuity of $\gamma$ and $\delta$.

Since $\delta_X(t) = \delta(X_t)$ increases with $t$, its discontinuity set $D = D(X)$ is at most countable. The set $D$ is divided into three parts:

$$D_+ = \{t\colon \delta_X(t-) = \delta_X(t) < \delta_X(t+)\}\,,$$
$$D_- = \{t\colon \delta_X(t-) < \delta_X(t) = \delta_X(t+)\}\,,$$
$$D_\pm = \{t\colon \delta_X(t-) < \delta_X(t) < \delta_X(t+)\}\,.$$

**Theorem 4.** *Let $X_t$ be a centralized additive process. Then $X_t$ is continuous i. p. at every point in $T - D$ ; discontinuous i. p. at every point in $D$; the right limit $X_{t+}$ and the left limit $X_{t-}$ both exist at every $t \in T$. We have*
*(i) $X_{t-} = X_t$ and $X_t \neq X_{t+}$ for $t \in D_+$,*
*(ii) $X_{t-} \neq X_t$ and $X_t = X_{t+}$ for $t \in D_-$,*
*(iii) $X_{t-} \neq X_t$, $X_t \neq X_{t+}$, and $X_{t-} \neq X_{t+}$ for $t \in D_\pm$.*

(Notice that both equality and inequality are to be understood in $\mathcal{R}$, so that $X_{t-} = X_t$ means $P(X_{t-} = X_t) = 1$ and $X_{t-} \neq X_t$ means $P(X_{t-} \neq X_t) > 0$.)

*Proof.* If $t \in T - D$, then

$$\delta(X_{t-} + (X_t - X_{t-})) = \delta(X_t) = \delta_X(t) = \delta_X(t-) = \delta(X_{t-})\,.$$

Since $\{X_{t-},\, X_t - X_{t-}\}$ is independent by Theorem 2, we can apply Theorem 0.3.1 to get $X_t - X_{t-} = c$ (constant) a. s. Then $\gamma(X_t) = \gamma(X_{t-}) + c$ by Theorem 0.2.1. Since $\gamma(X_t) = \gamma(X_{t-}) = 0$, we have $c = 0$ i. e. $X_t = X_{t-}$ a. s. Similarly $X_{t+} = X_t$ a. s. for $t \in T - D$. The rest of the proof can be carried out by the same argument.

The quantities $J_{t+} \equiv X_{t+} - X_t$ and $J_{t-} \equiv X_t - X_{t-}$ are called respectively the *right jump* and the *left jump* at $t$; both will vanish at $t \in T - D$. We will sum up the jumps to define the discontinuous part $\{Y_t\}$ of $\{X_t\}$, namely

$$Y_0 = 0$$

$$Y_t = J_{0+} \overset{\circ}{+} \sum_{0<s<t}^{\circ} (J_{s-} + J_{s+}) \overset{\circ}{+} J_{t-} \qquad (t > 0) ,$$

where the little circles in $\overset{\circ}{+}$ and $\sum^{\circ}$ indicate the *centralized sum* (see Section 0.3). The summands are independent and the dispersion of the finite partial sum is bounded by $\delta_X(t)$ by virtue of Theorem 2 and Theorem 0.3.1, so that $Y_t$ is well-defined.

**Theorem 5.** *$\mathcal{B}_{st}[\mathrm{d}Y] \subset \overline{\mathcal{B}}_{st}[\mathrm{d}X]$, $s < t$, and so $\{Y_t\}$ is a centralized additive process.*

*Proof.* Observe that

$$\mathcal{B}[J_{u+}] \subset \overline{\mathcal{B}}_{st}[\mathrm{d}X] \qquad \text{for } s \leq u < t$$

and

$$\mathcal{B}[J_{u-}] \subset \overline{\mathcal{B}}_{st}[\mathrm{d}X] \qquad \text{for } s < u \leq t .$$

The continuous part $\{Z_t\}$ of $\{X_t\}$ is defined by

$$Z_t = X_t - Y_t - \gamma(X_t - Y_t) .$$

**Theorem 6.** *The process $\{Z_t\}$ is a centralized additive process continuous i. p. and*

$$\mathcal{B}_{st}[\mathrm{d}Z] \subset \overline{\mathcal{B}}_{st}[\mathrm{d}X], \qquad s < t .$$

*Proof.* Everything except the continuity i. p. is obvious. By Theorem 0.3.4 we get

$$Y_{t-} = J_{0+} \overset{\circ}{+} \sum_{0<s<t}^{\circ} (J_{s-} + J_{s+}) ,$$

$$Y_{t+} = J_{0+} \overset{\circ}{+} \sum_{0<s\leq t}^{\circ} (J_{s-} + J_{s+}) .$$

Therefore

$$(1) \qquad Y_t = Y_{t-} \overset{\circ}{+} J_{t-} \sim Y_{t-} + (X_t - X_{t-}) ,$$

where $X \sim Y$ denotes $X = Y +$ const. Since $\{Z_t\}$ is a centralized additive process, $Z_{t-}$ is well-defined. By the definition of $\{Z_t\}$ we have, by (1),

$$Z_t - Z_{t-} \sim (X_t - Y_t) - (X_{t-} - Y_{t-}) \sim 0 ,$$

namely

$$Z_t = Z_{t-} + \text{const.}$$

Since $\{Z_t\}$ is centralized, we have $\gamma(Z_t) = \gamma(Z_{t-}) = 0$, so that the constant must be zero. This proves $Z_t = Z_{t-}$ (precisely $Z_t = Z_{t-}$ a. s.). Similarly $Z_t = Z_{t+}$ a. s.

**Theorem 7.** *The processes $\{Y_t\}$ and $\{Z_t\}$ are independent.*

*Proof.* Let $\{s_1, s_2, \ldots\}$ be an arrangement of $D$. We define $Y_t^{(n)}$ in the same way as $Y_t$, by using only the jumps corresponding to the points $\{s_1, s_2, \ldots, s_n\}$. Set

$$Z_t^{(n)} = X_t - Y_t^{(n)} - \gamma(X_t - Y_t^{(n)}) .$$

It is easy to see by Theorem 2 that $\{Y_t^{(n)}\}$ and $\{Z_t^{(n)}\}$ are independent for each $n$. Since $Y_t^{(n)} \to Y_t$ and $Z_t^{(n)} \to Z_t$ as $n \to \infty$, $\{Y_t\}$ and $\{Z_t\}$ are also independent.

Putting together all results, we have the decomposition of $\{X_t\}$ in three independent parts.

$$X_t = f(t) + Y_t + Z_t ,$$

where

$f(t) = \gamma(X_t)$ (deterministic),
$Y_t =$ the centralized sum of jumps of $\{X_t - \gamma(X_t)\}$,
$Z_t$ is a centralized additive process continuous i. p.

Since the structure of $f(t)$ is trivial and since that of $\{Y_t\}$ is rather simple, we are led to the study of additive processes continuous i. p. The rest of this chapter will be devoted to such processes.

## 1.3 The Lévy Modification of Additive Processes Continuous in Probability

A function $f\colon T = [0, \infty) \longrightarrow R^1$ is called a *D-function* or is said to *belong to the class D* if the following three conditions are satisfied:

(a) both $f(t+)$ and $f(t-)$ exist and are finite,
(b) $f(t+) = f(t)$,
(c) $f(0) = 0$.

An additive process $\{X_t\}_{t \in T}$ continuous i. p. is called a *Lévy process*[3] if almost all sample functions belong to $D$ namely if, for almost all $\omega$, $X_t(\omega)$ satisfies (a), (b), and (c) as a function of $t$.

The purpose of this section is to prove

**Theorem 1 (Doob).** [4] *For every additive process $\{X_t\}_{t \in T}$ continuous i. p. we can find a Lévy process $\{Y_t\}_{t \in T}$ such that $P(Y_t = X_t) = 1$ for every $t$. Such a Lévy process $\{Y_t\}$ is essentially unique in the sense that for two such processes $\{Y_t^i\}$, $i = 1, 2$, we have*

$$P(Y_t^1 = Y_t^2 \text{ for every } t) = 1 .$$

The Lévy process $\{Y_t\}$ obtained here is called the *Lévy modification* of $\{X_t\}$.

By virtue of this fact we can reduce the study of additive processes continuous i. p. to that of Lévy processes. The complete structure of Lévy processes will be described in the subsequent sections.

Let us start with some preliminary facts. Let $\{X_\alpha\}_{\alpha \in A}$ be a family of random variables indexed by a parameter $\alpha$ ranging over a *countable* linear set $A$. Fix $\omega$ for the moment and consider the sample process $\{X_\alpha(\omega)\}_\alpha$. For $a < b$ we will define the *down-crossing number* of $\{X_\alpha(\omega)\}_{\alpha \in A}$ over $[a, b]$. If we have $\alpha_1 < \alpha_2 < \cdots < \alpha_{2r}$ such that

$$X_{\alpha_i}(\omega) > b \text{ or } < a \text{ according as } i \text{ is odd or even,}$$

then we say that $\{X_\alpha(\omega)\}_\alpha$ has at least $r$ down-crossings. The least upper bound of such $r$ is called the down crossing number of $\{X_\alpha(\omega)\}_\alpha$ over $[a, b]$ and denoted by $N_{ab}(X_\alpha(\omega),\ \alpha \in A)$; it may be infinite. This number is measurable in $\omega$ because of the countability of $A$ and so is considered as a random variable taking values in $\{0, 1, 2, \ldots, \infty\}$.

Let $\{X_i\}_{i=1}^n$ be independent and set

$$S_0 = 0, \quad S_k = \sum_1^k X_i \quad (1 \le k \le n) .$$

Let $N$ be the down-crossing number of $\{S_i(\omega)\}_{i=0}^n$ over $[a, b]$ $(a < b)$. Then we have

---

[3] Recently the name "Lévy process" is usually used in a more restricted meaning. See the first footnote in Section 1.10.

[4] Original references are the following:

J. L. Doob, Stochastic processes depending on a continuous parameter, Trans. Amer. Math. Soc., **42**, 107–140 (1937).

J. L. Doob, Stochastic Processes, Wiley, New York, 1953.

**Lemma 1.** *If*

$$P\left(\max_{0\le p,q\le n} |S_p - S_q| > b - a\right) \le \delta < 1/2\,,$$

*then*

$$E(N) \le 2\delta\,.$$

*Proof.* Let us define a random sequence $\{\sigma_1(\omega), \sigma_2(\omega), \ldots\}$ as follows. Let $\sigma_1(\omega)$ be the minimum index $k$ for which $S_k(\omega) > b$ and $\sigma_2(\omega)$ the minimum index $k > \sigma_1(\omega)$ for which $S_k(\omega) < a$; if there exists no such $k$, we set $\sigma_i(\omega) = n$ ($i = 1$ or 2). Supposing that $\sigma_i(\omega)$, $i \le 2m$, are defined, we define $\sigma_{2m+1}(\omega)$ to be the minimum index $k > \sigma_{2m}(\omega)$ for which $S_k(\omega) > b$ and $\sigma_{2m+2}(\omega)$ to be the minimum index $k > \sigma_{2m+1}(\omega)$ for which $S_k(\omega) < a$; we set $\sigma_i(\omega) = n$ ($i = 2m+1$ or $2m+2$) as above in case there exists no such $k$. Then it is easy to see that

$$N \ge m \iff \sigma_1 < \sigma_2 < \cdots < \sigma_{2m} \text{ and } S_{\sigma_{2m}} < a\,.$$

Therefore

$$P(N \ge m) \le \sum_{(k)} P(A \cap B)\,,$$

where

$$A \equiv \{\sigma_1 = k_1,\, \sigma_2 = k_2,\, \ldots,\, \sigma_{2m-1} = k_{2m-1}\}\,,$$

$$B \equiv \left\{\sup_{p > k_{2m-1}} |S_p - S_{k_{2m-1}}| > b - a\right\}\,,$$

and the summation is carried out over all $(k) = (k_1, k_2, \ldots, k_{2m-1})$ such that $0 \le k_1 < k_2 < \cdots < k_{2m-1}$. Observing $A \in \mathcal{B}[X_i : i \le k_{2m-1}]$ and $B \in \mathcal{B}[X_i : i > k_{2m-1}]$, we get

$$P(A \cap B) = P(A)P(B) \le P(\sigma_1 = k_1, \ldots, \sigma_{2m-1} = k_{2m-1})\,\delta$$

by our assumption, so that

$$\begin{aligned} P(N \ge m) &\le \delta P(\sigma_1 < \sigma_2 < \cdots < \sigma_{2m-1}) \\ &\le \delta P(\sigma_1 < \sigma_2 < \cdots < \sigma_{2m-2} < n) \\ &\le \delta P(N \ge m - 1)\,. \end{aligned}$$

Therefore we have

$$P(N \ge m) \le \delta^m$$

and so

$$E(N) = \sum_{m \ge 1} P(N \ge m) \le \frac{\delta}{1-\delta} \le 2\delta\,.$$

**Lemma 2.** *If* $P(|S_k| > a) \le \delta < 1/2$ *for every* $k = 1, 2, \ldots, n$, *then*

$$P\left(\max_{0 \le p,q \le n} |S_p - S_q| > 4a\right) \le 4\delta \,.$$

*Proof.* It suffices to prove that

$$P\left(\max_{0 \le p \le n} |S_n - S_p| > 2a\right) \le 2\delta \,,$$

since $|S_p - S_q| \le |S_n - S_p| + |S_n - S_q|$. To prove this, we will use the same method as in the proof of the Kolmogorov inequality. Consider the events

$$A_p = \{|S_p| \le a\} \,,$$

$$B_p = \left\{\max_{p < i \le n} |S_n - S_i| \le 2a,\ |S_n - S_p| > 2a\right\} \quad (p = 0, 1, 2, \ldots, n) \,.$$

Then $\{B_p\}_p$ is disjoint, $\{A_p, B_p\}$ is independent for each $p$, and

$$E \equiv \left\{\max_{0 \le p \le n} |S_n - S_p| > 2a\right\} = \bigcup_p B_p \text{ (disjoint union).}$$

Since $|S_n - S_p| > 2a$ and $|S_p| \le a$ imply $|S_n| > a$, we have

$$\{|S_n| > a\} \supset \bigcup_p (A_p \cap B_p) \text{ (disjoint union).}$$

Then it holds that

$$\begin{aligned}\delta \ge P(|S_n| > a) &\ge \sum_p P(A_p \cap B_p) = \sum_p P(A_p) P(B_p) \\ &\ge (1 - \delta) \sum_p P(B_p) = (1 - \delta) P(E) \ge \frac{1}{2} P(E) \,,\end{aligned}$$

which completes the proof.

Combining Lemmas 1 and 2, we get

**Lemma 3.** *If* $P(|S_i| > (b - a)/4) < 1/8$ *for* $i = 1, 2, \ldots, n$, *then* $E(N) \le 1$.

Now we will prove Theorem 1. Let $Q$ be the set of all rational numbers and write $r$ for the generic element of $Q \cap T$.

Since $X_t$ is continuous i. p., the probability law $\mu_t$ of $X_t$ is continuous in $t$. Therefore $\{\mu_t\}_{t \in [0,n]}$ is compact, so that we can find, for every $\delta \in (0, 1/2)$, $a = a(\delta)$ such that

(1) $$P(|X_t| > a) < \delta \qquad \text{for } 0 \le t \le n\,.$$

Let $\{0 = r_0 < r_1 < r_2 < \cdots < r_m\} \subset Q \cap [0, n]$. Since $\{X_{r_i} - X_{r_{i-1}}\}_{i=1}^m$ is independent and since $X_{r_i}$ is its $i$-th partial sum, we can use Lemma 2 to get

$$P\left(\max_{1\le i\le m} |X_{r_i}| > 4a\right) \le 4\delta$$

by virtue of (1). It follows from this that

$$P\left(\sup_{r\le n} |X_r| > 4a\right) \le 4\delta\,,$$

a fortiori

$$P\left(\sup_{r\le n} |X_r| = \infty\right) \le 4\delta \to 0\,.$$

This implies that

(2) $$P\left(\sup_{r\le n} |X_r| < \infty \quad \text{for every } n\right) = 1\,.$$

Since $X_t$ is continuous i. p., it is uniformly continuous i. p. on every compact $t$-interval. Therefore for every $d > 0$ and $n > 0$ we can find $\varepsilon = \varepsilon(d, n)$ such that

(3) $$P(|X_v - X_u| > d/4) < 1/8$$

provided $|v - u| < \varepsilon$ and $u, v \in [0, n]$. Using the same argument as above, we can use Lemma 3 to obtain

(4) $$E(N(t, t+\varepsilon; a, b)) \le 1$$

provided $0 \le t < t + \varepsilon \le n$ and $b - a > d$, where $N(t, t+\varepsilon; a, b)$ is the down-crossing number of $X_r - X_t$, $r \in [t, t+\varepsilon] \cap Q$, over $[a, b]$. It follows from (4) that

$$P(N(t, t+\varepsilon; a, b) < \infty) = 1$$

provided $0 \le t < t + \varepsilon \le n$ and $b - a > d$. Therefore

$$P(N(t, t+\varepsilon; a, b) < \infty \text{ for every } (a, b) \in Q \times Q \text{ with } b - a > d) = 1\,.$$

This implies

$$P\left(\limsup_{r\to s+}(X_r - X_t) \le \liminf_{r\to s+}(X_r - X_t) + d \text{ for every } s \in [t, t+\varepsilon)\right) = 1\,,$$

namely

$$P\left(\limsup_{r\to s+} X_r \le \liminf_{r\to s+} X_r + d \text{ for every } s \in [t, t+\varepsilon)\right) = 1\,.$$

Recall that $r$ is always in $Q \cap T$. Since $[0, n)$ is covered by a finite number of intervals of the form $[t, t+\varepsilon)$, we have

$$P\left(\limsup_{r\to s+} X_r \le \liminf_{r\to s+} X_r + d \text{ for every } s \in [0, n)\right) = 1\,.$$

Letting $n \to \infty$ and $d \downarrow 0$, we have

$$P\left(\limsup_{r\to s+} X_r \le \liminf_{r\to s+} X_r \text{ for every } s \in T\right) = 1\,,$$

namely

$$P\left(\lim_{r\to t+} X_r \text{ exists for every } t\right) = 1\,. \tag{5}$$

Similarly

$$P\left(\lim_{r\to t-} X_r \text{ exists for every } t\right) = 1\,. \tag{6}$$

Combining (2), (5), and (6), we see that the set

$$\Omega_1 = \left\{\omega\colon \lim_{r\to t+} X_r \text{ and } \lim_{r\to t-} X_r \text{ exist and are finite for every } t\right\}$$

has probability 1. Now define $Y_t(\omega)$ by

$$Y_t(\omega) = \begin{cases} \lim_{r\to t+} X_r(\omega) & \text{for } \omega \in \Omega_1 \\ 0 & \text{otherwise.} \end{cases}$$

Then $Y_t$ is a Lévy process such that $P(Y_t = X_t) = 1$ for every $t$. If we have two such Lévy processes $\{Y_t^1\}$ and $\{Y_t^2\}$, then

$$P(Y_r^1 = Y_r^2 \quad \text{for } r \in Q \cap T) = 1\,.$$

By the right continuity of $Y_t^i$ $(i = 1, 2)$, we have

$$P(Y_t^1 = Y_t^2 \quad \text{for } t \in T) = 1\,.$$

## 1.4 Elementary Lévy Processes

Among the Lévy processes the following two types of processes are the simplest.

An *elementary Lévy process of type* 0 is a Lévy process whose sample function is continuous a. s. An *elementary Lévy process of type* 1 is a Lévy process whose sample function is step-wise with jumps = 1 a. s. By virtue of the following theorem the first one is also called a *Lévy process of Gauss type* and the second one a *Lévy process of Poisson type.*

**Theorem 1.** *(i) A Lévy process is an elementary Lévy process of type* 0 *if and only if every increment is Gauss distributed.*

*(ii) A Lévy process is an elementary Lévy process of type* 1 *if and only if every increment is Poisson distributed.*[5]

For a proof we need some preliminary facts.

**Lemma 1.** *Let* $z_{ni}$, $i = 1, 2, \ldots, p(n)$, $n = 1, 2, \ldots$, *be complex numbers satisfying the following three conditions:*

$$\max_{1 \le i \le p(n)} |z_{ni}| \to 0 \qquad \textit{as } n \to \infty , \tag{1}$$

$$\sup_n \sum_{i=1}^{p(n)} |z_{ni}| < \infty , \tag{2}$$

$$z_n \equiv \sum_{i=1}^{p(n)} z_{ni} \to z . \tag{3}$$

*Then we have*

$$\prod_{i=1}^{p(n)} (1 + z_{ni}) \to \mathrm{e}^z \qquad \textit{as } n \to \infty .$$

*Proof.* Use the fact

$$1 + a = \mathrm{e}^{a + O(|a|^2)} ,$$

where $a$ stands for a complex variable.

**Lemma 2.** *Let* $\{X_t\}$ *be a Lévy process and let* $u < v$. *If*

$$P(|X_t - X_u| > a) \le \delta < 1/2 \qquad \textit{for } u \le t \le v ,$$

*then*

$$P\left( \sup_{u \le t < s \le v} |X_s - X_t| > 4a \right) \le 4\delta .$$

*Proof.* Since $\{X_t\}$ is a Lévy process, the left hand side is

$$\sup_{(t)} P\left( \max_{1 \le i < j \le n} |X_{t_j} - X_{t_i}| > 4a \right) ,$$

$(t) = (t_1, t_2, \cdots, t_n)$ ranging over all $t_1 < t_2 < \cdots < t_n$ in $[u, v]$ and $n$ ranging over $\{2, 3, \ldots\}$. This is no more than $4\delta$ by Lemma 1.3.2.

[5] If $X = 0$ a.s., we say that $X$ is Gauss distributed with mean 0 and variance 0 as well as Poisson distributed with parameter 0.

Now we will prove Theorem 1.

*The "only if" part of (i).*
Let $\{X_t\}$ be a Lévy process whose sample function is continuous a. s. Let $I = [t_0, t_1]$. We want to prove that $X \equiv X(t_1) - X(t_0)$ is Gauss distributed. The proof is analogous to that of the central limit theorem. We assume that all time points $t, s$ etc. are in $I$ unless stated otherwise. By the continuity of the sample function we have $\delta = \delta(\varepsilon) > 0$ for every $\varepsilon > 0$ such that

$$P\left(\sup_{|t-s|<\delta} |X_t - X_s| < \varepsilon\right) > 1 - \varepsilon\,, \tag{4}$$

because of the uniform continuity of continuous functions on a compact set.

Take a sequence $\varepsilon_n \downarrow 0$ and a sequence of subdivisions of $[t_0, t_1]$,

$$t_0 = t_{n0} < t_{n1} < \cdots < t_{np(n)} = t_1,$$
$$0 < t_{nk} - t_{n,k-1} < \delta(\varepsilon_n) \quad \text{for } 1 \le k \le p(n) \qquad (n = 1, 2, \ldots)$$

and set

$$X_{nk} = \begin{cases} X(t_{nk}) - X(t_{n,k-1}) & \text{if this is less than } \varepsilon_n \text{ in modulus} \\ 0 & \text{otherwise} \end{cases}$$

and

$$S_n = \sum_{k=1}^{p(n)} X_{nk}\,.$$

It is obvious by (4) that $P(X = S_n) > 1 - \varepsilon_n$, so that

$$S_n \to X \quad \text{i. p.}$$

Let

$$m_{nk} = E(X_{nk}), \quad V_{nk} = V(X_{nk}), \quad m_n = \sum_{k=1}^{p(n)} m_{nk}, \quad V_n = \sum_{k=1}^{p(n)} V_{nk}\,.$$

Then we have

$$|m_{nk}| \le \varepsilon_n \to 0 \quad \text{and} \quad V_{nk} \le \varepsilon_n^2 \to 0$$

and we obtain for $\alpha$ real

$$\begin{aligned} E(\mathrm{e}^{\mathrm{i}\alpha X}) &= \lim_n E(\mathrm{e}^{\mathrm{i}\alpha S_n}) = \lim_n \prod_{k=1}^{p(n)} E(\mathrm{e}^{\mathrm{i}\alpha X_{nk}}) \\ &= \lim_n \mathrm{e}^{\mathrm{i}\alpha m_n} \prod_k E(\mathrm{e}^{\mathrm{i}\alpha(X_{nk}-m_{nk})}) \\ &= \lim_n \mathrm{e}^{\mathrm{i}\alpha m_n} \prod_k \left(1 - \frac{\alpha^2}{2} V_{nk}(1 + O(\varepsilon_n))\right)\,. \end{aligned} \tag{5}$$

Noticing that $1-\theta \le e^{-\theta/2}$ for small $\theta > 0$, we have

$$|E(e^{i\alpha X})| \le \liminf_n \prod_k e^{-(\alpha^2/4)V_{nk}} = \liminf_{n\to\infty} e^{-(\alpha^2/4)V_n} .$$

If $\{V_n\}$ is unbounded, then the right-hand side will vanish for $\alpha \ne 0$. This is a contradiction, because the left-hand side is $> 0$ for small $\alpha$. Thus $\{V_n\}$ must be bounded. By taking a subsequence we can assume that $V_n$ converges to a finite number $V \ge 0$.

It is easy to verify the conditions in Lemma 1 for

$$z_{nk} = -\frac{\alpha^2}{2} V_{nk}(1 + O(\varepsilon_n)), \qquad z = -\frac{\alpha^2}{2} V .$$

Therefore we have

$$\prod_k \left[1 - \frac{\alpha^2}{2} V_{nk}(1 + O(\varepsilon_n))\right] \to e^{-(\alpha^2/2)V} . \tag{6}$$

Now we want to prove that $\{m_n\}$ is bounded. By (5) and (6) we have

$$e^{i\alpha m_n} \to e^{(\alpha^2/2)V} E(e^{i\alpha X}) \equiv \varphi(\alpha) .$$

If $\{m_n\}$ is unbounded, then

$$\left|\int_0^\beta \varphi(\alpha)d\alpha\right| = \lim_n \left|\int_0^\beta e^{i\alpha m_n} d\alpha\right| \le \lim_n \left|\frac{e^{i\beta m_n} - 1}{i m_n}\right| \le \liminf_{n\to\infty} \frac{2}{|m_n|} = 0$$

for every $\beta > 0$ and so $\varphi(\alpha) \equiv 0$ by the continuity of $\varphi$, in contradiction with $\varphi(0) = 1$.

By taking a subsequence, if necessary, we can also assume that $m_n$ converges to some $m$. Then we get

$$E(e^{i\alpha X}) = e^{im\alpha - (V/2)\alpha^2} ,$$

which proves that $X$ is Gauss distributed.

*The "if" part of (i).*
Let $\{X_t\}$ be a Lévy process such that $X_t - X_s$ is Gauss distributed for $s < t$. Let $m(t) = E(X_t)$ and $V(t) = V(X_t)$. Then $E(X_t - X_s) = m(t) - m(s)$ and $V(X_t - X_s) = V(t) - V(s)$ by the additivity of $\{X_t\}$. The functions $m(t)$ and $V(t)$ are both continuous because of the continuity of $\{X_t\}$ i. p.[6] Let $Y_t = X_t - m(t)$. The continuity of the sample function of $X_t$ will follow from that of $Y_t$; $Y_t$ is also a Lévy process such that $Y_t - Y_s$ is $N(0, V(t) - V(s))$ distributed for $s < t$.

[6] Use that $Ee^{i\alpha X(t)} = e^{im(t)\alpha - (V(t)/2)\alpha^2}$.

Fix $m > 0$ for the moment. Since $Y_t - Y_s$ is $N(0, V(t) - V(s))$ distributed, we have

$$E[(Y_t - Y_s)^4] = (V(t) - V(s))^2 \frac{1}{\sqrt{2\pi}} \int_{-\infty}^{\infty} \mathrm{e}^{-y^2/2} y^4 \mathrm{d}y$$
$$= 3(V(t) - V(s))^2$$

and so

$$P(|Y_t - Y_s| > \varepsilon) \leq \frac{3(V(t) - V(s))^2}{\varepsilon^4}. \tag{7}$$

Since $V(t)$ is uniformly continuous on $0 \leq t \leq m$, we have $\delta = \delta(\varepsilon)$ such that

$$|V(t) - V(s)| < \varepsilon^5 \quad \text{for } 0 \leq s < t \leq m \text{ and } |t - s| < \delta .$$

Taking a subdivision $0 = t_0 < t_1 < \cdots < t_n = m$ of $[0, m]$ such that $\delta(\varepsilon)/2 < |t_i - t_{i-1}| < \delta(\varepsilon)$ for every $i$, we have

$$|V(t_i) - V(t_{i-1})| < \varepsilon^5 \quad \text{for } i = 1, 2, \ldots, n.$$

Then we get by (7)

$$P\left(\left|Y_t - Y_{t_{i-1}}\right| > \varepsilon\right) \leq \frac{3\varepsilon^5 (V(t) - V(t_{i-1}))}{\varepsilon^4} = 3\varepsilon(V(t) - V(t_{i-1})) \tag{8}$$

for $t_{i-1} \leq t \leq t_i$. Taking $\varepsilon$ small enough, we can assume that the right-hand side is $< 1/2$. It follows from (8) by Lemma 2 that

$$\begin{aligned} P\left(\max_{1 \leq i \leq n} \sup_{t_{i-1} \leq u < v \leq t_i} |Y_v - Y_u| > 4\varepsilon\right) &\leq \sum_{i=1}^{n} P\left(\sup_{t_{i-1} \leq u < v \leq t_i} |Y_v - Y_u| > 4\varepsilon\right) \\ &\leq 12\varepsilon \sum_{i=1}^{n} (V(t_i) - V(t_{i-1})) \\ &= 12\varepsilon\, V(m) . \end{aligned}$$

Therefore

$$P\left(\sup_{\substack{0 \leq s < t \leq m \\ |t-s| < \delta(\varepsilon)/2}} |Y_t - Y_s| > 8\varepsilon\right) \leq 12\varepsilon\, V(m) ,$$

$$P\left(\lim_{\delta \downarrow 0} \sup_{\substack{0 \leq s < t \leq m \\ |t-s| < \delta}} |Y_t - Y_s| > 8\varepsilon\right) \leq 12\varepsilon\, V(m) .$$

Letting $\varepsilon \downarrow 0$,

$$P\left(\lim_{\delta\downarrow 0} \sup_{\substack{0\le s<t\le m\\ |t-s|<\delta}} |Y_t - Y_s| > 0\right) = 0\,,$$

namely

$$P(Y_t \text{ is continuous on } 0 \le t \le m) = 1\,.$$

Since $m$ is arbitrary, $Y_t$ is continuous a. s.

*The "only if" part of (ii).*
Let $\{X_t\}$ be a Lévy process whose sample function is step-wise with jumps $= 1$ a. s. We want to prove that $X \equiv X(t_1) - X(t_0)$ is Poisson distributed. The proof is analogous to that of the *Poisson law of rare events.*

By the continuity i. p. of $X_t$ we have

$$\sup_{\substack{|t-s|\le 1/n\\ t_0\le s<t\le t_1}} P(|X_t - X_s| \ge 1) \to 0 \quad \text{as } n\to\infty\,. \tag{9}$$

For each $n$, let $t_0 = t_{n0} < t_{n1} < \cdots < t_{np_n} = t_1$, $t_{ni} - t_{n,i-1} \le 1/n$, be a subdivision of $[t_0, t_1]$ and

$$X_{nk} = \begin{cases} X(t_{nk}) - X(t_{n,k-1}) & \text{if this is 0 or 1}\\ 1 & \text{otherwise,}\end{cases}$$

$$X_n = \sum_k X_{nk}\,.$$

Since almost all sample functions of $\{X_t\}$ are step-wise with jumps 1, we have

$$P\left(X_n \to X\right) = 1\,.$$

Keeping (9) in mind and using Lemma 1, we can compute the Laplace transform of the probability law of $X$ as follows. Let $\lambda_{nk} = P(X_{nk} = 1)$ and $\lambda_n = \sum_k \lambda_{nk}$. Then

$$\begin{aligned} E(\mathrm{e}^{-\alpha X}) &= \lim_n E(\mathrm{e}^{-\alpha X_n}) = \lim_n \prod_{k=1}^{p_n} E(\mathrm{e}^{-\alpha X_{nk}})\\ &= \lim_n \prod_k [(1-\lambda_{nk}) + \lambda_{nk}\mathrm{e}^{-\alpha}]\\ &= \lim_n \prod_k [1 - \lambda_{nk}(1-\mathrm{e}^{-\alpha})]\\ &\le \liminf_{n\to\infty} \mathrm{e}^{-\lambda_n(1-\mathrm{e}^{-\alpha})}\,, \end{aligned}$$

since $1-\theta \le \mathrm{e}^{-\theta}$. If $\{\lambda_n\}$ is unbounded, then it turns out that

$$E(\mathrm{e}^{-\alpha X}) = 0 \quad \text{for } \alpha > 0$$

and, by the continuity of $E(\mathrm{e}^{-\alpha X})$ in $\alpha$, the left side is equal to 0 for $\alpha = 0$, which is a contradiction. Therefore $\{\lambda_n\}$ is bounded. By taking a subsequence we can assume that $\lambda_n \to \lambda$ for some $\lambda \in [0, \infty)$. Applying Lemma 1 to

$$z_{nk} = -\lambda_{nk}(1 - \mathrm{e}^{-\alpha}) ,$$

we have

$$E(\mathrm{e}^{-\alpha X}) = \mathrm{e}^{-\lambda(1-\mathrm{e}^{-\alpha})} ,$$

which shows that $X$ is Poisson distributed. Note that condition (1) of Lemma 1 is satisfied by virtue of (9).

*The "if" part of (ii).*
Let $\{X_t\}$ be a Lévy process such that $X_t - X_s$ is Poisson distributed for every $s < t$. Let $\lambda(t) = E(X_t)$. Then $E(X_t - X_s) = \lambda(t) - \lambda(s)$, $s < t$. Since $X_t - X_s$ is Poisson distributed with the parameter $= \lambda(t) - \lambda(s)$ for $s < t$, and since $\{X_t\}$ is continuous i. p., $\lambda(t)$ must be continuous. It is obvious that

$$P(X_t - X_s = 0, 1, 2, \ldots) = 1 \qquad \text{for } t > s \geq 0 .$$

Hence it follows that

$$P(X_t \text{ is increasing and takes only } 0, 1, 2, \ldots) = 1 ,$$

because almost all sample functions belong to $D$. This proves that almost all sample functions of $\{X_t\}$ are step-wise with jumps $= 1$. Indeed, consider the events

$$\begin{aligned} A &= \{X_t - X_{t-} \geq 2 \text{ for some } t\} , \\ A_n &= \{X_t - X_{t-} \geq 2 \text{ for some } t \leq n\} , \\ A_{n,m} &= \bigcup_{k=1}^{nm} \{X_{k/m} - X_{(k-1)/m} \geq 2\} . \end{aligned}$$

Then $A = \bigcup_n A_n$ and $A_n \subset A_{n,m}$ for every $m$. Hence

$$\begin{aligned} P(A_{n,m}) &\leq \sum_{k=1}^{nm} P(X_{k/m} - X_{(k-1)/m} \geq 2) \\ &= \sum_{k=1}^{nm} \left\{1 - \mathrm{e}^{-(\lambda(k/m) - \lambda((k-1)/m))} \left(1 + \lambda\left(\frac{k}{m}\right) - \lambda\left(\frac{k-1}{m}\right)\right)\right\} \\ &\leq \sum_{k=1}^{nm} \left[\lambda\left(\frac{k}{m}\right) - \lambda\left(\frac{k-1}{m}\right)\right]^2 \quad (\text{by } 1 - \mathrm{e}^{-\alpha}(1+\alpha) \leq \alpha^2,\ \alpha > 0) \\ &\leq \max_{1 \leq k \leq nm} \left[\lambda\left(\frac{k}{m}\right) - \lambda\left(\frac{k-1}{m}\right)\right] \lambda(n) \to 0 \qquad \text{as } m \to \infty . \end{aligned}$$

Then $P(A_n) = 0$ and $P(A) \le \sum_n P(A_n) = 0$. This completes the proof of the "if" part of (ii).

Let $\{X_t\}$ be an additive process and $\mu_{st}$ the probability law of $X_t - X_s$ ($s < t$). Then the joint distribution of $(X_{t_1}, X_{t_2}, \dots, X_{t_n})$, $t_1 < t_2 < \dots < t_n$, is given by

$$\begin{aligned} (10) \qquad & P((X_{t_1}, X_{t_2}, \dots, X_{t_n}) \in E) \\ & = \int \cdots \int_A \mu_{0t_1}(da_1)\mu_{t_1t_2}(da_2)\cdots\mu_{t_{n-1}t_n}(da_n)\,, \end{aligned}$$

where[7] $E \in \mathcal{B}(R^n)$ and

$$A = \{(a_1,\, a_1 + a_2,\, a_1 + a_2 + a_3, \dots, a_1 + \dots + a_n) \in E\}\,.$$

Since the Gauss distribution is determined by its mean and its variance and since the Poisson distribution is determined by its parameter (= mean), we have

**Theorem 2.** *(i) A Lévy process of Gauss type is determined in law by its mean function $m(t) = E(X(t))$ and its variance function $V(t) = V(X(t))$.*

*(ii) A Lévy process of Poisson type is determined in law by its mean function $\lambda(t) = E(X(t))$.*

**Theorem 3.** *(i) Functions $m(t)$ and $V(t)$ are respectively the mean function and covariance function of a Lévy process of Gauss type if and only if $m(t)$ is continuous, $V(t)$ is continuous and increasing (= non-decreasing), and $m(0) = V(0) = 0$.*

*(ii) A function $\lambda(t)$ is the mean function of a Lévy process of Poisson type if and only if $\lambda(t)$ is continuous and increasing and $\lambda(0) = 0$.*

*Proof.* The proof of the "only if" parts of (i) and (ii) is included in the proof of Theorem 1.

Suppose that $m(t)$ is continuous, $V(t)$ is continuous and increasing, and $m(0) = V(0) = 0$. Let $\mu_{st}$ denote the Gauss distribution $N(0, V(t) - V(s))$. Define $\nu_{t_1 \cdots t_n}(E)$ by the right-hand side of (10). Then it is easy to verify that the system $\{\nu_{t_1 \cdots t_n}\}$ satisfies Kolmogorov's consistency condition. Therefore there exists a stochastic process $\{Y_t\}$ such that

$$P((Y_{t_1}, Y_{t_2}, \dots, Y_{t_n}) \in E) = \nu_{t_1 \cdots t_n}(E), \qquad E \in \mathcal{B}(R^n)$$

for every $n$ and every $(t_1, \dots, t_n)$ with $t_1 < \dots < t_n$. It is also easy to see that $\{Y_t\}$ is an additive process and that $Y_t - Y_s$ ($s < t$) is $N(0, V(t) - V(s))$ distributed. Since $V(t)$ is continuous, $\{Y_t\}$ is continuous i.p. By taking the Lévy modification we can assume that $\{Y_t\}$ is a Lévy process. Therefore it

[7] $\mathcal{B}(R^n)$ is the class of all Borel sets in $R^n$.

is a Lévy process of Gauss type with $E(Y_t) = 0$ and $V(Y_t) = V(t)$. Now define $X_t$ to be $Y_t + m(t)$. Then $\{X_t\}$ is a Lévy process of Gauss type with $E(X_t) = m(t)$ and $V(X_t) = V(t)$. This completes the proof of the "if" part of (i).

The "if" part of (ii) can be proved the same way.

**Example 1.** The Wiener process (or the Brownian motion) is a Lévy process of Gauss type with $m(t) \equiv 0$ and $V(t) \equiv t$.

**Example 2.** The Poisson process is a Lévy process of Poisson type with $\lambda(t) \equiv \lambda t$, $\lambda$ being a positive constant.

## 1.5 Fundamental Lemma

In this section we will prove a lemma which will play a fundamental role in studying the structure of Lévy processes in the next section.

**Fundamental Lemma.** *Let $\mathfrak{F} = \{\mathcal{B}_{st} \colon s < t,\ s, t \in T\}$ be an additive family of $\sigma$-algebras, $X(t)$ a Lévy process, and $Y(t)$ a Lévy process of Poisson type, both $X$ and $Y$ being adapted to $\mathfrak{F}$. If with probability 1 there is no common jump point for $X(t)$ and $Y(t)$, then the two processes are independent; in other words if we have*

$$\text{(1)} \qquad P(\textit{either } X(t) = X(t-) \textit{ or } Y(t) = Y(t-) \textit{ for every } t) = 1\,,$$

*then $\mathcal{B}[X] \equiv \mathcal{B}[X(t) \colon t \in T]$ and $\mathcal{B}[Y] \equiv \mathcal{B}[Y(t) \colon t \in T]$ are independent.*

*Proof.* First we will prove that $X(t) - X(s)$ and $Y(t) - Y(s)$ are independent for every pair $s < t$. Divide the interval $(s, t]$ into $n$ disjoint subintervals $(t_{n,k-1}, t_{n,k}]$, $k = 1, 2, \ldots, n$, of equal length. Set

$$\begin{aligned} X &= X(t) - X(s), & Y &= Y(t) - Y(s), \\ X_{nk} &= X(t_{nk}) - X(t_{n,k-1}), & Y_{nk} &= Y(t_{nk}) - Y(t_{n,k-1}), \\ X_n &= \sum_{k=1}^{n} e_0(Y_{nk}) X_{nk}\,, \end{aligned}$$

where $e_0$ is the indicator function of a single point 0.

Below, for every $\varepsilon_n = (\varepsilon_{n1}, \varepsilon_{n2}, \ldots, \varepsilon_{nn})$ such that all $\varepsilon_{nk}$ are non-negative integers,

$$\sum_{k}^{\varepsilon_n +} \text{ denotes the summation over } k = 1, 2, \ldots, n \text{ with } \varepsilon_{nk} > 0$$

and

$$\sum_{k}^{\varepsilon_n 0} \text{ denotes the summation over } k = 1, 2, \ldots, n \text{ with } \varepsilon_{nk} = 0 .$$

It is obvious that

$$\sum_{k} = \sum_{k}^{\varepsilon_n +} + \sum_{k}^{\varepsilon_n 0} .$$

Similarly for the products $\prod_{k}^{\varepsilon_n +}$ and $\prod_{k}^{\varepsilon_n 0}$. Further, $|\varepsilon_n|$ denotes $\sum_{k=1}^{n} \varepsilon_{nk}$ and $\sum_{|\varepsilon_n|=p}$ denotes the summation over all $\varepsilon_n = (\varepsilon_{n1}, \ldots, \varepsilon_{nn})$ with $|\varepsilon_n| = p$.

Using the assumption that $\{X(t)\}$ and $\{Y(t)\}$ are adapted to $\mathfrak{F}$, we have

$$\begin{aligned}
&E(e^{izX_n}, Y = p)P(Y = 0) \\
&= \sum_{|\varepsilon_n|=p} E\left(\prod_{k}^{\varepsilon_n 0} e^{izX_{nk}}, Y_{n1} = \varepsilon_{n1}, \ldots, Y_{nn} = \varepsilon_{nn}\right) P(Y_{n1} = 0, \ldots, Y_{nn} = 0) \\
&= \sum_{|\varepsilon_n|=p} \prod_{k}^{\varepsilon_n 0} E(e^{izX_{nk}}, Y_{nk} = 0) \prod_{k}^{\varepsilon_n +} P(Y_{nk} = \varepsilon_{nk}) \prod_{k} P(Y_{nk} = 0) \\
&= \sum_{|\varepsilon_n|=p} \prod_{k}^{\varepsilon_n 0} E(e^{izX_{nk}}, Y_{nk} = 0) \prod_{k}^{\varepsilon_n +} P(Y_{nk} = 0) \prod_{k} P(Y_{nk} = \varepsilon_{nk}) ;
\end{aligned}$$

in the last step we used the relation

$$\prod_{k}^{\varepsilon_n 0} P(Y_{nk} = \varepsilon_{nk}) = \prod_{k}^{\varepsilon_n 0} P(Y_{nk} = 0) ,$$

which is obvious from the definition of $\prod_{k}^{\varepsilon_n 0}$. Similarly, using adaptedness, we have

$$E(e^{izX_n}, Y = 0)P(Y = p) = \sum_{|\varepsilon_n|=p} \prod_{k} E(e^{izX_{nk}}, Y_{nk} = 0) \prod_{k} P(Y_{nk} = \varepsilon_{nk}) .$$

Now observe

$$\begin{aligned}
&\left|E(e^{izX_n}, Y = p)P(Y = 0) - E(e^{izX_n}, Y = 0)P(Y = p)\right| \\
&\quad \le \sum_{|\varepsilon_n|=p} \left| \prod_{k}^{\varepsilon_n +} P(Y_{nk} = 0) - \prod_{k}^{\varepsilon_n +} E(e^{izX_{nk}}, Y_{nk} = 0) \right| \prod_{k} P(Y_{nk} = \varepsilon_{nk}) \\
&\quad \le \sum_{|\varepsilon_n|=p} \sum_{k}^{\varepsilon_n +} \left|P(Y_{nk} = 0) - E(e^{izX_{nk}}, Y_{nk} = 0)\right| \prod_{k} P(Y_{nk} = \varepsilon_{nk})
\end{aligned}$$

$$\begin{aligned}
&\leq \sum_{|\varepsilon_n|=p} \sum_k^{\varepsilon_n+} E(|1-\mathrm{e}^{\mathrm{i}zX_{nk}}|) \prod_k P(Y_{nk}=\varepsilon_{nk}) \\
&\leq p \sup_k E(|1-\mathrm{e}^{\mathrm{i}zX_{nk}}|) \\
&= p \sup_k E(|\mathrm{e}^{\mathrm{i}zX(t_{nk})} - \mathrm{e}^{\mathrm{i}zX(t_{n,k-1})}|) \,.
\end{aligned}$$

In the above the first and second inequalities are valid because all factors are $\leq 1$ in modulus.[8] Noticing that $\{X(t)\}_t$ is continuous in probability, we can easily see that $\{\mathrm{e}^{\mathrm{i}zX(t)}\}_t$ is continuous in the $L^1$-norm and so is uniformly continuous in the same norm on the interval $[s,t]$ for $z$ fixed. Therefore the last term tends to 0 as $n \to \infty$ for $z$ and $p$ fixed.

By our assumption of no common jump points we have $X_n \to X$ a. s. and so

$$E(\mathrm{e}^{\mathrm{i}zX_n}, Y=p) \to E(\mathrm{e}^{\mathrm{i}zX}, Y=p), \qquad p=0,1,2,\ldots\,.$$

Thus we have

$$E(\mathrm{e}^{\mathrm{i}zX}, Y=p)P(Y=0) = E(\mathrm{e}^{\mathrm{i}zX}, Y=0)P(Y=p)\,. \tag{2}$$

Adding both sides over $p=0,1,2,\ldots$, we get

$$E(\mathrm{e}^{\mathrm{i}zX})P(Y=0) = E(\mathrm{e}^{\mathrm{i}zX}, Y=0)\,.$$

Putting this in (2), we get

$$E(\mathrm{e}^{\mathrm{i}zX}, Y=p)P(Y=0) = E(\mathrm{e}^{\mathrm{i}zX})P(Y=0)P(Y=p)\,.$$

Since $Y$ is Poisson distributed, $P(Y=0)$ is positive and so

$$E(\mathrm{e}^{\mathrm{i}zX}, Y=p) = E(\mathrm{e}^{\mathrm{i}zX})P(Y=p)$$

namely

$$E(\mathrm{e}^{\mathrm{i}zX}\mathrm{e}^{\mathrm{i}wY}) = E(\mathrm{e}^{\mathrm{i}zX})E(\mathrm{e}^{\mathrm{i}wY})$$

for every $(z,w)$. This proves the independence of $X$ and $Y$.

Let $t_0 < t_1 < \cdots < t_n$. Then $X(t_k)-X(t_{k-1})$ and $Y(t_k)-Y(t_{k-1})$ are independent for each $k$. By our assumption

$$(X(t_k)-X(t_{k-1}), Y(t_k)-Y(t_{k-1})), \qquad k=1,2,\ldots n$$

[8] For the second inequality, observe that

$$|\alpha_1\cdots\alpha_n - \beta_1\cdots\beta_n| \leq |\alpha_1-\beta_1| + \cdots + |\alpha_n-\beta_n|$$

for complex numbers $\alpha_1,\ldots,\alpha_n,\beta_1,\ldots,\beta_n$ with moduli $\leq 1$. Indeed,

$$|\alpha_1\alpha_2 - \beta_1\beta_2| = |\alpha_1(\alpha_2-\beta_2) + (\alpha_1-\beta_1)\beta_2| \leq |\alpha_1-\alpha_2| + |\beta_1-\beta_2|$$

and similarly in general.

are also independent, since each random vector is measurable $(\mathcal{B}_{t_{k-1}t_k})$. Therefore

$$X(t_k) - X(t_{k-1}), Y(t_k) - Y(t_{k-1}), \qquad k = 1, 2, \dots n$$

are independent by Theorem 0.1.4 (ii). Therefore

$$\mathcal{B}[X(t_k) - X(t_{k-1}): k = 1, 2, \dots n] \text{ and } \mathcal{B}[Y(t_k) - Y(t_{k-1}): k = 1, 2, \dots n]$$

are independent by Theorem 0.1.4 (i). Since these $\sigma$-algebras are directed up as $\{t_0 < t_1 < \cdots < t_n\}$ ranges over all finite subsets of $T$ and generate $\mathcal{B}[X]$ and $\mathcal{B}[Y]$ respectively, $\mathcal{B}[X]$ and $\mathcal{B}[Y]$ are independent by Theorem 0.1.3. This completes the proof of the Fundamental Lemma.

## 1.6 Structure of Sample Functions of Lévy Processes (a) Number of Jumps

Let $X = X(t, \omega)$ be a Lévy process. The sample function of $X$ will be denoted by $X(\cdot, \omega)$ or simply by $X(\cdot)$. To avoid repetition of "a. s." we will assume, as we can, that $X(\cdot, \omega) \in D$ for *every* $\omega$. The set of all $t > 0$ for which

$$|X(t, \omega) - X(t-, \omega)| > 0$$

is denoted by $I(\omega)$. This set $I(\omega)$ consists of all discontinuity points of $X(\cdot, \omega)$. We also consider the two-dimensional set

$$J = J(\omega) = \{(t, X(t, \omega) - X(t-, \omega)): t \in I(\omega)\} .$$

Then $J(\omega)$ is a countable subset of $T_0 \times R_0$, where

$$T_0 = T - \{0\}, \qquad R_0 = R^1 - \{0\} ,$$

because $X(\cdot, \omega) \in D$.

Let $\mathcal{B}(T_0 \times R_0)$ denote the class of all Borel subsets of $T_0 \times R_0$, as usual, and let $\mathcal{B}^*(T_0 \times R_0)$ denote the class of all $A \in \mathcal{B}(T_0 \times R_0)$ such that

$$A \subset (0, a) \times \{u: |u| > 1/a\} \qquad \text{for some } a > 0 .$$

Then $\mathcal{B}^*(T_0 \times R_0)$ is closed under finite unions, countable intersections, and differences but neither under complements nor under countable unions in general.

By virtue of $X(\cdot, \omega) \in D$, $A \cap J(\omega)$ is a finite set depending on $\omega$ as long as $A \in \mathcal{B}^*(T_0 \times R_0)$. The number of points in this set $A \cap J(\omega)$ will be denoted by $N(A, \omega)$ and will play an important role in our discussion. First we will observe that $N(A, \omega)$ is measurable in $\omega$ and so is a random variable. More precisely we have

**Proposition 1.** *If* $A \in \mathcal{B}^*(T_0 \times R_0)$ *and* $A \subset (s,t] \times R_0$*, then* $N(A)$ *is measurable* ($\overline{\mathcal{B}}_{st}[\mathrm{d}X]$).

*Proof.* Let $E_{s,t,a}$ denote the set $(s,t] \times (a,\infty) \subset T_0 \times R_0$, where $0 \le s < t < \infty$ and $a > 0$, and let $Q$ be a countable dense subset of $(s,t]$ including the right end point $t$. It is easy to see that

$$\begin{aligned} \{N(E_{s,t,a}) \ge 1\} &= \{X(\tau) - X(\tau-) > a \quad \text{for some } \tau \in (s,t]\} \\ &= \bigcup_p \bigcap_q \bigcup_{\substack{s+1/p \le r < r' \le t \\ r,r' \in Q,\ r'-r<1/q}} \{X(r') - X(r) \ge a + 1/p\} \\ &\in \overline{\mathcal{B}}_{st}[\mathrm{d}X]\,. \end{aligned}$$

Observing

$$\{N(E_{s,t,a}) \ge k+1\} = \bigcup_{r \in Q \cap (s,t)} \{N(E_{s,r,a}) \ge k\} \cap \{N(E_{r,t,a}) \ge 1\}\,,$$

we get

$$\{N(E_{s,t,a}) \ge k\} \in \overline{\mathcal{B}}_{st}[\mathrm{d}X], \qquad k = 1,2,\dots\,,$$

which proves that $N(E_{s,t,a})$ is measurable ($\overline{\mathcal{B}}_{st}[\mathrm{d}X]$).

Let $\mathcal{D}$ denote the class of all sets $A \in \mathcal{B}(T_0 \times R_0)$ such that $N(A \cap E_{s,t,a})$ is measurable ($\overline{\mathcal{B}}_{st}[\mathrm{d}X]$) for $0 \le s < t < \infty$ and $a > 0$. Since $N(A \cap E_{s,t,a})$ is a bounded measure in $A$ for every $\omega$, $\mathcal{D}$ is a Dynkin class as introduced in Section 0.1. By the fact observed above, $\mathcal{D}$ includes the class

$$\mathcal{M} = \{(s,\infty) \times ((b,\infty) - \{0\})\colon 0 \le s < \infty,\ -\infty \le b < \infty\}\,,$$

which is obviously a multiplicative class in $T_0 \times R_0$ and generates the $\sigma$-algebra $\mathcal{B}(T_0 \times R_0)$. By Dynkin's theorem (Lemma 0.1.1) we have

$$\mathcal{D} \supset \mathcal{D}[\mathcal{M}] = \mathcal{B}[\mathcal{M}] = \mathcal{B}(T_0 \times R_0)\,.$$

Therefore $N(A \cap E_{s,t,a})$ is measurable ($\overline{\mathcal{B}}_{st}[\mathrm{d}X]$) for every $A \in \mathcal{B}(T_0 \times R_0)$.

Writing $E'_{s,t,a}$ for $(s,t] \times (-\infty,-a)$ $(a > 0)$ and using the same argument as above, we can see that $N(A \cap E'_{s,t,a})$ is also measurable ($\overline{\mathcal{B}}_{st}[\mathrm{d}X]$).

If $A \in \mathcal{B}^*(T_0 \times R_0)$ and $A \subset (s,t] \times R_0$, we have

$$A = (A \cap E_{s,t,a}) \cup (A \cap E'_{s,t,a})$$

and so

$$N(A) = N(A \cap E_{s,t,a}) + N(A \cap E'_{s,t,a})$$

by taking $a > 0$ sufficiently small. Therefore $N(A)$ is measurable ($\overline{\mathcal{B}}_{st}[\mathrm{d}X]$). This completes the proof.

**Proposition 2.** *For $A \in \mathcal{B}(T_0 \times R_0)$, $N(A)$ is either Poisson distributed or identically equal to $\infty$. (In the second case $N(A)$ is understood to be Poisson distributed with parameter $= \infty$ as a convention). If $A \in \mathcal{B}^*(T_0 \times R_0)$, then $N(A)$ is Poisson distributed with finite parameter.*

*Proof.* First we will discuss the case $A \in \mathcal{B}^*(T_0 \times R_0)$. Write $A(t)$ for the intersection of $A$ with $(0,t] \times R_0$ and consider the stochastic process

$$N(t) = N(t,\omega) = N(A(t),\omega) .$$

It is obvious that $N(t,\omega)$ is a right continuous step function in $t$ increasing with jumps $= 1$. Since

$$N(t) - N(s) = N(A(t) - A(s)) = N[A \cap ((s,t] \times R_0)] \quad \text{for } s < t ,$$

it is easy to see by Proposition 1 that $N(t)$ is an additive process. Since, for every $t$ fixed,

$$P(N(t) - N(t-) \neq 0) \leq P(X(t) - X(t-) \neq 0) = 0 ,$$

$N(t)$ is continuous i. p.. Therefore $N(t)$ is a Lévy process of Poisson type by Theorem 1.4.1 (ii). Since $A \in \mathcal{B}^*(T_0 \times R_0)$, we have $A = A(t)$ and so $N(A) \equiv N(A(t))$ for sufficiently large $t$. Therefore $N(A)$ is Poisson distributed with finite parameter.

Let $A \in \mathcal{B}(T_0 \times R_0)$. Then we have an increasing sequence $A_n \in \mathcal{B}^*(T_0 \times R_0)$, $n = 1, 2, \ldots$, such that $A_n \uparrow A$. Then

$$N(A) = \lim_{n\to\infty} N(A_n) .$$

Each $N(A_n)$ is Poisson distributed with the parameter $\lambda_n \equiv E(N(A_n))$, and $\lambda_n$ is increasing. If $\lambda = \lim_n \lambda_n < \infty$, then $N(A)$ is Poisson distributed with parameter $\lambda$. If $\lambda = \infty$, then

$$P(N(A) \leq k) \leq P(N(A_n) \leq k) = \mathrm{e}^{-\lambda_n} \sum_{j=0}^{k} \frac{\lambda_n{}^j}{j!} \to 0$$

as $n \to \infty$ for $k$ fixed. This proves $P(N(A) = \infty) = 1$, completing the proof.

It is obvious that $N(A,\omega)$ is a measure in $A \in \mathcal{B}(T_0 \times R_0)$ which takes the values $0, 1, 2, \ldots$ and $\infty$. Therefore

$$\begin{aligned} n(A) &= E(N(A)) \\ &= \text{the parameter of the Poisson variable } N(A) \end{aligned}$$

is also a measure in $A \in \mathcal{B}(T_0 \times R_0)$ which may take the value $\infty$. In Proposition 2 we saw that $n(A) < \infty$ if $A \in \mathcal{B}^*(T_0 \times R_0)$. The measure $n(A)$ is called the *Lévy measure* of the Lévy process $X$.

Now we consider

$$S(A) = S(A,\omega) = \sum_{(t,X(t,\omega)-X(t-,\omega))\in A} (X(t,\omega) - X(t-,\omega)) = \sum_{(t,u)\in A\cap J(\omega)} u$$

for $A \in \mathcal{B}^*(T_0 \times R_0)$. Since $A \cap J(\omega)$ is a finite set for such $A$, the sum is a finite sum and so there is no problem of convergence. This $S(A,\omega)$ is also measurable in $\omega$ and so a random variable, because

$$(1) \qquad S(A,\omega) = \lim_{n\to\infty} \sum_k \frac{k}{n} N\left(A \cap \left(T_0 \times \left(\frac{k-1}{n}, \frac{k}{n}\right]\right), \omega\right).$$

This relation is also expressed as

$$(1') \qquad S(A,\omega) = \iint_{(t,u)\in A} uN(\mathrm{d}t\mathrm{d}u, \omega).$$

In view of (1) it is easy to see

**Proposition 1′.** *Proposition 1 holds for $S(A)$ in place of $N(A)$.*

Let $A \in \mathcal{B}^*(T_0 \times R_0)$ and set

$$H(t;A) = X(t) - S(A(t)) \quad \text{with } A(t) = A \cap ([0,t] \times R_0).$$

**Proposition 3.** *The process $\{H(t;A)\}_t$ is a Lévy process independent of the process $\{N(A(t))\}_t$.*

*Proof.* It is easy to see by Proposition 1′ that $\{H(t;A)\}_t$ is a Lévy process adapted to $\{\overline{\mathcal{B}}_{st}[\mathrm{d}X]\}$. We have already seen that $\{N(A(t))\}_t$ is a Lévy process of Poisson type adapted to $\{\overline{\mathcal{B}}_{st}[\mathrm{d}X]\}$. It is obvious that the sample function of $\{H(t;A)\}_t$ and that of $\{N(A(t))\}_t$ have no common jump time for every $\omega$. Therefore they are independent by the Fundamental Lemma proved in the previous section.

The notions $N(A)$, $S(A)$, and $H(t;A)$ are defined with respect to the Lévy process $X$. To emphasize this point we often write $N_X$, $S_X$, and $H_X$ respectively for $N$, $S$, and $H$.

Now we will generalize Proposition 3.

**Proposition 4.** *Let $A_1, A_2, \ldots, A_n \in \mathcal{B}^*[T_0 \times R_0]$ be disjoint. Then the following processes are independent:*

$$\{N(A_1(t))\}_t, \ldots, \{N(A_n(t))\}_t, \{H(t; \textstyle\bigcup_{i=1}^n A_i)\}_t.$$

*Proof.* Set

$$N_k(t) = N(A_k(t)) \quad \text{and} \quad H_k(t) = H(t; \textstyle\bigcup_{i=1}^k A_i) .$$

Then it is easy to see that

$$N_{k+1}(t) = N_{H_k}(A_{k+1}(t)),$$
$$H_{k+1}(t) = H_{H_k}(t; A_{k+1}), \quad k = 0, 1, 2, \ldots, n-1,$$

where $H_0 \equiv X$. It follows at once from this that

$$\mathcal{B}[N_{k+1}] \vee \mathcal{B}[H_{k+1}] \subset \mathcal{B}[H_k], \quad k = 0, 1, 2, \ldots, n-1 ,$$

so that

$$\mathcal{B}[N_{k+1}] \vee \mathcal{B}[N_{k+2}] \vee \cdots \vee \mathcal{B}[N_n] \vee \mathcal{B}[H_n] \subset \mathcal{B}[H_k], \quad k = 0, 1, 2, \ldots, n-1$$

Applying Proposition 3 to the Lévy process $H_k$ and the set $A_{k+1} \in \mathcal{B}^*(T_0 \times R_0)$, we see that $\mathcal{B}[N_{k+1}]$ is independent of $\mathcal{B}[H_{k+1}]$. Therefore $\mathcal{B}[N_1], \mathcal{B}[N_2], \ldots, \mathcal{B}[N_n], \mathcal{B}[H_n]$ are independent. Indeed, if $B_i \in \mathcal{B}[N_i]$, $i = 1, 2, \ldots, n$, and $C \in \mathcal{B}[H_n]$, then $B_{k+1} \cap \cdots \cap B_n \cap C \in \mathcal{B}[H_k]$, $k = 0, 1, \ldots, n-1$, and so

$$\begin{aligned} P(B_1 \cap B_2 \cap \cdots \cap B_n \cap C) &= P(B_1)P(B_2 \cap \cdots \cap B_n \cap C) \\ &= P(B_1)P(B_2)P(B_3 \cap \cdots \cap B_n \cap C) \\ &= \cdots \\ &= P(B_1)P(B_2)P(B_3) \cdots P(B_n)P(C) . \end{aligned}$$

This completes the proof.

In view of (1) we can derive the following from Proposition 4.

**Proposition 4′.** *Proposition 4 holds for $S$ in place of $N$.*

As a corollary of Proposition 4 we get

**Proposition 5.** *Let $A_1, A_2, \ldots, A_n \in \mathcal{B}(T_0 \times R_0)$ be disjoint. Then $N(A_1), N(A_2), \ldots, N(A_n)$ are independent.*

*Proof.* Since every $A \in \mathcal{B}(T_0 \times R_0)$ can be expressed as the limit of an increasing sequence of sets in $\mathcal{B}^*(T_0 \times R_0)$, we can assume that

$$A \in \mathcal{B}^*(T_0 \times R_0), \quad i = 1, 2, \ldots, n .$$

Then we can take $t_0$ so large that $A_i = A_i(t_0)$ for every $i$. Therefore $N(A_i)$, $i = 1, 2, \ldots, n$, are independent by Proposition 4.

For $A \in \mathcal{B}^*(T_0 \times R_0)$ we set

$$A_{mk} = \left\{(s,u) \in A \colon \frac{k}{m} < u \le \frac{k+1}{m}\right\} .$$

Then

$$S(A,\omega) = \iint_A uN(\mathrm{d}s\mathrm{d}u,\omega) = \lim_{m\to\infty} \sum_k \frac{k}{m} N(A_{mk},\omega) . \tag{2}$$

**Proposition 6.** *If $A \in \mathcal{B}^*(T_0 \times R_0)$, then*

$$E[\mathrm{e}^{\mathrm{i}zS(A)}] = \exp\left\{\iint_A (\mathrm{e}^{\mathrm{i}zu} - 1)n(\mathrm{d}s\mathrm{d}u)\right\} .$$

*Proof.* Since $\{A_{mk}\}$ are disjoint for each $m$, $\{N(A_{mk})\}_k$ are independent. Since $N(A_{mk})$ is Poisson distributed with parameter $n(A_{mk})$, we have

$$E\left[\mathrm{e}^{\mathrm{i}z(k/m)N(A_{mk})}\right] = \exp\{(\mathrm{e}^{\mathrm{i}zk/m} - 1)n(A_{mk})\} .$$

Therefore we have

$$\begin{aligned}
E[\mathrm{e}^{\mathrm{i}zS(A)}] &= \lim_{m\to\infty} E\left[\mathrm{e}^{\mathrm{i}z\sum_k (k/m)N(A_{mk})}\right] \\
&= \lim_{m\to\infty} \prod_k E\left[\mathrm{e}^{\mathrm{i}z(k/m)N(A_{mk})}\right] \\
&= \lim_{m\to\infty} \prod_k \exp\{(\mathrm{e}^{\mathrm{i}zk/m} - 1)n(A_{mk})\} \\
&= \lim_{m\to\infty} \exp\left\{\sum_k (\mathrm{e}^{\mathrm{i}zk/m} - 1)n(A_{mk})\right\} \\
&= \exp\left\{\iint_A (\mathrm{e}^{\mathrm{i}zu} - 1)n(\mathrm{d}s\mathrm{d}u)\right\} ,
\end{aligned}$$

noticing that $n(A) < \infty$.

Similarly we have

**Proposition 7.** *If $A \in \mathcal{B}^*(T_0 \times R_0)$ is included in $\{(s,u) \colon |u| < m\}$ for some $m < \infty$, then*

$$E[S(A)] = \iint_A u\, n(\mathrm{d}s\mathrm{d}u) ,$$

$$V[S(A)] = \iint_A u^2 n(\mathrm{d}s\mathrm{d}u) .$$

Since $(0,t] \times \{u\colon |u| > \varepsilon\}$ $(\varepsilon > 0)$ belongs to $\mathcal{B}^*(T_0 \times R_0)$, we have

$$\iint\limits_{\substack{0<s\le t\\ |u|>\varepsilon}} n(\mathrm{d}s\mathrm{d}u) < \infty\,.$$

From the behavior of the measure $n(\mathrm{d}s\mathrm{d}u)$ near $u = 0$ we have

**Proposition 8.** $$\iint\limits_{\substack{0<s\le t\\ |u|<1}} u^2 n(\mathrm{d}s\mathrm{d}u) < \infty\,.$$

*Proof.* Set

$$E_\varepsilon = \{(s,u)\colon 0 < s \le t,\ \varepsilon < |u| < 1\}, \qquad \varepsilon > 0\,.$$

Then $E_\varepsilon \in \mathcal{B}^*(T_0 \times R_0)$ and we have

$$E\left[\mathrm{e}^{\mathrm{i}zS(E_\varepsilon)}\right] = \exp\left\{\iint\limits_{E_\varepsilon} (\mathrm{e}^{\mathrm{i}zu} - 1) n(\mathrm{d}s\mathrm{d}u)\right\}$$

by Proposition 6. By Proposition 4′ the two processes

$$\{S(E_\varepsilon(s))\}_s \quad \text{and} \quad \{X(s) - S(E_\varepsilon(s))\}_s$$

are independent and so $S(E_\varepsilon(t))$ and $X(t) - S(E_\varepsilon(t))$ are independent. Since $E_\varepsilon(t) = E_\varepsilon$, we see that $S(E_\varepsilon)$ and $X(t) - S(E_\varepsilon)$ are independent. Therefore we have

$$E[\mathrm{e}^{\mathrm{i}zX(t)}] = E\left[\mathrm{e}^{\mathrm{i}zS(E_\varepsilon)}\right] E\left[\mathrm{e}^{\mathrm{i}z(X(t)-S(E_\varepsilon))}\right]$$

and so

$$(3) \qquad |E\{\mathrm{e}^{\mathrm{i}zX(t)}\}| \le |E[\mathrm{e}^{\mathrm{i}zS(E_\varepsilon)}]| = \exp\left\{\iint\limits_{E_\varepsilon} (\cos zu - 1)\, n(\mathrm{d}s\mathrm{d}u)\right\}$$

$$\le \exp\left\{-\frac{z^2}{4}\iint\limits_{E_\varepsilon} u^2 n(\mathrm{d}s\mathrm{d}u)\right\} \qquad \text{for } |z| < 1,$$

noticing that $\cos x \le 1 - x^2/4$ for $|x| \le 1$.

Suppose that

$$(4) \qquad \iint\limits_{\substack{0<s\le t\\ 0<|u|<1}} u^2 n(\mathrm{d}s\mathrm{d}u) = \infty\,.$$

Then, letting $\varepsilon \downarrow 0$ in (3), we have

$$|E[\mathrm{e}^{\mathrm{i}zX(t)}]| \le 0 \qquad \text{for } 0 < |z| < 1\,.$$

Letting $z \to 0$, we get $1 \le 0$, which is absurd. Therefore the integral in the left side of (4) must be finite.

Since $X(t) = X(t-)$ a. s. for each $t$, we have

$$N(\{t\} \times R_0) = 0 \quad \text{a. s. for each } t\,.$$

Therefore we get

**Proposition 9.** $n(\{t\} \times R_0) = 0\,.$

## 1.7 Structure of Sample Functions of Lévy Processes (b) Lévy's Decomposition Theorem

We will use the same notation as in the previous section. Set

$$\begin{aligned} S_1(t,\omega) &= \iint_{\substack{0<s\leq t \\ u\geq 1}} uN(\mathrm{d}s\mathrm{d}u,\omega) \\ &= S((0,t] \times [1,\infty),\omega) \\ &= \sum_{\substack{0<s\leq t \\ X(s,\omega)-X(s-,\omega)\geq 1}} (X(s,\omega) - X(s-,\omega)) \end{aligned}$$

and, for $k = 2, 3, \dots,$

$$\begin{aligned} S_k(t,\omega) &= \iint_{\substack{0<s\leq t \\ 1/k\leq u<1/(k-1)}} uN(\mathrm{d}s\mathrm{d}u,\omega) \\ &= S\left((0,t] \times \left[\frac{1}{k},\frac{1}{k-1}\right),\omega\right) \\ &= \sum_{\substack{0<s\leq t \\ 1/k\leq X(s,\omega)-X(s-,\omega)<1/(k-1)}} (X(s,\omega) - X(s-,\omega))\,, \end{aligned}$$

$$T_k(t,\omega) = S_1(t,\omega) + \sum_{j=2}^{k} (S_j(t,\omega) - E(S_j(t)))\,.$$

Using the results obtained in the previous section we can easily see that $\{S_n(t,\omega)\}_t$, $n = 1, 2, \dots,$ are independent Lévy processes with step-wise sample functions. Since

$$E[S_k(t)] = \iint_{\substack{0<s\leq t \\ 1/k\leq u<1/(k-1)}} u\,n(\mathrm{d}s\mathrm{d}u)\,,$$

this is continuous in $t$ by Proposition 1.6.9. Therefore, $\{T_k(t,\omega)\}_t$ is also a Lévy process for every $k$ and

$$T_k(t,\omega) = \iint_{\substack{0<s\leq t\\ u\geq 1/k}} uN(\mathrm{d}s\mathrm{d}u,\omega) - \iint_{\substack{0<s\leq t\\ 1>u\geq 1/k}} u\, n(\mathrm{d}s\mathrm{d}u)\,.$$

Setting

$$R_{km}(t,\omega) = \sum_{\nu=k+1}^{m} (S_\nu(t,\omega) - E(S_\nu(t))) \qquad (k<m)\,,$$

we get

**Proposition 1 (Continuous version of Kolmogorov's inequality).**

$$P\left(\sup_{0\leq t\leq a} |R_{km}(t,\omega)| > b\right) \leq \frac{1}{b^2}E(R_{km}(a)^2) = \frac{1}{b^2}\iint_{\substack{0<s\leq a\\ 1/m\leq u<1/k}} u^2 n(\mathrm{d}s\mathrm{d}u)$$

*for $k<m$ and $b>0$.*

*Proof.* It is easy to see that $R_{km}(t)$ is a Lévy process. Therefore the probability in question is

$$\lim_{p\to\infty} P\left(\max_{1\leq i\leq p}\left|R_{km}\left(\frac{i}{p}a,\omega\right)\right| > b\right).$$

Now set

$$X_i = R_{km}\left(\frac{i}{p}a,\omega\right) - R_{km}\left(\frac{i-1}{p}a,\omega\right), \qquad i=1,2,\ldots,p\,.$$

Then it is also easy to see that $\{X_i\}_i$ are independent random variables with $E(X_i)=0$. By Kolmogorov's inequality we get

$$P\left(\max_{1\leq i\leq p}|X_1+X_2+\cdots+X_i| > b\right) \leq \frac{1}{b^2}E[(X_1+X_2+\cdots+X_p)^2]\,,$$

which means

$$\begin{aligned} P\left(\max_{1\leq i\leq p}\left|R_{km}\left(\frac{i}{p}a,\omega\right)\right| > b\right) &\leq \frac{1}{b^2}E[R_{km}(a)^2] \\ &= \frac{1}{b^2}V\left[S\left((0,a]\times\left[\frac{1}{m},\frac{1}{k}\right)\right)\right] \\ &= \frac{1}{b^2}\iint_{\substack{0<s\leq a\\ 1/m\leq u<1/k}} u^2 n(\mathrm{d}s\mathrm{d}u)\,. \end{aligned}$$

This completes the proof.

Now set

$$\theta(k,m,a,b) = \frac{1}{b^2} \iint_{\substack{0<s\le a \\ 1/m\le u<1/k}} u^2 n(dsdu)$$

for $a > 0$ and $b > 0$. Since

$$\iint_{\substack{0<s\le a \\ |u|\le 1}} u^2 n(dsdu) < \infty$$

by Proposition 1.6.8, it is easy to see that

$$\lim_{k,m\to\infty} \theta(k,m,a,b) = 0\,.$$

**Proposition 2.** *If* $\theta(k,m,a,b) < 1/2$ *, then*

*(i)* $\displaystyle P\left(\max_{k<\nu\le m} \sup_{0\le t\le a} |R_{k\nu}(t,\omega)| > 2b\right) \le 2\theta(k,m,a,b)\,,$

*(ii)* $\displaystyle P\left(\max_{k<\nu<\mu\le m} \sup_{0\le t\le a} |R_{\nu\mu}(t,\omega)| > 4b\right) \le 4\theta(k,m,a,b)\,.$

*Proof.* Since (ii) follows from (i) at once, it is enough to prove (i). Set

$$Y_{\nu\mu} = \sup_{0\le t\le a} |R_{\nu\mu}(t,\omega)|\,.$$

It is clear that $\{Y_{j,j+1}\}_j$ are independent and that, if $\nu < \mu < \lambda$, then

$$Y_{\nu\lambda} \ge Y_{\nu\mu} - Y_{\mu\lambda}\,.$$

What we have to prove is that

$$P\left(\max_{k<\nu\le m} Y_{k\nu} > 2b\right) \le 2\theta(k,m,a,b)\,.$$

Set

$$\begin{aligned} B_\nu &= \{Y_{k,k+1} \le 2b,\ \dots,\ Y_{k,\nu-1} \le 2b,\ Y_{k\nu} > 2b\}\,, \\ C_\nu &= \{Y_{\nu m} \le b\} \end{aligned}$$

for $\nu = k+1, \dots, m$ with the understanding that $C_m = \Omega$. Then, by Proposition 1,

$$\begin{aligned} P(C_\nu) &= 1 - P(Y_{\nu m} > b) \ge 1 - \theta(\nu,m,a,b) \ge 1 - \theta(k,m,a,b) \\ &> 1/2 \end{aligned}$$

and, since $Y_{km} \geq Y_{k\nu} - Y_{\nu m}$,

$$\bigcup_{\nu=k+1}^{m} (B_\nu \cap C_\nu) \subset \{Y_{km} > b\} .$$

Since the left side is a disjoint union, we have

$$\sum_\nu P(B_\nu \cap C_\nu) \leq \theta(k, m, a, b)$$

by Proposition 1. Since $B_\nu$ and $C_\nu$ are independent by the independence of the family of processes $\{\{R_{j,j+1}(t) \colon 0 \leq t \leq a\}\}_j$, we have

$$\sum_\nu P(B_\nu)P(C_\nu) \leq \theta(k, m, a, b) ,$$

and, since $P(C_\nu) > 1/2$,

$$\sum_\nu P(B_\nu) \leq 2\theta(k, m, a, b) .$$

Therefore

$$P\left(\bigcup_\nu B_\nu\right) \leq 2\theta(k, m, a, b) ,$$

that is,

$$P\left(\max_{k<\nu\leq m} Y_{n\nu} > 2b\right) \leq 2\theta(k, m, a, b) .$$

**Proposition 3.** *With probability* 1, $T_k(t, \omega)$ *is convergent uniformly on every bounded* $t$*-interval, as* $k \to \infty$.

*Proof.* It is enough to prove that given $a > 0$, with probability 1, $T_k(t, \omega)$ is convergent uniformly on $0 \leq t \leq a$ as $k \to \infty$. Using Proposition 2, we have

$$\begin{aligned}
&P\left(\lim_{k\to\infty} \lim_{m\to\infty} \max_{k<\nu<\mu\leq m} \sup_{0\leq t\leq a} |R_{\nu\mu}(t,\omega)| > 4b\right) \\
&\quad \leq \lim_{k\to\infty} \lim_{m\to\infty} P\left(\max_{k<\nu<\mu\leq m} \sup_{0\leq t\leq a} |R_{\nu\mu}(t,\omega)| > 4b\right) \\
&\quad \leq 4 \lim_{k\to\infty} \lim_{m\to\infty} \theta(k, m, a, b) = 0 .
\end{aligned}$$

Letting $b \downarrow 0$, we have

$$P\left(\lim_{k\to\infty} \lim_{m\to\infty} \max_{k<\nu<\mu\leq m} \sup_{0\leq t\leq a} |R_{\nu\mu}(t,\omega)| > 0\right) = 0 .$$

This proves that with probability 1, $T_k(t, \omega)$ is convergent uniformly in $0 \leq t \leq a$ as $k \to \infty$.

Now consider

$$X_k^+(t,\omega) = \iint_{\substack{0<s\le t\\ u\ge 1/k}} uN(\mathrm{d}s\mathrm{d}u,\omega) - \iint_{\substack{0<s\le t\\ u\ge 1/k}} \frac{u}{1+u^2}\, n(\mathrm{d}s\mathrm{d}u)\,.$$

Then

$$\begin{aligned} X_k^+(t,\omega) - T_k(t,\omega) &= -\iint_{\substack{0<s\le t\\ u\ge 1}} \frac{u}{1+u^2}\, n(\mathrm{d}s\mathrm{d}u) + \iint_{\substack{0<s\le t\\ 1>u\ge 1/k}} \frac{u^3}{1+u^2}\, n(\mathrm{d}s\mathrm{d}u) \\ &\rightrightarrows -\iint_{\substack{0<s\le t\\ u\ge 1}} \frac{u}{1+u^2}\, n(\mathrm{d}s\mathrm{d}u) + \iint_{\substack{0<s\le t\\ 1>u>0}} \frac{u^3}{1+u^2}\, n(\mathrm{d}s\mathrm{d}u) \end{aligned}$$

on every bounded $t$-interval, because

$$\iint_{\substack{0<s<a\\ u\ge 1}} n(\mathrm{d}s\mathrm{d}u) < \infty, \qquad \iint_{\substack{0<s<a\\ 0<u<1}} u^2 n(\mathrm{d}s\mathrm{d}u) < \infty\,.$$

Here the symbol $\rightrightarrows$ denotes uniform convergence. Therefore

**Proposition 4.** *With probability* 1, $X_k^+(t,\omega)$ *converges to a Lévy process, (say* $X^+(t,\omega)$*) uniformly on every bounded* $t$*-interval as* $k \to \infty$.

*Proof.* It is enough to observe that if a sequence of functions in $D$ converges uniformly in every bounded $t$-interval, then the limit function is also in $D$.

Similarly consider

$$X_k^-(t,\omega) = \iint_{\substack{0<s\le t\\ u\le -1/k}} uN(\mathrm{d}s\mathrm{d}u,\omega) - \iint_{\substack{0<s\le t\\ u\le -1/k}} \frac{u}{1+u^2}\, n(\mathrm{d}s\mathrm{d}u)\,.$$

Then we have

**Proposition 4′.** *Proposition 4 holds for* $X_k^-(t,\omega)$ *in place of* $X_k^+(t,\omega)$. *(Write the limit as* $X^-(t,\omega)$.*)*

Now set

$$X^0(t,\omega) = X(t,\omega) - X^+(t,\omega) - X^-(t,\omega)\,.$$

Then we have

**Proposition 5.** *The process* $X^0(t,\omega)$ *is a Lévy process of Gauss type.*

*Proof.* Since $X^0(t)$ is adapted to $\{\mathcal{B}_{st}[\mathrm{d}X]\}_{st}$ and since $X^0(0) \equiv 0$ is obvious, $X^0(t)$ is an additive process. It remains only to prove that its sample function is continuous a. s. With probability 1,

$$X_k^0(t,\omega) = X(t,\omega) - X_k^+(t,\omega) - X_k^-(t,\omega)$$

converges to $X^0(t,\omega)$ uniformly in every bounded $t$-interval. Since

$$X_k^+(t,\omega) + X_k^-(t,\omega) = \iint_{\substack{0<s\le t\\ |u|\ge 1/k}} uN(\mathrm{d}s\mathrm{d}u,\omega) - \iint_{\substack{0<s\le t\\ |u|\ge 1/k}} \frac{u}{1+u^2}\, n(\mathrm{d}s\mathrm{d}u)$$

and the second term in the right-hand side is a continuous function of $t$ by Proposition 1.6.9, $X_k^0(t,\omega)$ has no jump with the absolute size $> 1/m$ a. s. as long as $k \ge m$. Therefore $X^0(t,\omega)$ is continuous a. s. because of the uniform convergence on every bounded $t$-interval.

**Proposition 6.** *The family $\{N(A,\omega)\colon A \in \mathcal{B}^*(T_0 \times R_0)\}$ and the process $X^0(t,\omega)$ are independent.*

*Proof.* Since $X_k^0(t,\omega)$ is different from

$$X(t,\omega) - S((0,t] \times \{u\colon |u| \ge 1/k\})$$

by a (continuous) function independent of $\omega$, $\{X_k^0(t,\omega)\}_t$ is independent of the family

$$\{N(A,\omega)\colon A \in \mathcal{B}^*(T_0 \times R_0),\ A \subset \{(t,u)\colon |u| \ge 1/m\}\}$$

as far as $k > m$ (by Proposition 1.6.4) and so is $\{X^0(t,\omega)\}_t$. It is now easy to complete the proof.

Summarizing the results obtained above and writing $X^0(t)$ as $X_0(t)$, we have

**Theorem 1 (Lévy's decomposition theorem).** [9] *Let $X(t,\omega)$ be a Lévy process. Set*

[9] This result is usually called the Lévy–Itô decomposition of sample functions of a Lévy process. Original references are the following:

P. Lévy, Sur les intégrales dont les éléments sont des variables aléatoires indépendantes, Ann. Scuola Norm. Sup. Pisa, Sér. 2, **3**, 337–366; **4**, 217–218 (1934) (see Œuvre de Paul Lévy, Volume 4, Gauthier-Villars, 1980).

P. Lévy, Théorie de l'Addition des Variables Aléatoires, Gauthier-Villars, Paris, 1937.

K. Itô, On stochastic processes, I (Infinitely divisible laws of probability), Japan. J. Math., **18**, 261–301 (1942) (see Kiyosi Itô, Selected Papers, Springer, 1987.

$N(A, \omega) =$ *the number of jumps of* $X(t, \omega)$ *whose pairs of time and size are in* $A$ ,

*and*

$$n(A) = E[N(A)]$$

*for* $A \in \mathcal{B}(T_0 \times R_0)$. *Then we have the following.*

*(i)* $\displaystyle\iint_{0<s\le t} \frac{u^2}{1+u^2}\, n(dsdu)$ *is finite for every* $t < \infty$ *and continuous in* $t$.

*(ii)* $N(A, \omega)$ *is a measure in* $A$ *a. s.,* $N(A, \omega)$ *is Poisson distributed with parameter* $n(A)$, *and if* $A_1, A_2, \ldots, A_n$ *are disjoint, then* $N(A_1, \omega), \ldots, N(A_n, \omega)$ *are independent. (Such a family* $\{N(A, \omega)\}_A$ *is called a* Poisson random measure *on* $T_0 \times R_0$ *with mean measure* $\{n(A)\}_A$.*)*

*(iii)* $X(t, \omega)$ *is expressed as*

$$X(t, \omega) = X_0(t, \omega) + \lim_{k \to \infty} \left( \iint_{\substack{0<s\le t \\ |u|\ge 1/k}} uN(dsdu, \omega) - \iint_{\substack{0<s\le t \\ |u|\ge 1/k}} \frac{u}{1+u^2}\, n(dsdu) \right),$$

*where* $\{X_0(t, \omega)\}_t$ *is a Lévy process of Gauss type independent of the Poisson random measure* $\{N(A, \omega)\}_A$.

## 1.8 Three Components of Lévy Processes

Let $X = X(t)$ be a Lévy process. We have derived the Lévy decomposition of the sample function of $X$:

$$X(t) = X_0(t) + \lim_{m \to \infty} \left( \iint_{\substack{0<s\le t \\ |u|\ge 1/m}} uN(dsdu) - \iint_{\substack{0<s\le t \\ |u|\ge 1/m}} \frac{u}{1+u^2}\, n(dsdu) \right).$$

Consider

$$m(t) = E[X_0(t)] \quad \text{and} \quad V(t) = V[X_0(t)]\,.$$

The functions $m(t)$ and $V(t)$ and the measure $n(dsdu)$ are called the *three components* of $X$. It is obvious that these three components satisfy

(m) $m(t)$ is continuous and $m(0) = 0$,

(V) $V(t)$ is continuous and increasing and $V(0) = 0$,

(n) $n(dsdu)$ is a measure on $T_0 \times R_0$ such that $\displaystyle\iint_{\substack{0<s\le t \\ u\in R_0}} \frac{u^2}{1+u^2}\, n(dsdu) < \infty$ and $n(\{t\} \times R_0) = 0$ for every $t \in T_0$.

In this section we will prove the following

**Theorem 1.** *Given $m(t)$, $V(t)$, and $n(dsdu)$ satisfying* (m), (V), *and* (n), *we can construct a Lévy process $X$ whose three components are $m(t)$, $V(t)$, and $n(dsdu)$. Such a Lévy process $X$ is uniquely determined in law by $m$, $V$, and $n$.*

*Proof of uniqueness.* Let $X = X(t)$ be a Lévy process with the three components $m(t)$, $V(t)$, and $n(dsdu)$. Consider the Lévy decomposition of the sample function of $X$:

$$\begin{aligned} X(t) &= X_0(t) + \lim_{m\to\infty}\left(\iint_{\substack{0<s\le t\\ |u|\ge 1/m}} uN(dsdu) - \iint_{\substack{0<s\le t\\ |u|\ge 1/m}} \frac{u}{1+u^2}\,n(dsdu)\right) \\ &= X_0(t,\omega) + \lim_{m\to\infty}(S_m(t) - \alpha_m(t)), \end{aligned}$$

where $S_m(t) = S((0,t] \times \{u\colon |u| \ge 1/m\})$ and $\alpha_m(t)$ is a deterministic function.

Since $X_0(t)$ is a Lévy process of Gauss type, it is determined by its mean function $m(t)$ and its variance function $V(t)$.

The distribution of $S_m(t_2) - S_m(t_1)$ $(t_2 > t_1)$ is determined by $n(dsdu)$ because

$$E\left[e^{iz(S_m(t_2)-S_m(t_1))}\right] = \exp\left\{\iint_{\substack{t_1<s\le t_2\\ |u|\ge 1/m}} (e^{izu} - 1)\,n(dsdu)\right\}.$$

Since $S_m(t)$ is a Lévy process, $\{S_m(t)\}_t$ is determined in law by $n(dsdu)$.

*Proof of existence* (construction). Suppose that we are given $m(t)$, $V(t)$, and $n(dsdu)$ with the conditions (m), (V), and (n). Let $X_0(t)$ be a Lévy process of Gauss type with the mean function $m(t)$ and the variance function $V(t)$; $X_0(t)$ exists by Theorem 1.4.3. Let $N(A,\omega)$ be a Poisson random measure on $T_0 \times R_0$ with mean measure $n(A)$, whose existence will be proved in the next section. We can assume that $\{X_0(t)\}_t$ and $\{N(A)\}_A$ are independent. We can apply the arguments used in the previous section to prove that

$$X(t) \equiv X_0(t) + \lim_{m\to\infty}\left(\iint_{\substack{0<s\le t\\ |u|\ge 1/m}} uN(dsdu) - \iint_{\substack{0<s\le t\\ |u|\ge 1/m}} \frac{u}{1+u^2}\,n(dsdu)\right)$$

is a Lévy process whose three components are the given $m(t)$, $V(t)$, and $n(dsdu)$.

**Theorem 2.** *Let $\{X(t)\}$ be a Lévy process with the three components $m(t)$, $V(t)$, and $n(dsdu)$. Then the characteristic function of $X(t_2) - X(t_1)$ for $t_1 < t_2$ is*

$$\exp\left\{ \mathrm{i}z(m(t_2) - m(t_1)) - \frac{z^2}{2}(V(t_2) - V(t_1)) + \iint\limits_{\substack{t_1 < s \le t_2 \\ u \in R_0}} \left( \mathrm{e}^{\mathrm{i}zu} - 1 - \frac{\mathrm{i}zu}{1+u^2} \right) n(\mathrm{d}s\mathrm{d}u) \right\}$$

*and is subject to an infinitely divisible distribution.*

*Proof.* By Theorem 1.7.1 we have

$$\begin{aligned} X(t_2) - X(t_1) &= X_0(t_2) - X_0(t_1) + \lim_{k\to\infty} \left( \iint\limits_{\substack{t_1 < s \le t_2 \\ |u| \ge 1/k}} uN(\mathrm{d}s\mathrm{d}u) - \iint\limits_{\substack{t_1 < s \le t_2 \\ |u| \ge 1/k}} \frac{u}{1+u^2} n(\mathrm{d}s\mathrm{d}u) \right) \\ &\equiv (X_0(t_2) - X_0(t_1)) + \lim_{k\to\infty} S_k(t_1, t_2) , \end{aligned}$$

where these two terms are independent. Therefore

$$\begin{aligned} E\left[\mathrm{e}^{\mathrm{i}z(X(t_2)-X(t_1))}\right] &= E\left[\mathrm{e}^{\mathrm{i}z(X_0(t_2)-X_0(t_1))}\right] E\left[\lim_{k\to\infty} \mathrm{e}^{\mathrm{i}zS_k(t_1,t_2)}\right] \\ &= \mathrm{e}^{\mathrm{i}z(m(t_2)-m(t_1))-(z^2/2)(V(t_2)-V(t_1))} \lim_{k\to\infty} E\left[\mathrm{e}^{\mathrm{i}zS_k(t_1,t_2)}\right] . \end{aligned}$$

But

$$E\left[\mathrm{e}^{\mathrm{i}zS_k(t_1,t_2)}\right] = \exp\left\{ \iint\limits_{\substack{t_1 < s \le t_2 \\ |u| \ge 1/k}} \left( \mathrm{e}^{\mathrm{i}zu} - 1 - \frac{\mathrm{i}zu}{1+u^2} \right) n(\mathrm{d}s\mathrm{d}u) \right\} .$$

This completes the proof.[10]

Since every additive process continuous in probability has a Lévy modification, we have

**Theorem 3.** *Every increment of an additive process continuous in probability is subject to an infinitely divisible distribution.*

[10] The infinite divisibility of the distribution of $X(t_2) - X(t_1)$ follows from the expression of its characteristic function, by Theorem 0.4.2.

## 1.9 Random Point Measures

Let $(S, \mathcal{B}(S))$ be a measurable space. In this section we assume

(A) there exists a countable subfamily $\{A_n\}_{n\geq 1}$ of $\mathcal{B}(S)$ which generates $\mathcal{B}(S)$ and separates points in $S$.

An important consequence of this assumption is that every single point set belongs to $\mathcal{B}(S)$. Indeed, for every $a \in S$, set $B_n = A_n$ or $A_n^{\mathrm{C}}$ according as $a \in A_n$ or not.[11] Then it is obvious by (A) that $\{a\} = \bigcap B_n \in \mathcal{B}(S)$.

**Definition 1.** A *point measure* $N(A)$ on $(S, \mathcal{B}(S))$ is defined to be a $\sigma$-finite non-negative integer-valued (admitting $\infty$) measure. It is called *proper* if $N(\{a\}) = 0$ or $1$ for every $a \in S$. The set $\sigma(N) = \{a\colon N(\{a\}) > 0\}$ is called the *support* of $N$ by virtue of the following proposition.

**Proposition 1.** *For any point measure $N$, $\sigma(N)$ is a countable set and $N(\sigma(N)^{\mathrm{C}}) = 0$.*

*Proof.* The set $\sigma(N)$ is countable, because $N$ is $\sigma$-finite.

Suppose $N(\sigma(N)^{\mathrm{C}}) > 0$. Using the $\sigma$-finiteness of $N$, choose $D \in \mathcal{B}(S)$ such that $D \subset \sigma(N)^{\mathrm{C}}$ and $0 < N(D) < \infty$. Then set

$$B_1 = A_1 \text{ or } A_1^{\mathrm{C}} \text{ according as } N(D \cap A_1) > 0 \text{ or not,}$$

$$B_2 = A_2 \text{ or } A_2^{\mathrm{C}} \text{ according as } N(D \cap B_1 \cap A_2) > 0 \text{ or not,}$$

and so on, where $\{A_n\}$ is the subfamily of $\mathcal{B}(S)$ in (A). Then $N(D \cap B_1 \cap B_2 \cap \cdots \cap B_n) > 0$, i.e. $\geq 1$, for $n = 1, 2, \ldots$. Therefore we can conclude that

$$N\left(D \cap \bigcap_n B_n\right) \geq 1\,,$$

since $N(D) < \infty$. Since $\{A_n\}$ separates points, $\bigcap_n B_n$ contains at most one point. Therefore $D \cap \bigcap_n B_n$ must be a single-point set, say $\{a\}$. Then $a \in \sigma(N)^{\mathrm{C}}$ and $N(\{a\}) > 0$, which is a contradiction.

Consider the function

$$f(a) = N(\{a\})\,.$$

If $a \in \sigma(N)$, then $f(a) = 1, 2, 3, \ldots$ (note that $f(a) \neq \infty$ by the $\sigma$-finiteness of $N$). If $a \notin \sigma(N)$, then $f(a) = 0$. It is clear that $N$ is completely determined by $f$ because

$$N(A) = \sum_{a\in A} f(a)\,.$$

*If $N$ is proper*, then $f(a) = 1$ or $0$ according as $a \in \sigma(N)$ or not. Thus $N$ is determined by the set $\sigma(N)$. In fact $N$ is the *counting measure* of the set $\sigma(N)$ namely

$$N(A) = \text{ the number of elements of } \sigma(N) \cap A\,.$$

[11] The complement of $A_n$ is denoted by $A_n^{\mathrm{C}}$.

**Definition 2.** A family $\{N(B,\omega)\}_{B\in\mathcal{B}(S)}$ is called a *random point measure* if

(a) $N(B,\omega)$ is a point measure in $B$ for almost every $\omega$,
(b) $N(B,\omega)$ is measurable in $\omega$ for every $B$,
(c) there exists $\{B_n\}_{n\geq 1}$ such that $S=\bigcup_n B_n$ and $P(N(B_n,\omega)<\infty) = 1, \quad n=1,2,3,\dots$.

**Definition 3.** Let $n = 1,2,\dots$ and let $\nu$ be a probability measure on $(S,\mathcal{B}(S))$. A random point measure $\{N(B,\omega)\}_{B\in\mathcal{B}(S)}$ is called a *polynomial random (point) measure* of type $(n,\nu)$ if for every decomposition

$$S = A_1 \cup A_2 \cup \cdots \cup A_r \text{ (disjoint)}$$

we have for $k_1+\cdots+k_r = n$

$$(1)\quad P(N(A_1)=k_1,\dots,N(A_r)=k_r) = \frac{n!}{k_1!k_2!\cdots k_r!}\nu(A_1)^{k_1}\cdots\nu(A_r)^{k_r}$$

namely

$$(1')\qquad E\left(t_1^{N(A_1)}\cdots t_r^{N(A_r)}\right) = \left(\sum_{i=1}^r t_i\nu(A_i)\right)^n .$$

**Theorem 1.** *There exists one and only one (up to equivalence in law) polynomial random measure of type* $(n,\nu)$.

*Proof.* The uniqueness follows from the definitions. Indeed, if we have two such random measures $N_1$ and $N_2$, then

$$P(N_1(A_1)=k_1,\dots,N_1(A_r)=k_r) = P(N_2(A_1)=k_1,\dots,N_2(A_r)=k_r)$$

for every decomposition $\{A_i\}$ of $S$. It is easy to derive from this that this identity holds for every family $\{A_i\}$ by the additivity of $N_1$, $N_2$, and $P$. To construct such a random point measure, set

$$(\Omega,\mathcal{B},P) = (S,\mathcal{B}(S),\nu)^n,$$
$$X_j(\omega) = s_j \quad \text{for } \omega=(s_1,s_2,\dots,s_n)\in\Omega,$$

and

$$N(A,\omega) = \text{the number of } X_j(\omega)\in A,\ j=1,2,\dots,n.$$

Then $X_1,X_2,\dots,X_n$ are independent and identically $\nu$-distributed. Let $S=\bigcup_{i=1}^r A_i$ be a decomposition of $S$ and write $e_i$ for the indicator of $A_i$. Then

$$N(A_i,\omega) = \sum_{j=1}^n e_i(X_j(\omega)) .$$

Therefore we have

$$
\begin{aligned}
E\left(\prod_i t_i^{N(A_i,\omega)}\right) &= E\left(\prod_i \prod_j t_i^{e_i(X_j)}\right) \\
&= E\left(\prod_j \prod_i t_i^{e_i(X_j)}\right) \\
&= \prod_j E\left(\prod_i t_i^{e_i(X_j)}\right) \quad \text{(by the independence of } \{X_j\}) \\
&= \prod_j \sum_i t_i P(X_j \in A_i) \quad \text{(because } \{A_i\} \text{ are disjoint)} \\
&= \left(\sum_i t_i \nu(A_i)\right)^n .
\end{aligned}
$$

This proves that $N$ is a polynomial random measure of type $(n, \nu)$.

**Definition 4.** Let $\mu$ be a $\sigma$-finite measure on $(S, \mathcal{B}(S))$. A random point measure $\{N(B,\omega)\}_{B\in\mathcal{B}(S)}$ is called a *Poisson random point measure with mean measure* $\mu$ if for any disjoint $A_1, A_2, \ldots, A_r$ with $\mu(A_i) < \infty$, $i = 1, 2, \ldots, r$, we have

$$
\begin{aligned}
(2) \qquad & P(N(A_1) = k_1, \ldots, N(A_r) = k_r) \\
&= \frac{1}{k_1! \cdots k_r!} \mathrm{e}^{-\mu(A_1 \cup A_2 \cup \cdots \cup A_r)} \mu(A_1)^{k_1} \cdots \mu(A_r)^{k_r}
\end{aligned}
$$

namely

$$
(2') \qquad E\left(t_1^{N(A_1)} \cdots t_r^{N(A_r)}\right) = \exp\left\{\sum_{i=1}^r (t_i - 1)\mu(A_i)\right\}.
$$

The condition (2) or (2′) is equivalent to the following pair of conditions:

(2a) $N(A)$ is Poisson distributed with mean $\mu(A)$,

(2b) $N(A_1), \ldots, N(A_r)$ are independent if $A_1, \ldots, A_r$ are disjoint.

(Note that the finiteness of $\mu(A)$ and $\mu(A_i)$ is not assumed in these conditions).[12]

**Theorem 2.** *Let $\{N(B,\omega)\}_{B\in\mathcal{B}(S)}$ be a Poisson random measure with mean measure $\mu$. Let $A$ be a set in $\mathcal{B}(S)$ with $\mu(A) < \infty$. Then*

[12] For the definition of "Poisson distributed with mean $\infty$" see Proposition 1.6.2. It follows from the condition (2) that if $\mu(A) = \infty$, then $P(N(A) = \infty) = 1$. This is proved by using a sequence $A_n \uparrow A$ satisfying $\mu(A_n) < \infty$.

$$\{N(B,\omega)\colon B \in \mathcal{B}(S),\ B \subset A\}$$

*is a polynomial random measure of type* $(n, \mu(\cdot)/\mu(A))$ *with respect to the conditional probability measure* $P(\cdot \mid N(A) = n)$.

*Proof.* Let $A = B_1 \cup B_2 \cup \cdots \cup B_r$ (disjoint) and $k_1 + k_2 + \cdots + k_r = n$. Then

$$\begin{aligned}
&P(N(B_1) = k_1, \ldots, N(B_r) = k_r \mid N(A) = n) \\
&\quad = P(N(B_1) = k_1, \ldots, N(B_r) = k_r, N(A) = n)/P(N(A) = n) \\
&\quad = P(N(B_1) = k_1, \ldots, N(B_r) = k_r)/P(N(A) = n) \\
&\qquad\qquad\qquad\qquad \text{(because } N(A) = \textstyle\sum N(B_i) \text{ and } n = \sum k_i) \\
&\quad = \frac{n!}{k_1! k_2! \cdots k_r!} \left(\frac{\mu(B_1)}{\mu(A)}\right)^{k_1} \cdots \left(\frac{\mu(B_r)}{\mu(A)}\right)^{k_r}.
\end{aligned}$$

by (2).

**Theorem 3.** *There exists one and only one (up to equivalence in law) Poisson random measure with mean measure* $\mu$.

*Proof.* First consider the case $0 < \mu(S) < \infty$. Keeping Theorem 2 in mind, we will construct such a Poisson random measure as follows. Consider a sequence of independent random variables

$$Y, X_1, X_2, \ldots$$

such that $Y$ is Poisson distributed with mean $\mu(S)$ and every $X_i$ has distribution $\nu(\cdot) = \mu(\cdot)/\mu(S)$. The existence of such a random sequence is obvious. Now set

$$N(A,\omega) = \text{ the number of } X_j \in A,\ j = 1, 2, \ldots, Y.$$

Then

$$N(S,\omega) = Y\ .$$

Consider

$$N_n(A,\omega) = \text{ the number of } X_i \in A,\ i = 1, 2, \ldots, n.$$

Then for $k_1 + k_2 + \cdots + k_r = n$ and $S = \bigcup_i A_i$ (disjoint) we have

$$P(N_n(A_1) = k_1, \ldots,\ N_n(A_r) = k_r) = \frac{n!}{k_1! \cdots k_r!} \nu(A_1)^{k_1} \cdots \nu(A_r)^{k_r}$$

(see the proof of Theorem 1), and so

$$\begin{aligned}
&P(N(A_1) = k_1, \ldots, N(A_r) = k_r) \\
&\quad = P(N(A_1) = k_1, \ldots, N(A_r) = k_r, N(S) = n) \text{ (by } S = \textstyle\bigcup A_i \text{ (disjoint))} \\
&\quad = P(N(A_1) = k_1, \ldots, N(A_r) = k_r\,, Y = n)
\end{aligned}$$

$$
\begin{aligned}
&= P(N_n(A_1) = k_1, \ldots, N_n(A_r) = k_r\,, Y = n) \\
&= P(N_n(A_1) = k_1, \ldots, N_n(A_r) = k_r) P(Y = n) \\
&\qquad\qquad \text{(by independence of } Y, X_1, \ldots, X_n) \\
&= \frac{n!}{k_1! k_2! \cdots k_r!} \nu(A_1)^{k_1} \cdots \nu(A_r)^{k_r} \mathrm{e}^{-\mu(S)} \frac{\mu(S)^n}{n!} \\
&= \frac{1}{k_1! \cdots k_r!} \mathrm{e}^{-\mu(S)} \mu(A_1)^{k_1} \cdots \mu(A_r)^{k_r}.
\end{aligned}
$$

Let $A_1, A_2, \ldots, A_r$ be disjoint. Set $A = S - \bigcup_{i=1}^r A_i$. Then

$$
\begin{aligned}
&P(N(A_1) = k_1, \ldots, N(A_r) = k_r) \\
&\quad = \sum_{k=0}^{\infty} P(N(A_1) = k_1, \ldots, N(A_r) = k_r, N(A) = k) \\
&\quad = \sum_{k=0}^{\infty} \frac{1}{k_1! \cdots k_r! k!} \mathrm{e}^{-\mu(\bigcup_1^r A_i)} \mathrm{e}^{-\mu(A)} \mu(A_1)^{k_1} \cdots \mu(A_r)^{k_r} \mu(A)^k \\
&\quad = \frac{1}{k_1! \cdots k_r!} \mathrm{e}^{-\mu(\bigcup_1^r A_i)} \mu(A_1)^{k_1} \cdots \mu(A_r)^{k_r}.
\end{aligned}
$$

Therefore $N$ is a Poisson random measure with mean measure $\mu$.

If $\mu(S) = \infty$, then $S$ is decomposed as

$$
S = \bigcup_n S_n \text{ (disjoint)}, \qquad 0 < \mu(S_n) < \infty\,.
$$

Let $\mathcal{B}(S_n)$ be the restriction of $\mathcal{B}(S)$ to $S_n$. Define a Poisson random measure $N_n$ on $(S_n, \mathcal{B}(S_n))$ for every $n = 1, 2, \ldots$ such that $N_1, N_2, \ldots$ are independent. Now set

$$
N(A, \omega) = \sum_n N(A \cap S_n, \omega)\,.
$$

Then $N$ is a Poisson random measure with mean measure $\mu$, as we can easily verify.

The uniqueness can be proved in the same way as for the polynomial random measure.

**Proposition 2.** *Let $\mu$ be a measure on $(S, \mathcal{B}(S))$. If $\mu(\{a\}) = 0$ for every $a \in S$, then, for $A \in \mathcal{B}(S)$ with $\mu(A) < \infty$ and $\delta > 0$, we can find a decomposition of $A$,*

$$
A = \bigcup_{i=1}^m B_i \text{ (disjoint)},
$$

*such that*

$$
\mu(B_i) < \delta, \qquad i = 1, 2, \ldots, m\,.
$$

*Proof.* It is enough to give a proof under the assumption $\mu(S) < \infty$. Let $\{A_m\}$ be the family in the condition (A) and set

$$B_{\varepsilon_1, \varepsilon_2, \ldots, \varepsilon_m} = A_1^{\varepsilon_1} \cap A_2^{\varepsilon_2} \cap \cdots \cap A_m^{\varepsilon_m} ,$$

where $\varepsilon_i = 1$ or $0$ and $A_i^1 = A_i$, $A_i^0 = A_i^{\mathrm{C}}$. Then we have a sequence of decompositions of $S$,

$$S = \bigcup_{(\varepsilon_1, \ldots, \varepsilon_m)} B_{\varepsilon_1, \ldots, \varepsilon_m}, \qquad m = 1, 2, \ldots .$$

For any given $\delta > 0$ we can find $m$ such that

$$\mu(B_{\varepsilon_1, \ldots, \varepsilon_m}) < \delta \quad \text{for every } (\varepsilon_1, \ldots, \varepsilon_m) ,$$

which will complete the proof. Suppose that such $m$ does not exist. Namely we have

$$\max_{(\varepsilon_1, \ldots, \varepsilon_m)} \mu(B_{\varepsilon_1, \ldots, \varepsilon_m}) \geq \delta, \qquad m = 1, 2, \ldots .$$

Then there exists[13] $\varepsilon_1^0$ such that

$$\max_{(\varepsilon_2, \ldots, \varepsilon_m)} \mu(B_{\varepsilon_1^0, \varepsilon_2, \ldots, \varepsilon_m}) \geq \delta, \qquad m = 2, 3, \ldots .$$

Then there exists $\varepsilon_2^0$ such that

$$\max_{(\varepsilon_3, \ldots, \varepsilon_m)} \mu(B_{\varepsilon_1^0, \varepsilon_2^0, \varepsilon_3, \ldots, \varepsilon_m}) \geq \delta, \qquad m = 3, 4, \ldots .$$

Similarly we can determine $\varepsilon_3^0, \varepsilon_4^0, \ldots$ . Then

$$\mu(B_{\varepsilon_1^0, \varepsilon_2^0, \ldots, \varepsilon_m^0}) \geq \delta, \qquad m = 1, 2, \ldots .$$

It follows that

$$\mu\left(\bigcap_n A_n^{\varepsilon_n^0}\right) \geq \delta$$

since $\mu(S) < \infty$. But this intersection is one point or empty and so the left measure must be 0 by our assumption. This is a contradiction.

**Theorem 4.** *If $\nu(\{a\}) = 0$ for every $a \in S$, then every polynomial random measure of type $(n, \nu)$ is proper. A similar fact holds for Poisson random measures.*

[13] Because, otherwise there exist $m(1)$ and $m(0)$ such that

$$\max_{(\varepsilon_2, \ldots, \varepsilon_{m(1)})} \mu(B_{1, \varepsilon_2, \ldots, \varepsilon_{m(1)}}) < \delta \quad \text{and} \quad \max_{(\varepsilon_2, \ldots, \varepsilon_{m(0)})} \mu(B_{0, \varepsilon_2, \ldots, \varepsilon_{m(0)}}) < \delta$$

and it follows that $\max_{(\varepsilon_1, \ldots, \varepsilon_m)} \mu(B_{\varepsilon_1, \ldots, \varepsilon_m}) < \delta$ for $m = m(1) \vee m(0)$.

*Proof.* Let $N$ be a polynomial random measure of type $(n, \nu)$. Then

$$P(N(B) \geq 2) = \sum_{k=2}^{n} \binom{n}{k} \nu(B)^k \nu(S-B)^{n-k} \leq 2^n \nu(B)^2 .$$

For every $\delta > 0$ we have a decomposition

$$S = \bigcup_{i=1}^{m} B_i \quad \text{with } \nu(B_i) < \delta,\ i = 1, 2, \dots m .$$

Then

$$\begin{aligned} P(N(\{a\}) \geq 2 \text{ for some } a) &\leq \sum_i P(N(B_i) \geq 2) \leq \sum_i 2^n \nu(B_i)^2 \\ &\leq \delta\, 2^n \nu(S) . \end{aligned}$$

Since $\delta$ is arbitrary,

$$\begin{aligned} P(N(\{a\}) \geq 2 \text{ for some } a) &= 0 , \\ P(N(\{a\}) = 0 \text{ or } 1 \text{ for every } a) &= 1 . \end{aligned}$$

This proves that $N$ is proper. The proof for Poisson random measures is similar.

## 1.10 Homogeneous Additive Processes and Homogeneous Lévy Processes

An additive process $\{X(t)\}$ is called *homogeneous* if the probability law of $X(t_2) - X(t_1)$ depends only on $t_2 - t_1$. A homogeneous additive process is often called a *process with stationary independent increments.* A *homogeneous Lévy process* is defined in the same way.[14]

**Theorem 1.** *Let $\{X(t)\}$ be a homogeneous Lévy process. Then, for any $t_0 \in T$, $\{Y(t) \equiv X(t_0 + t) - X(t_0)\colon t \in T\}$ is a Lévy process having the same probability law as $\{X(t)\}_t$.*

*Proof.* Obvious by the definition.

**Theorem 2.** *Let $X(t)$ be a Lévy process. Then $X(t)$ is homogeneous if and only if its three components are of the form*

$$m(t) = mt, \quad V(t) = Vt, \quad n(\mathrm{d}s\mathrm{d}u) = \mathrm{d}s\, n(\mathrm{d}u),$$

[14] In the recent usual terminology, a homogeneous Lévy process in the sense of this book is called a Lévy process.

*where $m$ is real, $V \geq 0$, and $n(\mathrm{d}u)$ is a measure*[15] *on $R_0$ such that*

$$\int_{R_0} \frac{u^2}{1+u^2} n(\mathrm{d}u) < \infty .$$

*Proof.* Suppose that $X(t)$ is a homogeneous Lévy process with three components $m(t)$, $V(t)$, and $n(\mathrm{d}s\mathrm{d}u)$. Let $E$ be a real Borel set separated from 0 by a positive distance. Then $N_E(t) \equiv N((0,t] \times E)$, $t \in T$, is a Poisson process with mean function

$$n_E(t) = n((0,t] \times E) .$$

It is easy by Theorem 1 to see that $N_E(t+s) - N_E(t)$ has the same probability law as $N_E(s)$. Therefore

$$n_E(t+s) - n_E(t) = n_E(s)$$

i. e.

$$n_E(t+s) = n_E(t) + n_E(s) .$$

It is obvious that $n_E(t)$ is continuous in $t$. Therefore

$$n_E(t) = n_E(1)\, t .$$

Here $n(E) \equiv n_E(1)$ is a measure on $R_0$ with

$$\iint\limits_{\substack{0<s\leq t\\ u \in R_0}} \frac{u^2}{1+u^2} n(\mathrm{d}u)\mathrm{d}s < \infty \quad \text{i. e.} \quad \int_{R_0} \frac{u^2}{1+u^2} n(\mathrm{d}u) < \infty ,$$

since

$$n(\mathrm{d}s\mathrm{d}u) = \mathrm{d}s\, n(\mathrm{d}u) .$$

Writing $X(t)$ as

$$\begin{aligned} X(t) &= X_0(t) + \lim_{k\to\infty} \Bigg( \iint\limits_{\substack{0<s\leq t\\ |u|\geq 1/k}} uN(\mathrm{d}s\mathrm{d}u) - \iint\limits_{\substack{0<s\leq t\\ |u|\geq 1/k}} \frac{u}{1+u^2} n(\mathrm{d}s\mathrm{d}u) \Bigg) \\ &= X_0(t) + \lim_{k\to\infty} S_k(t) , \end{aligned}$$

we can easily see that $S_k(t)$ is a homogeneous Lévy process independent of $X_0(t)$. Therefore $X_0(t)$ is a homogeneous Lévy process (of Gauss type). Then we have

$$m(t+s) - m(t) = E(X(t+s) - X(t)) = E(X(s)) = m(s)$$

[15] This measure $n$ is called the *Lévy measure* of the homogeneous Lévy process.

namely

$$m(t+s) = m(t) + m(s) .$$

It follows from this that $m(t) = t\,m(1)$, because $m(t)$ is continuous in $t$. Similarly $V(t) = t\,V(1)$. This completes the "only if" part. The proof of the "if" part is easy and so is omitted.

Observing that the $m$, $V$, and $n(\mathrm{d}u)$ in Theorem 2 are the three components of the probability law of $X(1)$ and recalling that the three components $m(t)$, $V(t)$, and $n(\mathrm{d}s\mathrm{d}u)$ of the Lévy process $X(t)$ determine the probability law governing the Lévy process, we obtain

**Theorem 3.** *The probability law governing a homogeneous Lévy process is determined by the probability law of the value at the time point* 1.

Since the Lévy modification of a homogeneous additive process continuous in probability is also homogeneous, we obtain

**Theorem 4.** *Theorem 3 holds for homogeneous additive processes continuous in probability.*[16]

## 1.11 Lévy Processes with Increasing Paths

The decomposition formula of a Lévy process takes a simpler form in the case where its sample function increases a. s.

Suppose that $X(t,\omega)$ be a Lévy process whose sample function is increasing a. s. It is often called simply a *Lévy process with increasing paths.*[17] Let $\{N(E)\}_E$ be the Poisson random measure expressing the counting measure of the set of pairs of jump time and jump size as before.

Since every increasing right continuous function is expressed as its continuous part plus the sum of jumps, we have

$$X(t,\omega) = X_0(t,\omega) + X_1(t,\omega) ,$$

where $X_0$ is continuous and increasing for a. e. $\omega$ and

$$X_1(t,\omega) = \iint_{\substack{0<s\le t \\ 0<u<\infty}} uN(\mathrm{d}s\mathrm{d}u), \quad t \in T \qquad \text{a.s.}$$

[16] The restriction "continuous in probability" is added by the Editors. A counter example without this restriction is given by $X(t) \equiv f(t)$ with a function $f(t)$ which satisfies $f(t+s) = f(t) + f(s)$ and which is not Borel measurable. Such a function exists and is everywhere discontinuous.

[17] A homogeneous Lévy process with increasing paths is called a *subordinator.*

As $X_1(t,\omega)$ is adapted to $\{\mathcal{B}_{st}[\mathrm{d}X]\}$, so is $X_0(t,\omega)$ and so $X_0(t,\omega)$ is a Lévy process of Gauss type. Let $m(t)$ and $V(t)$ be its mean function and variance function. Then $X_0(t)$ is $N(m(t),V(t))$ distributed. If $V(t)>0$, then

$$P(X_0(t)<0)>0\,,$$

in contradiction with

$$X_0(t)\geq X_0(0)=0 \qquad \text{a. s.}$$

Therefore

$$X_0(t,\omega)\equiv m(t) \qquad \text{a. s.,}$$

where $m(t)$ should be continuous and increasing. Notice here that the continuous part $X_0(t,\omega)$ is deterministic (i. e. does not depend on $\omega$).

Writing $n(E)$ for the parameter (mean) of the Poisson variable $N(E)$ as before, we have

$$\begin{aligned} 0<E\left[\mathrm{e}^{-X(t)}\right] &= \mathrm{e}^{-m(t)}E\left[\mathrm{e}^{-X_1(t)}\right] \\ &= \mathrm{e}^{-m(t)}\exp\Bigg\{-\iint\limits_{\substack{0<s\leq t\\0<u<\infty}}(1-\mathrm{e}^{-u})\,n(\mathrm{d}s\mathrm{d}u)\Bigg\}, \end{aligned}$$

as is easily verified. Therefore

$$\iint\limits_{\substack{0<s\leq t\\0<u<\infty}}(1-\mathrm{e}^{-u})\,n(\mathrm{d}s\mathrm{d}u)<\infty\,.$$

Recalling that

$$0<a<(1-\mathrm{e}^{-u})\frac{1+u}{u}<b<\infty \qquad (u>0)$$

for $a,b$ independent of $u$, we have

(n′) $n$ is a measure on $(0,\infty)\times(0,\infty)$ satisfying $\displaystyle\iint\limits_{\substack{0<s\leq t\\0<u<\infty}}\frac{u}{1+u}\,n(\mathrm{d}s\mathrm{d}u)<\infty\,.$

As we mentioned above, we have

(m′) $m(t)$ is continuous and increasing and $m(0)=0$.

Thus we have

**Theorem 1.** *Any Lévy process $X(t,\omega)$ with increasing paths is expressed as*

$$X(t,\omega)=m(t)+\iint\limits_{\substack{0<s\leq t\\0<u<\infty}}uN(\mathrm{d}s\mathrm{d}u)\,,$$

*where $m(t)$ and the mean measure $n$ of $N$ satisfy* (m′) *and* (n′) *and $N$ is a Poisson random measure representing the counting measure of pairs of jump time and jump size.*

Conversely we have

**Theorem 2.** *Given $m(t)$ and $n(\mathrm{d}s\mathrm{d}u)$ satisfying* (m′) *and* (n′), *we can construct one and only one (up to equivalence in law) Lévy process with increasing paths.*

The proof is similar to the corresponding theorem for general Lévy processes; in fact it is simpler and so is omitted.

In view of these theorems a Lévy process with increasing paths is completely determined by $m(t)$ and $n(\mathrm{d}s\mathrm{d}u)$. These are called the *two components* of the process.

Now we will consider special cases.

**Theorem 3.** *Let $X(t,\omega)$ be a Lévy process with increasing paths.*

*(i) If the sample function increases only by jumps, then $m(t) \equiv 0$.*

*(ii) If $X(t)$ is homogeneous, then*

$$m(t) = mt, \qquad n(\mathrm{d}s\mathrm{d}u) = \mathrm{d}s\, n(\mathrm{d}u)\ ,$$

*where $m \geq 0$ and $n(\mathrm{d}u)$ is a measure on $(0,\infty)$ with*

$$\int_0^\infty \frac{u}{1+u}\, n(\mathrm{d}u) < \infty\ .$$

Let $\mu$ be an infinitely divisible distribution concentrated on $[0,\infty)$. Then its Laplace transform has the form

$$L(\alpha,\mu) = \exp\left\{-m\alpha - \int_0^\infty (1-\mathrm{e}^{-\alpha u})n(\mathrm{d}u)\right\}$$

with $m \geq 0$ and

$$\int_0^\infty \frac{u}{1+u}\, n(\mathrm{d}u) < \infty$$

(see Problem 0.21 in Exercises). The homogeneous Lévy process whose value at $t = 1$ is governed by this $\mu$ is a Lévy process with increasing paths determined by the two components $m(t) = mt$ and $n(\mathrm{d}s\mathrm{d}u) = \mathrm{d}s\, n(\mathrm{d}u)$.

**Theorem 4.** *Let $X(t)$ be a Lévy process with increasing paths and two components $m(t)$ and $n(\mathrm{d}s\mathrm{d}u)$.*

*(i) The number of jumps on the time interval $(t_1, t_2]$ is finite a. s. or infinite a. s. according as*

$$\iint_{\substack{t_1<s\le t_2\\0<u<\infty}} n(\mathrm{d}s\mathrm{d}u) < \infty \text{ or } = \infty$$

*($\int_0^\infty n(\mathrm{d}u) < \infty$ or $= \infty$ in the homogeneous case).*

*(ii) The Lebesgue measure of the values of $X(t,\omega)$ taken on $t_1 < t \le t_2$ is $m(t_2) - m(t_1)$ a. s. ($(t_2 - t_1)m$ in the homogeneous case).*

*Proof.* The assertion (i) is obvious. To prove (ii), let $A(\omega)$ denote the set of values of $X(t,\omega)$ on $(t_1, t_2]$. Let $|A(\omega)|$ denote the Lebesgue measure of $A(\omega)$. Then either the interval $(X(t_1,\omega), X(t_2,\omega)]$ or the interval $[X(t_1,\omega), X(t_2,\omega)]$ is decomposed into the following disjoint parts:

$$\bigcup_k (X(S_k-,\omega), X(S_k,\omega)) \cup A(\omega)\,,$$

where $\{S_k\}$ are the discontinuity points in the interval $(t_1, t_2]$ and depend obviously on $\omega$. We take the Lebesgue measures of both expressions to get

$$\begin{aligned} X(t_2,\omega) - X(t_1,\omega) &= \sum_k (X(S_k,\omega) - X(S_k-,\omega)) + |A(\omega)| \\ &= \iint_{\substack{t_1<s\le t_2\\0<u<\infty}} uN(\mathrm{d}s\mathrm{d}u) + |A(\omega)|\,, \end{aligned}$$

namely

$$m(t_2) - m(t_1) = |A(\omega)|\,.$$

## 1.12 Stable Processes

**Definition.** A homogeneous Lévy process $\{X(t)\}$ is called a *stable process* if for every $a > 0$ we can find $b > 0$ and $c$ real such that the following two processes are equivalent in law:

$$X_1(t) = X(at), \quad X_2(t) = bX(t) + ct\,.$$

This property is roughly stated as follows. Every time-changed process $\{X_1(t)\}$ is equivalent in law with a certain space-changed process $\{X_2(t)\}$.

A deterministic process $X(t) \equiv mt$ is a stable process according to the definition, but we will exclude this trivial case.

Since $\{X(t)\}_t$ is a homogeneous Lévy process, it is expressed as

$$(1)\quad X(t) = mt + \sqrt{V}B(t) + \lim_{k\to\infty} \iint_{\substack{0<a\le t\\|u|\ge 1/k}} \left(uN(\mathrm{d}s\mathrm{d}u) - \frac{u}{1+u^2}\mathrm{d}s\, n(\mathrm{d}u)\right)\,.$$

Notice that if $X(t)$ is stable then the jump part (integral part) in (1) and the rest (continuous part) are both stable having the same $b$ and possibly different $c$ for a given $a$.

Let $\{X(t)\}_t$ be a stable process. As $\{X(t)\}_t$ is a homogeneous Lévy process, so are both $\{X_1(t)\}_t$ and $\{X_2(t)\}_t$. Let $N_1(E,\omega)$ and $N_2(E,\omega)$ be the $N$-measures for $\{X_1(t)\}$ and $\{X_2(t)\}$, respectively. Then[18] for $u > 0$

$$\begin{aligned} N_1(t,u) &\equiv N_1((0,t] \times (u,\infty),\omega) \\ &= \#\{s \le t\colon X(as,\omega) - X(as-,\omega) > u\} \\ &= \#\{s \le at\colon X(s,\omega) - X(s-,\omega) > u\} \\ &= N((0,at] \times (u,\infty),\omega) \,. \end{aligned}$$

Similarly

$$N_2(t,u) \equiv N_2((0,t] \times (u,\infty),\omega) = N((0,t] \times (u/b,\infty),\omega) \,.$$

Since $\{X_1(t)\}$ and $\{X_2(t)\}$ have the same probability law, $N_1(t,u)$ and $N_2(t,u)$ have the same distribution. Therefore their means are the same, namely[19]

$$at\, n(u,\infty) = t\, n(u/b,\infty) \,,$$

that is,

$$a\, n(u,\infty) = n(u/b,\infty), \qquad u > 0 \,.$$

Unless $n(du) \equiv 0$ on $u > 0$, the $b$ is uniquely determined by $a$ and so we write $b = b(a)$. Indeed, suppose that we have two $b$ with this property, say $b_1 > b_2$. Then

$$n(u/b_1,\infty) = n(u/b_2,\infty) \,.$$

Therefore

$$n(u/b_1, u/b_2] = 0 \quad \text{for every } u \,.$$

For every point $v \in (0,\infty)$ we can find $u \in (0,\infty)$ such that

$$u/b_1 < v < u/b_2 \,,$$

and thus $v$ does not belong to the support of $n$. Therefore the support of $n$ on $(0,\infty)$ should be empty, namely $n(du) \equiv 0$ on $u > 0$, in contradiction with our assumption.

Observing that

$$\begin{aligned} n(u/b(a_1a_2),\infty) &= a_1a_2n(u,\infty) = a_1n(u/b(a_2),\infty) \\ &= n(u/(b(a_1)b(a_2)),\infty) \,, \end{aligned}$$

[18] Notation: $\#A$ is the number of elements of the set $A$.
[19] $n(u,\infty) \equiv n((u,\infty))$.

we have $b(a_1a_2) = b(a_1)b(a_2)$. Since $a\,n(u,\infty)$ is increasing in $a$, $b(a)$ is increasing in $a$. Therefore

$$b(a) = a^\beta \quad \text{with } \beta > 0\,.$$

Writing $\beta = 1/\alpha$ ($\alpha > 0$), we get

$$b(a) = a^{1/\alpha}.$$

Thus we have

$$a\,n(u,\infty) = n(a^{-1/\alpha}u,\infty), \qquad u, a > 0\,.$$

Setting $a = u^\alpha$, we get

$$u^\alpha n(u,\infty) = n(1,\infty) \quad \text{i.e.} \quad n(u,\infty) = \frac{n(1,\infty)}{u^\alpha}$$

and so $n(\mathrm{d}u)$, $u > 0$, is of the following form:

$$n(\mathrm{d}u) = \frac{\gamma_+}{u^{\alpha+1}}\mathrm{d}u \quad (\gamma_+ > 0), \qquad u > 0$$

so long as $n(\mathrm{d}u) \not\equiv 0$ on $(0,\infty)$. But admitting $\gamma_+ = 0$, this formula takes care of the case $n(\mathrm{d}u) \equiv 0$ on $(0,\infty)$.

By the same argument, we have

$$n(\mathrm{d}u) = \frac{\gamma_-}{|u|^{\alpha+1}}\mathrm{d}u \quad (\gamma_- \geq 0), \qquad u < 0$$

but we are not sure that this $\alpha$ is the same $\alpha$ as above. In fact it is the same, as we can easily see by comparing

$$\widetilde{N}_1(t,u) = N_1((0,t] \times (-\infty,-u) \cup (u,\infty))$$

and

$$\widetilde{N}_2(t,u) = N_2((0,t] \times (-\infty,-u) \cup (u,\infty))$$

in the same way. Since $\int_{|u|\leq 1} u^2 n(\mathrm{d}u) < \infty$ and $\int_{|u|>1} n(\mathrm{d}u) < \infty$, we have $0 < \alpha < 2$ unless $\gamma_+ = \gamma_- = 0$. This $\alpha$ is called the exponent.

*Case 1.* $\gamma_+ = \gamma_- = 0$. Then $N(E) \equiv 0$ and

$$X(t) = mt + \sqrt{V}B(t) \quad (V > 0),$$

since $X(t) \equiv mt$ is excluded. Then

$$X_1(t) = mat + \sqrt{V}B(at)\,,$$
$$X_2(t) = (b(a)m + c(a))t + b(a)\sqrt{V}B(t)\,.$$

Since these have the same distribution, we get

$$ma = b(a)m + c(a), \qquad Va = b(a)^2 V,$$

comparing the means and the variances at $t = 1$. Thus

$$b(a) = a^{1/2}, \qquad c(a) = m(a - a^{1/2}).$$

This is the case of the exponent $= 2$.

*Case 2.* $\gamma_+ + \gamma_- > 0$. Then $0 < \alpha < 2$. We divide this case into three subcases.

(i) $0 < \alpha < 1$. Then

$$\int_{|u|\leq 1} |u|\, n(\mathrm{d}u) < \infty\,.$$

Therefore we can express $X(t)$ as

$$X(t) = mt + \sqrt{V}B(t) + \iint_{\substack{0<s\leq t\\ u\in R_0}} uN(\mathrm{d}s\mathrm{d}u)\,,$$

where $m$ is different from the $m$ in (1) and

$$E(N(\mathrm{d}s\mathrm{d}u)) = \gamma_\pm \mathrm{d}s\frac{\mathrm{d}u}{|u|^{\alpha+1}} \quad \text{for } u \gtrless 0\,.$$

Comparing $X_1(t)$ and $X_2(t)$ as before, we have

$$\begin{aligned} ma &= a^{1/\alpha}m + c(a)\,, \\ Va &= a^{2/\alpha}V\,. \end{aligned}$$

Therefore $V = 0$ and $c(a) = m(a - a^{1/\alpha})$.

(ii) $1 < \alpha < 2$. In this case

$$\int_{|u|>1} |u|\, n(\mathrm{d}u) < \infty\,.$$

Therefore the Lévy decomposition (1) is written as

$$\begin{aligned} X(t) =& mt + \sqrt{V}B(t) + \lim_{k\to\infty} \iint_{\substack{0<s\leq t\\ u\geq 1/k}} u\left[N(\mathrm{d}s\mathrm{d}u) - \mathrm{d}s\frac{\gamma_+\mathrm{d}u}{u^{\alpha+1}}\right] \\ &+ \lim_{k\to\infty} \iint_{\substack{0<s\leq t\\ u\leq -1/k}} u\left[N(\mathrm{d}s\mathrm{d}u) - \mathrm{d}s\frac{\gamma_-\mathrm{d}u}{|u|^{\alpha+1}}\right]\,. \end{aligned}$$

By the same argument as in (i) we can verify $V = 0$.

(iii) $\alpha = 1$. In this case we have $V = 0$ again and

$$X(t) = mt + \lim_{k\to\infty} \iint_{\substack{0<s\le t\\ u\ge 1/k}} u\left[N(\mathrm{d}s\mathrm{d}u) - \frac{1}{1+u^2}\mathrm{d}s\frac{\gamma_+\mathrm{d}u}{u^{\alpha+1}}\right]$$

$$+ \lim_{k\to\infty} \iint_{\substack{0<s\le t\\ u\le -1/k}} u\left[N(\mathrm{d}s\mathrm{d}u) - \frac{1}{1+u^2}\mathrm{d}s\frac{\gamma_-\mathrm{d}u}{|u|^{\alpha+1}}\right].$$

In particular if $\gamma_+ = \gamma_-$, then $X(t)$ is governed by a Cauchy distribution[20] for each $t$ and so it is called a *Cauchy process.*

[20] See Problem 0.30 in Exercises.

# 2 Markov Processes

## 2.1 Transition Probabilities and Transition Operators on Compact Metrizable Spaces

Let us consider a particle moving in a space $S$, called the *state space.* We assume the *Markovian character* of the motion that the particle that starts at $x$ at present will move into $B \subset S$ with probability $p_t(x, B)$ after time $t$ *irrespectively of its past motion*; $\{p_t(x, B)\}_{t,x,B}$ are called the *transition probabilities* of the motion. The time parameter moves in $T = [0, \infty)$.

From now on we will assume that the state space $S$ is a *compact Hausdorff space with a countable open base*, so that it is homeomorphic with a compact separable metric space by Urysohn's metrization theorem. Let $\mathcal{B}(S)$ stand for the topological $\sigma$-algebra on $S$, the $\sigma$-algebra generated by the open sets. A set $A \in \mathcal{B}(S)$ is called a Borel set.

We will assume the following conditions on the transition probabilities $\{p_t(x, B)\}_{t \in T,\, x \in S,\, B \in \mathcal{B}(S)}$ .

(T.0) $p_t(x, B)$ is Borel measurable in $x$ for $t$ and $B$ fixed.

(T.1) $p_t(x, B)$ is a probability measure in $B$ for $t$ and $x$ fixed.

(T.2) $p_0(x, B) = \delta_x(B)$ (= the $\delta$-measure concentrated at $x$).

(T.3) For $t$ fixed, $p_t(x, \cdot) \underset{\text{weak}}{\to} p_t(x_0, \cdot)$ as $x \to x_0$ .

(T.4) For every neighborhood $U(x)$ of $x$, $p_t(x, U(x)) \to 1$ as $t \downarrow 0$ .

(T.5) Chapman–Kolmogorov equation:

$$p_{t+s}(x, B) = \int_S p_t(x, \mathrm{d}y) p_s(y, B) .$$

The assumptions (T.0) and (T.1) are made for convenience of mathematical treatment; (T.2) is an obvious assumption; (T.3)[1] claims that the transition probability measure changes continuously as the starting point changes; (T.4) means that the particle will stay near the starting point for a short

[1] The meaning of (T.3) is that $\lim_{x \to x_0} \int f(y) p_t(x, \mathrm{d}y) = \int f(y) p_t(x_0, \mathrm{d}y)$ for all continuous functions $f$ on $S$.

while with high probability; (T.5) follows from the Markovian character. The assumption (T.0) follows from the assumption (T.3) and so can be omitted.

Let $C = C(S)$ be the space of all continuous functions. It is a separable Banach space with the supremum norm. Consider the operators $p_t$, called *transition operators*:

$$(p_t f)(x) = \int_S p_t(x, \mathrm{d}y) f(y), \qquad f \in C\,.$$

The assumptions mentioned above for the transition probabilities can be restated in terms of the transition operators as follows.[2]

(p.1) $p_t \colon C \longrightarrow C$ and linear ((T.1) and (T.3)).

(p.2) $p_0 = I$ (= identity operator) ((T.2)).

(p.3) $(p_t f)(x) \to f(x)$ as $t \downarrow 0$ for $f \in C$ and $x \in S$ ((T.4)).

(p.4) $p_{t+s} = p_t p_s$ ((T.4)).

(p.5) $p_t \cdot 1 = 1$, $p_t \geq 0$ (i.e. $p_t f \geq 0$ for $f \geq 0$).

The family $\{p_t\}_{t \in T}$ is called a *semi-group* of transition operators by virtue of (p.4). Each $p_t$ is a bounded operator by (p.1) and (p.5); in fact $\|p_t\| = 1$.

Suppose conversely that we are given $\{p_t\}$ with the properties (p.1)–(p.5). It follows from (p.1) and (p.5) by the Riesz representation theorem that we can find a unique measure $p_t(x, \cdot\,)$ depending on $(t, x)$ such that

$$(p_t f)(x) = \int_S p_t(x, \mathrm{d}y) f(y) \qquad \text{for } f \in C\,.$$

Then $\{p_t(x, B)\}$ satisfies the conditions (T.1)–(T.5). Thus we have a one-to-one correspondence between the transition probabilities with (T.1)–(T.5) and the transition semi-groups with (p.1)–(p.5).

The following two examples satisfy (T.1)–(T.5), where $R^1 \cup \{\infty\}$ is the one-point compactification of $R^1$.

**Example 1 (Brownian transition probabilities).** Let $S = R^1 \cup \{\infty\}$. Define

$$p_t(x, \mathrm{d}y) = \frac{1}{\sqrt{2\pi t}} \mathrm{e}^{-(y-x)^2/(2t)} \mathrm{d}y \qquad \text{in } R^1,$$
$$p_t(\infty, B) = \delta_\infty(B)\,.$$

**Example 2 (Infinitely divisible transition probabilities).** Let $\mu_t$ have characteristic function $\exp[t\psi(z)]$ with

[2] In (p.5), 1 is the constant function with value 1. When $f$ is this function, we sometimes write $p_t f = p_t \cdot 1$.

$$\psi(z) = \mathrm{i}\, mz - \frac{v}{2}z^2 + \int_{-\infty}^{\infty} \left( \mathrm{e}^{\mathrm{i}zu} - 1 - \frac{\mathrm{i}zu}{1+u^2} \right) n(\mathrm{d}u) .$$

Let $S = R^1 \cup \{\infty\}$ and define[3]

$$\begin{aligned} p_t(x, B) &= \mu_t(B - x) \qquad \text{in } R^1, \\ p_t(\infty, B) &= \delta_\infty(B) . \end{aligned}$$

The Chapman–Kolmogorov equation follows from the fact

$$\mu_{t+s} = \mu_t * \mu_s .$$

## 2.2 Summary of the Hille–Yosida Theory of Semi-Groups

Let $E$ be a separable Banach space with real coefficients and norm $\|\cdot\|$ and let $L(E, E)$ be the space of all bounded linear operators: $E \longrightarrow E$. $L(E, E)$ is also a linear space. We have three different topologies in $L(E, E)$. $E^*$ denotes the dual space i. e. $L(E, R^1)$.

(i) *Uniform topology.* Let $\|A\|$ be defined by

$$\|A\| = \sup_{f \in E,\, \|f\| \le 1} \|Af\| .$$

The uniform topology in $L(E, E)$ is given by the following neighborhoods:

$$U_\varepsilon(A_0) = \{A \in L(E, E) \colon \|A - A_0\| < \varepsilon\}, \qquad \varepsilon > 0 .$$

(ii) *Strong topology.* This is given by the neighborhoods:

$$\begin{aligned} U_{f_1, f_2, \ldots, f_n, \varepsilon}(A_0) = \{A \in L(E, E) \colon \|Af_i - A_0 f_i\| < \varepsilon,\ i = 1, 2, \ldots, n\} \\ f_1, \ldots, f_n \in E,\ \varepsilon > 0 . \end{aligned}$$

(iii) *Weak topology.* This is given by the neighborhoods:

$$\begin{aligned} &U_{f_1, \ldots, f_n, \mu_1, \ldots, \mu_n, \varepsilon}(A_0) \\ &\quad = \{A \in L(E.E) \colon |(Af_i, \mu_i) - (A_0 f_i, \mu_i)| < \varepsilon,\ i = 1, 2, \ldots, n\}, \\ &\qquad f_1, f_2, \ldots, f_n \in E,\ \mu_1, \mu_2, \ldots, \mu_n \in E^*,\ \varepsilon > 0 . \end{aligned}$$

Among these topologies the uniform topology is the strongest and the weak topology is the weakest.

---

[3] $B - x = \{y - x \colon y \in B\}$.

Let $A(\alpha)$ be a function of $\alpha$ with values in $L(E,E)$. The continuity of $A(\alpha)$ in $\alpha$, the differential coefficients of $A(\alpha)$ in $\alpha$, and the integral of $A(\alpha)$ over $\alpha_0 \le \alpha \le \alpha_1$ are understood in three ways according to the topology we are referring to. There are close relations among those concepts. For example, the *strong integral* of $A(\alpha)$ over $\alpha_0 \le \alpha \le \alpha_1$ is defined to be

$$\left(\int_{\alpha_0}^{\alpha_1} A(\alpha)\mathrm{d}\alpha\right) f = \int_{\alpha_0}^{\alpha_1} A(\alpha) f \mathrm{d}\alpha \qquad \text{(Bochner integral)}^4$$

and the *weak integral* is defined to be

$$\left(\left(\int_{\alpha_0}^{\alpha_1} A(\alpha)\mathrm{d}\alpha\right) f, \mu\right) = \int_{\alpha_0}^{\alpha_1} (A(\alpha)f, \mu)\mathrm{d}\alpha\,.$$

If the strong integral exists, then the weak integral exists and the two values coincide. The strong integral exists if $A(\alpha)$ is strongly continuous, while the weak integral exists if $A(\alpha)$ is weakly integrable i. e. if $(A(\alpha)f, \mu)$ is integrable in $\alpha$ for every $f \in E$ and $\mu \in E^*$. The weak integral is more convenient.

**Definition 1.** A one-parameter family $\{H_t : t \in T\}$ of bounded linear operators in $E$ is called a *semi-group* (in the sense of Hille–Yosida) if the following conditions are satisfied:

(H.1) $\qquad \|H_t\| \le 1 \qquad (\|\ \| = \text{uniform norm}).$

(H.2) $\qquad H_{t+s} = H_t H_s\,.$

(H.3) $\qquad \operatorname*{s-lim}_{t \downarrow 0} H_t = I \qquad (\text{s-lim} = \text{strong limit}).$

*Remark.* (H.3) is equivalent to the following apparently weaker condition under the other conditions:

(H.3′) $\qquad \operatorname*{w-lim}_{t \downarrow 0} H_t = I \qquad (\text{w-lim} = \text{weak limit})$

by virtue of a theorem due to Dunford (K. Yosida: Functional Analysis, Springer, p. 233).

By (H.2) it is clear that $H_t H_s = H_s H_t$. We can easily verify the strong continuity of $H_t$ ,

(H.4) $$\operatorname*{s-lim}_{s \to t} H_s = H_t\,.$$

**Definition 2.** Let $\{H_t : t \in T\}$ be a semi-group. The operator $R_\alpha$ defined by

$$R_\alpha = \int_T \mathrm{e}^{-\alpha t} H_t \mathrm{d}t \qquad \text{(integration in the strong topology)}$$

is called the *resolvent operator* with index $\alpha > 0$ or the *Green operator* of order $\alpha > 0$.

[4] See K. Yosida: Functional Analysis, Springer, p. 132.

**Theorem 1.**

(R.1) $\|R_\alpha\| \leq 1/\alpha$ .

(R.2) *(Resolvent equation)* $R_\alpha - R_\beta + (\alpha - \beta)R_\alpha R_\beta = 0$ *(and so* $R_\alpha R_\beta = R_\beta R_\alpha$*)* .

(R.3) $\operatorname*{s-lim}_{\alpha \to \infty} \alpha R_\alpha = I$ .

(R.4) $\dfrac{\mathrm{d}^n R_\alpha}{\mathrm{d}\alpha^n} = (-1)^n n!\, R_\alpha^{n+1}$ *(differentiation in the uniform topology).*

(R.5) $R_\alpha = \sum_{n=0}^{\infty} (-1)^n (\alpha - \alpha_0)^n R_{\alpha_0}^{n+1}$ *for* $0 < \alpha < 2\alpha_0$ .

(R.6) *The range* $\mathfrak{R}_\alpha$ *of* $R_\alpha$ *is independent of* $\alpha$ *and dense in* $E$*, so that we will write* $\mathfrak{R}$ *for* $\mathfrak{R}_\alpha$ .

(R.7) $R_\alpha$ *is one-to-one:* $E \longrightarrow \mathfrak{R}$ .

(R.8) *The map:* $u \in \mathfrak{R} \longrightarrow \alpha u - R_\alpha^{-1} u$ *is independent of* $\alpha$ .

*Proof.* (R.1) follows from (H.1), (R.2) from (H.2), and (R.3) from (H.3). Using (R.1) and (R.2), we obtain (R.4) and so (R.5). (R.6) follows from (R.2) and (R.3). To prove (R.7), observe that $\Pi = R_\alpha^{-1}\{0\}$ is independent of $\alpha$ by (R.2). Let $f \in \Pi$. Then

$$R_\alpha f = 0 \qquad \text{for every } \alpha = 0 ,$$

which implies $f = 0$ by (R.3). Therefore $\Pi = \{0\}$. This means (R.7). For the proof of (R.8), it is enough to observe that

$$\begin{aligned} &R_\alpha R_\beta[(\alpha u - R_\alpha^{-1} u) - (\beta u - R_\beta^{-1} u)] \\ &\quad = \alpha R_\alpha R_\beta u - R_\beta u - \beta R_\alpha R_\beta u + R_\alpha u = 0 , \end{aligned}$$

which implies $\alpha u - R_\alpha^{-1} u = \beta u - R_\beta^{-1} u$ by (R.7).

**Theorem 2.** *The whole family* $\{R_\alpha\}$ *is determined by any single* $R_{\alpha_0}$.

*Proof.* If we are given $R_{\alpha_0}$, then $R_\alpha$ is determined by (R.5) for $0 < \alpha < 2\alpha_0$. Using (R.5) again, we can determine $R_\alpha$ for $0 < \alpha < 4\alpha_0$. Repeating this procedure, we can get $R_\alpha$ for every $\alpha > 0$.

**Theorem 3.** *A semi-group is determined by its resolvent operators* $R_\alpha$, $\alpha > 0$*, and so by any single* $R_{\alpha_0}$ *by virtue of the previous theorem.*

*Proof.* Use the uniqueness theorem for the inverse Laplace transform.

**Definition 3.** Let $\{H_t\}$ be a semi-group. The operator[5]

$$Au = \lim_{t \downarrow 0} \frac{H_t u - u}{t}, \qquad \mathfrak{D}(A) = \{u\colon \text{ this limit exists}\}^6$$

is called the *(infinitesimal) generator* of the semi-group.

[5] By $\mathfrak{D}(A)$ we denote the domain of definition of $A$.

[6] The limit is in the sense of the norm in $E$.

**Theorem 4.**

(A.1) $\mathfrak{D}(A) = \mathfrak{R}$, $u \in \mathfrak{R} \longrightarrow Au = \alpha u - R_\alpha^{-1} u \quad (\alpha > 0)$.

(A.2) *$A$ is a closed linear operator.*

(A.3) *$\alpha - A$ has an inverse and $\|(\alpha - A)^{-1}\| \le 1/\alpha \quad (\alpha > 0)$.*

(A.4) *$u \in \mathfrak{D}(A)$ implies $H_t u \in \mathfrak{D}(A)$ and*
$$\frac{\mathrm{d}H_t u}{\mathrm{d}t} = AH_t u = H_t A u$$
*(differentiation in the sense of norm convergence in $E$).*

*Proof.* (A.1): Define $\widetilde{A}$ by $\widetilde{A}u = \alpha u - R_\alpha^{-1} u$ for $u \in \mathfrak{R}$. This is well-defined by Theorem 1 (R.8). It is easy to see that $A$ is an extension of $\widetilde{A}$ since

$$\begin{aligned} \frac{1}{t}(H_t R_\alpha f - R_\alpha f) &= \frac{1}{t}(\mathrm{e}^{\alpha t} - 1) R_\alpha f - \frac{1}{t}\int_0^t \mathrm{e}^{\alpha(t-s)} H_s f \mathrm{d}s \\ &\to \alpha R_\alpha f - f \qquad (t \downarrow 0) . \end{aligned}$$

Suppose $u \in \mathfrak{D}(A)$ and set $f = \alpha u - Au$ and $v = R_\alpha f$. Then $v \in \mathfrak{D}(A)$,

$$\begin{aligned} Av = \alpha v - f &= \alpha v - (\alpha u - Au) , \\ A(u - v) &= \alpha(u - v) , \\ H_t(u - v) &= (u - v) + t\alpha(u - v) + o(t) . \end{aligned}$$

Since $\|H_t\| \le 1$, we have

$$\begin{aligned} \|u - v\| &\ge (1 + t\alpha)\|u - v\| + o(t) , \\ t\alpha\|u - v\| + o(t) &\le 0 , \\ \|u - v\| + o(1) &\le 0 , \\ \|u - v\| &\le 0 , \end{aligned}$$

and so $u = v \in \mathfrak{D}(\widetilde{A}) = \mathfrak{R}$.

(A.2): $\alpha - A$ is a closed linear operator as an inverse of $\mathfrak{R}_\alpha$, and so is $A$.

(A.3) follows at once from (A.1). (A.4) follows from (H.2) and the definition of $A$.

**Theorem 5.** *A linear operator $A$ is the generator of a semi-group (uniquely determined by $A$), if and only if*

(a.1) *$\mathfrak{D}(A)$ is dense in $E$.*

(a.2) *For any $\alpha > 0$, $\alpha - A$ has an inverse defined on $E$ and $\|(\alpha - A)^{-1}\| \le 1/\alpha$.*

*Proof.* It is obvious by Theorems 1 and 4 that these conditions are necessary. Suppose (a.1) and (a.2) hold for $A$. Let $I_\alpha$ denote $\alpha(\alpha - A)^{-1}$. Then we can verify

$$\operatorname*{s-lim}_{\alpha\to\infty} I_\alpha = I\ .$$

Furthermore, $A_\alpha \equiv AI_\alpha$ is a bounded operator since

$$A_\alpha = \alpha(I_\alpha - I)\ .$$

Then the semi-group can be constructed by

$$H_t = \operatorname*{s-lim}_{\alpha\to\infty} \sum_{n=0}^{\infty} \frac{t^n A_\alpha^n}{n!}\ .$$

Here the infinite sum is also understood in the strong topology.[7]

**Example 1.** The resolvent operator of the *Brownian semi-group* (= the semi-group corresponding to the Brownian transition probabilities) is as follows.[8]

$$(R_\alpha f)(x) = \frac{1}{\sqrt{2\alpha}} \int_{-\infty}^{\infty} \mathrm{e}^{-\sqrt{2\alpha}|y-x|} f(y)\mathrm{d}y\, e_{R^1}(x) + \frac{f(\infty)}{\alpha}\, e_{\{\infty\}}(x)\ .$$

*Proof.* It is enough to prove

$$\int_0^\infty \mathrm{e}^{-\alpha t} \frac{1}{\sqrt{2\pi t}} \mathrm{e}^{-x^2/(2t)}\mathrm{d}t = \frac{1}{\sqrt{2\alpha}} \mathrm{e}^{-\sqrt{2\alpha}|x|}\ .$$

The left-hand side is

$$\begin{aligned}
&= 2\int_0^\infty \mathrm{e}^{-\alpha s^2} \frac{1}{\sqrt{2\pi}} \mathrm{e}^{-x^2/(2s^2)}\mathrm{d}s \qquad (t = s^2) \\
&= \frac{2\sqrt{c}}{\sqrt{2\pi}} \int_0^\infty \mathrm{e}^{-\alpha c u^2 - x^2/(2cu^2)}\mathrm{d}u \qquad (s = \sqrt{c}u) \\
&= \frac{2\sqrt{c}}{\sqrt{2\pi}} \int_0^\infty \mathrm{e}^{-\beta(u^2+1/u^2)}\mathrm{d}u \qquad (c = |x|/\sqrt{2\alpha},\ \beta = \sqrt{\alpha/2}|x|)
\end{aligned}$$

---

[7] Let $H_t^\alpha \equiv \sum_{n=0}^\infty t^n A_\alpha^n/n! \equiv \mathrm{e}^{tA_\alpha}$. Then $H_t^\alpha = \mathrm{e}^{-t\alpha}\mathrm{e}^{t\alpha I_\alpha}$, $\|H_t^\alpha\| \le \mathrm{e}^{-t\alpha}\mathrm{e}^{t\alpha\|I_\alpha\|} \le 1$, $H_{t+s}^\alpha = H_t^\alpha H_s^\alpha$, and $\operatorname{s-lim}_{t\downarrow 0} H_t^\alpha = I$. We have

$$\begin{aligned}
\|H_t^\alpha u - H_t^\beta u\| &= \|\mathrm{e}^{tA_\alpha}u - \mathrm{e}^{tA_\beta}u\| \\
&\le n\|\mathrm{e}^{(t/n)A_\alpha}u - \mathrm{e}^{(t/n)A_\beta}u\| \to t\|A_\alpha u - A_\beta u\|
\end{aligned}$$

as $n \to \infty$, and hence $\|H_t^\alpha u - H_t^\beta u\| \le t\|A_\alpha u - A_\beta u\|$. Since $\operatorname{s-lim}_{\alpha\to\infty} A_\alpha u = Au$ for $u \in \mathfrak{D}(A)$, we can prove that $H_t^\alpha$ converges to a semi-group $H_t$ in the strong topology and that $\{H_t\}$ has generator $A$.

[8] $e_{R^1}(x)$ and $e_{\{\infty\}}(x)$ are the indicators of $R^1$ and $\{\infty\}$, respectively.

$$= \frac{2\sqrt{c}}{\sqrt{2\pi}} \mathrm{e}^{-2\beta} \int_0^\infty \mathrm{e}^{-\beta(u-1/u)^2} du \, .$$

Further,

$$\begin{aligned}
\int_0^\infty \mathrm{e}^{-\beta(u-1/u)^2} du &= \int_0^1 + \int_0^\infty \\
&= \int_0^1 \mathrm{e}^{-\beta(u-1/u)^2} \left(1 + \frac{1}{u^2}\right) du \quad (u \longrightarrow 1/u \text{ in the second integral}) \\
&= \int_{-\infty}^0 \mathrm{e}^{-\beta v^2} dv \qquad (v = u - 1/u) \\
&= \frac{1}{2}\sqrt{\frac{\pi}{\beta}}
\end{aligned}$$

Put this in the previous formula and replace $c$ and $\beta$ respectively with $|x|/\sqrt{2\alpha}$ and $\sqrt{\alpha/2}\,|x|$.

**Example 2.** The generator $A$ of the Brownian semi-group is given by[9]

(a) $$\mathfrak{D}(A) = \left\{ u \in C(R^1 \cup \{\infty\}) \colon u \in C^2 \text{ in } R^1, \lim_{x \in R^1,\, x \to \infty} u''(x) = 0 \right\},$$

(b) $$Au(x) = \begin{cases} (1/2)u''(x) & x \in R^1, \\ 0 & x = \infty \, . \end{cases}$$

*Proof.* Let $\mathfrak{D}$ denote the family on the right-hand side of (a). Let $\alpha > 0$. We have

$$\mathfrak{D}(A) = \{R_\alpha f \colon f \in C(S)\}$$

and

$$Au = \alpha u - R_\alpha^{-1} u \qquad \text{for } u \in \mathfrak{D}(A) \, .$$

Remember the following formula in Example 1

$$R_\alpha f(x) = \begin{cases} \int_{-\infty}^\infty \frac{1}{\sqrt{2\alpha}} \mathrm{e}^{-\sqrt{2\alpha}|y-x|} f(y) dy & x \in R^1 \, , \\ f(\infty)/\alpha & x = \infty \, . \end{cases}$$

By dividing the integral into two parts $\int_{-\infty}^x$ and $\int_x^\infty$, we can easily see that $u \equiv R_\alpha f \in C^2$ in $R^1$ and

$$\alpha u - (1/2)u'' = f \qquad \text{in } R^1 \, .$$

---

[9] $C^2$ is the class of functions which are twice differentiable with continuous second derivatives.

This differential equation implies that $u''(x)$ tends to 0 as $x\,(\in R^1) \to \infty$, because both $f$ and $R_\alpha f$ are in $C(R^1 \cup \{\infty\})$ and $R_\alpha f(\infty) = f(\infty)/\alpha$. Therefore $u \in \mathfrak{D}$. Thus we have proved $\mathfrak{D}(A) \subset \mathfrak{D}$.

Let $u \in \mathfrak{D}$. Define $f$ by

$$f(x) = \begin{cases} \alpha u(x) - (1/2)u''(x) & x \in R^1\,, \\ \alpha u(\infty) & x = \infty\,. \end{cases}$$

Then $f \in C(R^1 \cup \{\infty\})$ and $v \equiv R_\alpha f\ (\in \mathfrak{D}(A))$ satisfies

$$\alpha v - (1/2)v'' = f \qquad \text{in } R^1\,,$$
$$v(\infty) = f(\infty)/\alpha = u(\infty)$$

as above. Let $w = u - v$. Then

$$\alpha w - (1/2)w'' = 0\,, \qquad w(\infty) = 0\,,$$

which implies $w \equiv 0$ i.e. $u = v \in \mathfrak{D}(A)$. Thus we have $\mathfrak{D}(A) = \mathfrak{D}$.

If $u \in \mathfrak{D}(A)$ then $u = R_\alpha f$ for some $f \in C(R^1 \cup \{\infty\})$. Then

$$Au = \alpha u - f = (1/2)u'' \qquad \text{in } R^1\,,$$
$$Au(\infty) = \alpha u(\infty) - f(\infty) = 0\,.$$

This completes the proof.

## 2.3 Transition Semi-Group

The semi-group of transition operators on $C(S)$ introduced in Section 2.1 is called a *transition semi-group*. It is easy to see that a transition semi-group is a semi-group in the sense of Hille–Yosida. To characterize transition semi-groups, we have

**Theorem 1.** *A semi-group of operators $\{H_t\}$ on $C(S)$ in the Hille–Yosida sense ($S$ being compact and metrizable) is a transition semi-group if and only if*

$$H_t \geq 0 \quad \textit{and} \quad H_t \cdot 1 = 1\,. \tag{1}$$

*Proof.* The "only if" part is obvious. The "if" part follows from the Riesz representation theorem.

**Theorem 2.** *In order for $A$ to be the generator of a transition semi-group, the following four conditions are jointly necessary and sufficient.*[10]

[10] $\overline{\mathfrak{D}(A)}$ denotes the closure of $\mathfrak{D}(A)$.

(a) *$A$ is a linear operator on $C(S)$ and $\overline{\mathfrak{D}(A)} = C(S)$.*

(b) $A \cdot 1 = 0$.

(c) *If $u \in \mathfrak{D}(A)$ and $u(x_0) = \max_x u(x)$, then $Au(x_0) \leq 0$.*

(d) *There exists $\alpha_0 > 0$ such that, for every $f \in C(S)$, $(\alpha_0 - A)u = f$ has at least one solution.*

*Proof.* The necessity of these conditions is obvious. Suppose these conditions are satisfied. First we will prove that the solution of $(\alpha_0 - A)u = f$ is unique. Let $u_1$ and $u_2$ be two solutions. Then $u = u_1 - u_2$ satisfies

$$(\alpha_0 - A)\, u = 0 \,.$$

Let $u(x_0) = \max_x u(x)$. Then

$$\alpha_0 u(x_0) = Au(x_0) \leq 0$$

by (c). Therefore $u(x) \leq 0$. Since

$$(\alpha_0 - A)(-u) = -(\alpha_0 - A)u = 0 \,,$$

we will get $-u(x) \leq 0$ and so $u \equiv 0$, i. e. $u_1 = u_2$ .

Write $Rf$ for the unique solution of

$$(\alpha_0 - A)u = f \,.$$

We shall prove that $R$ is a linear operator satisfying

$$\|R\| \leq 1/\alpha_0 \,.$$

The linearity is obvious. Let $u = Rf$ and $u(x_0) = \max_x u(x)$. Then

$$\alpha_0 u(x_0) = Au(x_0) + f(x_0) \leq 0 + f(x_0) \leq \|f\|$$

by (c). Therefore

$$\alpha_0 u(x) \leq \|f\| \,.$$

Since $(\alpha_0 - A)(-u) = -f$, we have

$$\alpha_0(-u(x)) \leq \|-f\| = \|f\| \,.$$

Therefore $\|u\| \leq \|f\|/\alpha_0$, i. e. $\|R\| \leq 1/\alpha_0$, i. e.

$$\|(\alpha_0 - A)^{-1}\| \leq 1/\alpha_0 \,.$$

Now we will prove the existence of a solution of $(\alpha - A)u = f$ for every $\alpha > 0$. For $0 < \alpha < 2\alpha_0$, set

$$R_\alpha = \sum_{n=0}^{\infty} (-1)^n (\alpha - \alpha_0)^n R^{n+1} \,. \tag{2}$$

This is convergent in the uniform norm because of $\|R\| \le 1/\alpha_0$. Noticing

$$R_\alpha = R\left(\sum_{n=0}^{\infty}(-1)^n(\alpha-\alpha_0)^n R^n\right),$$

we have

$$\begin{aligned}(\alpha_0 - A)R_\alpha f &= \sum_{n=0}^{\infty}(-1)^n(\alpha-\alpha_0)^n R^n f \\ &= f - (\alpha-\alpha_0)R_\alpha f\end{aligned}$$

and so

$$(\alpha - A)R_\alpha f = f,$$

which shows that $u = R_\alpha f$ is a solution. The solution is unique and $\|(\alpha - A)^{-1}\| = \|R_\alpha\| \le 1/\alpha$ in the same way as in case $\alpha = \alpha_0$. Replacing $R$ with $R_{3\alpha_0/2}$ in (2), we can define $R'_\alpha$ for $0 < \alpha < 3\alpha_0$ such that $u = R'_\alpha f$ is also the unique solution of $(\alpha - A)u = f$ for $0 < \alpha < 3\alpha_0$. Replacing $R$ with $R_{5\alpha_0/2}$ in (2), we can define $R''_\alpha$ for $0 < \alpha < 5\alpha_0$ such that $u = R''_\alpha f$ is the unique solution of $(\alpha - A)u = f$ for $0 < \alpha < 5\alpha_0$. Repeating this procedure, we can find a unique solution of $(\alpha - A)u = f$ for every $\alpha > 0$.

Using Theorem 5 in the previous section we can get the semi-group $\{H_t\}$ whose generator is $A$. Using the condition (c) as above, we can prove that $R_\alpha \ge 0$, i.e. $R_\alpha f \ge 0$ for $f \ge 0$. Therefore $I_\alpha = \alpha R_\alpha \ge 0$. Thus the semi-group constructed in Theorem 5 in the previous section should satisfy $H_t \ge 0$, because

$$\sum_{n=0}^{\infty}\frac{t^n A_\alpha^n}{n!} = \mathrm{e}^{-\alpha t}\sum_{n=0}^{\infty}\frac{(\alpha t)^n}{n!}(\alpha R_\alpha)^n \ge 0.$$

Since $A \cdot 1 = 0$, we have $(\alpha - A)\cdot 1 = \alpha \cdot 1$ and so $R_\alpha \cdot 1 = \alpha^{-1}\cdot 1$, i.e.

$$\int_0^\infty \mathrm{e}^{-\alpha t}H_t \cdot 1\,\mathrm{d}t = \alpha^{-1}\cdot 1,$$

which implies $H_t \cdot 1 = 1$ for almost all $t$. But $H_t \cdot 1$ is continuous in $t$ and so $H_t \cdot 1 \equiv 1$. This completes the proof that $\{H_t\}$ is a transition semi-group.

## 2.4 Probability Law of the Path

Let $X(t,\omega)$ be a motion governed by the transition probabilities mentioned in Section 2.1. Then we have

$$\begin{aligned}(1)\quad & P(X(t_1)\in B_1,\ X(t_2)\in B_2,\ldots,\ X(t_n)\in B_n) \\ &= \int_{x\in S}\int_{x_1\in B_1}\cdots\int_{x_n\in B_n}\mu(\mathrm{d}x)p(t_1,x,\mathrm{d}x_1)p(t_2-t_1,x_1,\mathrm{d}x_2) \\ &\qquad\cdots p(t_n-t_{n-1},x_{n-1},\mathrm{d}x_n)\end{aligned}$$

for $0 < t_1 < t_2 < \cdots < t_n$, where $\mu$ is the probability law of $X(0)$, namely the *initial distribution.* Such a stochastic process can be constructed by Kolmogorov's extension theorem.

However, the properties of the sample function (or often called the *sample path*) of the process such as the continuity of the path cannot be discussed in this model, because such properties depend on the values of the process at uncountably many time points. Therefore we have to take a reasonable modification of the process. The following theorem will provide such a modification.

**Theorem 1.** *For the process $X(t)$ introduced above we have a stochastic process $Y(t)$ such that*

*(a) the sample path $Y(t,\omega)$, $t \in T$ is right continuous and has left limit at every $t$ for every $\omega$ (or we simply say that $Y(\cdot,\omega) \in D(S)$ for every $\omega$) and that*

*(b) $Y(t)$ is a modification of $X(t)$, i. e.*

$$P(X(t) = Y(t)) = 1 \qquad \textit{for every } t\,.$$

*Such $Y(t)$ is unique in the sense that if we have two such $Y$'s, say $Y_1, Y_2$, then*

$$P(Y_1(t) = Y_2(t) \textit{ for every } t) = 1\,.$$

*Proof.* We divide the proof into several steps.

*Step 1.* If $f \in C(S)$, then (for $\mathcal{B}_t = \mathcal{B}[X_\tau : \tau \le t]$)

$$E(f(X_{t+s}) \,|\, \mathcal{B}_t) = H_s f(X_t)$$

i. e.

$$E(f(X_{t+s}), B) = E(H_s f(X_t), B) \qquad \text{for } B \in \mathcal{B}_t\,. \tag{2}$$

By Dynkin's Theorem (Lemma 0.1.1) it is enough to prove this for

$$B = \{X_{t_1} \in B_1\} \cap \{X_{t_2} \in B_2\} \cap \cdots \cap \{X_{t_n} \in B_n\}$$

with $0 \le t_1 < t_2 < \cdots < t_n \le t$. In this case (2) follows at once from (1).

*Step 2.* If $f \in C(S)$ and if $f \ge 0$, then $\mathrm{e}^{-\alpha t} R_\alpha f(X_t)$ is a supermartingale and continuous in probability.

The proof is as follows. Multiplying $\mathrm{e}^{-\alpha(t+s)}$ on both sides of (2) and integrating them on $0 \le s < \infty$, we have, for $B \in \mathcal{B}_t$,

$$\begin{aligned} E(\mathrm{e}^{-\alpha t} R_\alpha f(X_t), B) &= E\left(\int_0^\infty \mathrm{e}^{-\alpha(t+s)} f(X_{t+s}) \mathrm{d}s, B\right) \\ &= E\left(\int_t^\infty \mathrm{e}^{-\alpha s} f(X_s) \mathrm{d}s, B\right)\,. \end{aligned}$$

Since $B \in \mathcal{B}_t \subset \mathcal{B}_{t+u}$ for $u > 0$, we have

$$E(\mathrm{e}^{-\alpha(t+u)} R_\alpha f(X_{t+u}), B) = E\left(\int_{t+u}^{\infty} \mathrm{e}^{-\alpha s} f(X_s)\mathrm{d}s, B\right)$$

$$\leq E\left(\int_{t}^{\infty} \mathrm{e}^{-\alpha s} f(X_s)\mathrm{d}s, B\right)$$

and so

$$E(\mathrm{e}^{-\alpha(t+u)} R_\alpha f(X_{t+u}), B) \leq E(\mathrm{e}^{-\alpha t} R_\alpha f(X_t), B) .$$

This is true for every $B \in \mathcal{B}_t$. Therefore

$$E(\mathrm{e}^{-\alpha(t+u)} R_\alpha f(X_{t+u}) \,|\, \mathcal{B}_t) \leq \mathrm{e}^{-\alpha t} R_\alpha f(X_t) \qquad \text{a. s.}$$

Since

$$p_\varepsilon(x, \cdot\,) \underset{\text{weak}}{\rightarrow} \delta_x(\cdot\,) \qquad (\varepsilon \downarrow 0) ,$$

$\mathrm{e}^{-\alpha t} R_\alpha f(X_t)$ is continuous (in $t$) in probability. (See Remark at the end of this section).

*Step 3.* Let $\{a_n\}_n$ be a countable dense set in $S$ and set $f_n(x) = \rho(x, a_n)$, $\rho$ being a metric consistent with the topology in $S$. Let $\{g_k\}_k$ be an enumeration of $\{R_m f_n\}$, $m, n = 1, 2, \ldots$. Then the map $g\colon S \longrightarrow R^\infty(= R^1 \times R^1 \times \cdots)$ defined by

$$g(x) = (g_1(x), g_2(x), \ldots)$$

is continuous and one-to-one. As every $f_n$ is continuous, so is $g$. If $g(x) = g(y)$, then

$$f_n(x) = \lim_{m\to\infty} mR_m f_n(x) = \lim_{m\to\infty} mR_m f_n(y) = f_n(y)$$

and so $\rho(x, a_n) = \rho(y, a_n)$ and

$$\rho(x, y) \leq \rho(x, a_n) + \rho(y, a_n) = 2\rho(x, a_n) .$$

Since $\{a_n\}$ is dense in $S$, $\inf_n \rho(x, a_n) = 0$ and so $\rho(x, y) = 0$, i.e. $x = y$. Therefore $g$ is one-to-one.

*Step 4.* Using the notation in Step 3, we see that $g$ defines a homeomorphism from $S$ onto $K \equiv g(S)$ and both $g$ and $h \equiv g^{-1}\colon K \longrightarrow S$ are uniformly continuous, because $S$ is compact; it is obvious that both $S$ and $K$ are compact and metrizable.

*Step 5.* Let $\xi_t$ be a stochastic process with values in a metric space, $R^1$, $S$, or $K$ for example. A modification of $\xi_t(\omega)$ whose sample path is right continuous and has left limit at every $t$, for every $\omega$, is called a $D$-modification for the moment. Since $\mathrm{e}^{-\alpha t} R_\alpha f(X_t)$ $(f \in C(S), f \geq 0)$ is a supermartingale continuous in probability, we can find a $D$-modification of this process by Doob's

Theorem 0.7.7. Since $e^{\alpha t}$ is continuous, $R_\alpha f(X_t)$ has also a $D$-modification. Now we will apply this fact to a $D$-modification $Z_k(t)$ of $g_k(X_t)$. Then

$$Z(t) \equiv (Z_1(t), Z_2(t), \ldots)$$

is a $D$-modification of

$$g(X_t) \equiv (g_1(X_t), g_2(X_t), \ldots) .$$

Let $Q$ be the set of all nonnegative rational numbers in $R^1$. Then

$$P(Z(r) = g(X_r) \in K \text{ for every } r \in Q) = 1 .$$

As $Z(t, \omega)$ is right continuous in $t$ for every $\omega$, we have

$$P(Z(t) \in K \text{ for every } t) = 1 .$$

Therefore $Y(t) \equiv h(Z(t))$ is right continuous and has left limits for every $\omega$, because $h\colon K \longrightarrow S$ is a homeomorphism. It is obvious that

$$P(X(t) = Y(t)) = P(g(X(t)) = g(Y(t))) = P(g(X(t)) = Z(t)) = 1 .$$

Therefore $Y(t)$ is a $D$-modification of $X(t)$. The uniqueness is easy to see. This completes the proof of Theorem 1.

Let $D(S)$ denote the space of all functions that are right continuous and have left limit at every time point and let $\mathcal{B}_{D(S)}$ be the $\sigma$-algebra generated by the sets

$$\{f \in D(S)\colon f(t) \in B\}, \qquad B \in \mathcal{B}(S),\, t \in T .$$

The *probability law* $P_\mu$ of the sample path of the $Y(t)$ process determined in Theorem 1 is given by

$$P_\mu(\Lambda) = P(Y(\,\cdot\,, \omega) \in \Lambda), \quad \Lambda \in \mathcal{B}_{D(S)} ;$$

we put the suffix $\mu$ because this probability law depends not only on the transition probabilities but also on the initial distribution $\mu$.

Let $P_a$ denote $P_{\delta_a}$. It is the probability law governing the motion starting at $a$. Since

$$P_\mu(\Lambda) = \int_S \mu(\mathrm{d}a) P_a(\Lambda) ,$$

it is enough to consider the family $\{P_a\}_{a \in S}$ instead of $\{P_\mu\}$. Note that $\{P_a\}_{a \in S}$ is completely determined by the transition semi-group $\{p_t\}$ and therefore by a single operator $A$, the generator of the semi-group. Thus we have the following correspondence:

$$\text{transition probability } \{p_t(x, B)\} \longleftrightarrow \text{transition semi-group } \{H_t \equiv p_t\}$$
$$\longleftrightarrow \text{generator } A \longleftrightarrow \text{probability law of the path } \{P_a\} .$$

**Example.** Let the space $S$ be a finite set endowed with discrete topology. We denote the points in $S$ by $1, 2, \ldots, n$. Then we have an isomorphism

$$E \equiv C(S) \simeq R^n, \quad f \longleftrightarrow \begin{pmatrix} f(1) \\ \vdots \\ f(n) \end{pmatrix},$$

$$L(E, E) \simeq \text{the space of } n \times n \text{ matrices}, \quad \Phi \longleftrightarrow \begin{pmatrix} \varphi(1,1) & \cdots & \varphi(1,n) \\ \vdots & & \vdots \\ \varphi(n,1) & \cdots & \varphi(n,n) \end{pmatrix}.$$

In particular a transition operator $p$ corresponds to a matrix $(p(i,j))$ where

$$\sum_j p(i,j) = 1, \qquad p(i,j) \geq 0 .$$

Such a matrix is called a *stochastic matrix*.

The transition matrices are represented by a family of stochastic matrices $(p_t(i,j))$ satisfying[11]

(a) $\quad p_{t+s} = p_t p_s$ (matrix multiplication),

(b) $\quad \lim_{t \downarrow 0} p_t(i,j) = \delta(i,j)$ for each $(i,j)$.

Now we will determine the generator $A$ of this semi-group. The general theory shows that $A = (a(i,j))$ is a linear operator with $\overline{\mathfrak{D}(A)} = R^n$. In fact $\mathfrak{D}(A) = R^n$ by finite dimensionality, so that $\overline{\mathfrak{D}(A)} = R^n$ is a vacuous condition. Therefore

$$a(i,j) = \lim_{t \downarrow 0} \frac{p_t(i,j) - \delta(i,j)}{t} \qquad \text{for every pair } (i,j) .$$

Since $p_t(i,j)$ is a stochastic matrix,

$$a(i,j) \geq 0 \text{ for } i \neq j \text{ and } \sum_j a(i,j) = 0 ,$$

so that

$$a(i,i) = -\sum_{j \neq i} a(i,j) .$$

Thus $A$ must be of the form

$$A = \begin{pmatrix} -a(1) & a(1,2) & \ldots & a(1,n) \\ a(2,1) & -a(2) & \ldots & a(2,n) \\ \vdots & \vdots & & \vdots \\ a(n,1) & a(n,2) & \ldots & -a(n) \end{pmatrix}, \quad a(i,j) \geq 0, \ a(i) = \sum_{j \neq i} a(i,j) .$$

[11] $\delta(i,j) = 1$ or $0$ according as $i = j$ or $i \neq j$.

Now we want to prove that such $A$ generates a transition semi-group $\{p_t(i,j)\}$. For this purpose it is enough to check the four conditions in Theorem 2.3.2.

(a) is obvious.

(b) follows from $(A \cdot 1)(i) = -a(i) + \sum_{j \neq i} a(i,j) = 0$.

(c) Let $u = (u(i))$ and suppose that $u(i_0) \geq u(j)$ for every $j$. Then

$$\begin{aligned}(Au)(i_0) &= -a(i_0)u(i_0) + \sum_{j \neq i_0} a(i_0, j)u(j) \\ &\leq -a(i_0)u(i_0) + \sum_{j \neq i_0} a(i_0, j)u(i_0) \\ &= 0\,.\end{aligned}$$

(d) It is enough to prove that the linear equation

$$(\alpha\delta - A)u = f$$

has a solution $u$ for some $\alpha$, namely that

$$\Delta(\alpha) \equiv \det(\alpha\delta - A) \neq 0$$

for some $\alpha$. But

$$\Delta(\alpha) = \alpha^n + g_{n-1}(\alpha), \qquad g_{n-1} = \text{ polynomial of degree } \leq n-1\,.$$

Therefore $\Delta(\alpha) > 0$ for $\alpha$ big enough. Hence all conditions are satisfied.

Let us consider the probability law $P_a$ of the path starting at $a$. $P_a$ is a probability measure on the space $D$ that consists of all right continuous functions having left limits. Since every function $\omega \in D$ takes only values $1, 2, 3, \ldots, n$, $\omega$ should be a step function.

By the correspondence $A \longleftrightarrow \{P_a\}$, $P_a$ is determined by the constants $a_{ij}$ $(i \neq j)$ (and $a(i) = \sum_{j \neq i} a(i,j)$).

The relationship between these two concepts will be discussed later.

*Remark.* The statement at the end of Step 2 in the proof of Theorem 1 may need some explanation. Since $\mathrm{e}^{-\alpha t}$ is continuous in $t$ and since $R_\alpha f \in C(S)$, it is enough to prove that $X_t$ is continuous in probability. In fact we can prove that $X_t$ is uniformly continuous in probability. It is obvious by the definition of $P_\mu$ that[12]

$$(3) \quad P_\mu(d(X(t), X(s)) > \varepsilon) = \int_S \int_S \mu(\mathrm{d}a) p_s(a, \mathrm{d}b) p_{t-s}(b, U_\varepsilon(b)^{\mathrm{C}}), \quad t > s.$$

[12] Note that $p_h(b, U_\varepsilon(b)^{\mathrm{C}})$ is Borel measurable in $b$ for $h > 0$ and $\varepsilon > 0$ fixed. To prove this, it is enough to show that if $f(x,y)$ is bounded and Borel measurable in $(x,y)$, then $\int p_h(x, \mathrm{d}y)\, f(x,y)$ is Borel measurable in $x$. When $f$ is the indicator of $C \in \mathcal{B}(S \times S)$, we can show this by starting from the case that $C = C_1 \times C_2$ and by using Lemma 0.1.1. The case of a general $f$ is handled by approximation.

For the proof[13] that this probability tends to 0 as $t-s \to 0$, it is enough to use the following

**Lemma 1.** *For* $\varepsilon > 0$, $p_h(b, U_\varepsilon(b)^{\mathrm{C}}) \to 0$ *uniformly in* $b$ *as* $h \downarrow 0$.

*Proof.* First we will find $\eta = \eta(a, \varepsilon, \delta)$ for $\varepsilon, \delta > 0$ and $a \in S$ such that

$$p_h(b, U(b)^{\mathrm{C}}) < \delta \qquad \text{for } 0 \le h \le \eta \text{ and } b \in U_{\varepsilon/3}(a)\,. \tag{4}$$

Take a continuous function $f$ such that

$$f = \begin{cases} 0 & \text{on } \overline{U_{\varepsilon/3}(a)}, \\ 1 & \text{on } U_{2\varepsilon/3}(a)^{\mathrm{C}}, \end{cases}$$
$$0 \le f \le 1 \qquad \text{elsewhere.}$$

Since $\|H_h f - f\| \to 0$ as $h \downarrow 0$, we have $\eta = \eta(f, \delta) = \eta(a, \varepsilon, \delta)$ such that

$$\|H_h f - f\| < \delta \qquad \text{for } 0 \le h \le \eta\,.$$

As $f = 0$ on $\overline{U_{\varepsilon/3}(a)}$, we have $H_h f(b) < \delta$ for $0 \le h \le \eta$ and $b \in \overline{U_{\varepsilon/3}(a)}$. But

$$H_h f(b) = \int_S p_h(b, \mathrm{d}c) f(c) \ge p_h(b, U_\varepsilon(b)^{\mathrm{C}})$$

for such $b$ and $h$, because

$$\begin{aligned} b \subset \overline{U_{\varepsilon/3}(a)},\ c \in U_\varepsilon(b)^{\mathrm{C}} \quad &\Longrightarrow \quad d(a,b) \le \varepsilon/3,\ d(b,c) \ge \varepsilon \\ \Longrightarrow \quad d(a,c) \ge 2\varepsilon/3 \quad &\Longrightarrow \quad c \in U_{2\varepsilon/3}(a)^{\mathrm{C}} \quad \Longrightarrow \quad f(c) = 1\,. \end{aligned}$$

Thus (4) is proved.

Since $S$ is compact, we have $S = \bigcup_{i=1}^n U_{\varepsilon/3}(a_i)$ for some $a_1, \ldots, a_n$. Setting $\eta_1 = \eta_1(\varepsilon, \eta) = \min_i \eta(a_i, \varepsilon, \delta)$, we get

$$p_h(b, U_\varepsilon(b)^{\mathrm{C}}) < \delta \qquad \text{for } 0 \le h \le \eta_1 \text{ and } b \in S\,.$$

This completes the proof.

[13] Another proof of (3) is to use Lebesgue's bounded convergence theorem, since $p_{t-s}(b, U_\varepsilon(b)^{\mathrm{C}}) \to 0$ for each $b$ as $t - s \to 0$. However, we will use Lemma 1 in Section 2.9.

## 2.5 Markov Property

We will use the following notation.
$S$: state space, a compact Hausdorff space with a countable open base,
$T$: time parameter space $= [0, \infty)$,
$p(t, a, B)$: transition probabilities,
$\{H_t\}$: transition semi-group,
$R_\alpha$: resolvent operator of $\{H_t\}$,
$A$: generator of $\{H_t\}$,
$\Omega = D(T, S) =$ the space of all right continuous functions: $T \longrightarrow S$ with left limits,
$\mathcal{B}$: the $\sigma$-algebra on $\Omega$ generated by the cylinder sets,[14]
$P_a$ $(P_\mu)$: the probability law of the path starting at $a$ (having the initial distribution $\mu$) introduced in Section 2.4,
$X_t(\omega) = \omega(t)$ for $\omega \in \Omega$,
$\mathcal{B}_t$: the sub-$\sigma$-algebra of $\mathcal{B}$ generated by $X_s$, $s \leq t$ .

**Definition 1.** The system of stochastic processes[15]

$$\{X_t(\omega),\ t \in T,\ \omega \in (\Omega, \mathcal{B}, P_a)\}_{a \in S}$$

is called a *Markov process* with transition probabilities $\{p(t, a, B)\}$.

We will express $p(t, a, B)$, $H_t$ and $R_\alpha$ in terms of the process.

**Theorem 1.** *Let $f \in C(S)$.*
*(i)* $p(t, a, B) = P_a(X_t \in B)$ .
*(ii)* $H_t f(a) = E_a(f(X_t))$ *where* $E_a(\cdot) = \int_\Omega \cdot\, P_a(\mathrm{d}\omega)$ .
*(iii)* $R_\alpha f(a) = E_a\left(\int_0^\infty \mathrm{e}^{-\alpha t} f(X_t)\mathrm{d}t\right)$ .

*Proof.* (i) and (ii) are obvious. (iii):

$$R_\alpha f(a) = \int_0^\infty \mathrm{e}^{-\alpha t} H_t f(a)\mathrm{d}t = \int_0^\infty \mathrm{e}^{-\alpha t} E_a(f(X_t))\mathrm{d}t\ .$$

Since $f(X_t(\omega))$ is right continuous in $t$ for $\omega$ fixed and measurable in $\omega$ for $t$ fixed, it is measurable in the pair $(t, \omega)$. Therefore, we can use Fubini's theorem to get

$$R_\alpha f(a) = E_a\left(\int_0^\infty \mathrm{e}^{-\alpha t} f(X_t)\mathrm{d}t\right)\ .$$

The operator $\theta_t \colon \Omega \longrightarrow \Omega$ defined by

$$(\theta_t\omega)(s) = \omega(s + t) \qquad \text{for every } s \in T$$

[14] That is, $\mathcal{B}$ is the $\sigma$-algebra generated by $\{X_t \in A\}$, $t \in T$, $A \in \mathcal{B}(S)$.
[15] For each $a \in S$, $\{X_t\}_{t \in T}$ is a stochastic process defined on the probability space $(\Omega, \mathcal{B}, P_a)$.

is called a shift operator. It is obvious that

$$\theta_{t+s} = \theta_t \theta_s \qquad \text{(semi-group property)} .$$

Let $\mathcal{C}$ be a $\sigma$-algebra on $\Omega$. The space of all bounded $\mathcal{C}$-measurable functions is denoted by $\mathbf{B}(\Omega, \mathcal{C})$ or simply by $\mathbf{B}(\mathcal{C})$.

**Theorem 2 (Markov property).** *For $\Lambda \in \mathcal{B}$, we have*

$$P_a(\theta_t \omega \in \Lambda \mid \mathcal{B}_t) = P_{X_t(\omega)}(\Lambda) \qquad a.\ s.\ (P_a) ,$$

*that is,* $P_a(\theta_t^{-1} \Lambda \mid \mathcal{B}_t) = P_{X_t(\omega)}(\Lambda)$ .

*Remark.* $P_{X_t(\omega)}(\Lambda) = P_b(\Lambda)|_{b=X_t(\omega)}$ .

*Proof.* It is enough to prove that

$$(1) \qquad P_a(\theta_t^{-1} \Lambda \cap M) = E_a(P_{X_t}(\Lambda), M) \qquad \text{for } \Lambda \in \mathcal{B},\ M \in \mathcal{B}_t .$$

*Case 1.* Let

$$\begin{aligned} \Lambda &= \{X_{s_1} \in B_1\} \cap \{X_{s_2} \in B_2\} \cap \cdots \cap \{X_{s_n} \in B_n\} , \\ M &= \{X_{t_1} \in A_1\} \cap \{X_{t_2} \in A_2\} \cap \cdots \cap \{X_{t_m} \in A_m\} \end{aligned}$$

with $0 \le s_1 < s_2 < \cdots < s_n$, $0 \le t_1 < t_2 < \cdots < t_m \le t$, and $B_i, A_j \in \mathcal{B}(S)$. Both sides of (1) are expressed as integrals on $S^{m+n}$ in terms of transition probabilities. It is easy to see that they are equal.

*Case 2.* Let $\Lambda$ be the same as in Case 1. Let $M$ be a general member of $\mathcal{B}_t$. With $\Lambda$ fixed, the family $\mathcal{D}$ of all $M$'s that satisfy (1) is a Dynkin class. The family $\mathcal{M}$ of all $M$'s in Case 1 is multiplicative and $\mathcal{M} \subset \mathcal{D}$. Therefore

$$\mathcal{D} \supset \mathcal{D}[\mathcal{M}] = \mathcal{B}[\mathcal{M}] = \mathcal{B}_t .$$

This proves that (1) holds for $\Lambda$ in Case 1 and for $M$ general in $\mathcal{B}_t$.

*Case 3.* General case. Fixing an arbitrary $M \in \mathcal{B}_t$ we can carry out the same argument to derive this case from Case 2.[16]

The intuitive meaning of Theorem 2 is that if we know the behavior of the path up to time $t$, then the future motion is as if it started at the point $X_t(\omega)$ i.e. the position at $t$.

**Corollary to Theorem 2.**

$$\begin{aligned} E_a(G \circ \theta_t, M) &= E_a(E_{X_t}(G), M) \quad \textit{for } G \in \mathbf{B}(\mathcal{B}),\ M \in \mathcal{B}_t , \\ E_a(F \cdot (G \circ \theta_t)) &= E_a(F \cdot E_{X_t}(G)) \quad \textit{for } G \in \mathbf{B}(\mathcal{B}),\ F \in \mathbf{B}(\mathcal{B}_t) \\ E_a(G \circ \theta_t \mid \mathcal{B}_t) &= E_{X_t}(G) \quad a.\ s.\ (P_a)\ \textit{for } G \in \mathbf{B}(\mathcal{B}) . \end{aligned}$$

[16] Notice that, for any $\Lambda \in \mathcal{B}$, $P_a(\Lambda)$ is Borel measurable in $a$. This fact is also proved by Dynkin's theorem.

We will now extend Markov property slightly. Let $\mathcal{B}_{t+} = \bigcap_{s>t} \mathcal{B}_s$. Since $\mathcal{B}_t$ increases with $t$,

$$\mathcal{B}_{t+} = \bigcap_n \mathcal{B}_{t+1/n} \,.$$

It is obvious that $s < t$ implies

$$\mathcal{B}_s \subset \mathcal{B}_{s+} \subset \mathcal{B}_t \subset \mathcal{B}_{t+} \,.$$

More precisely $\mathcal{B}_t \subsetneqq \mathcal{B}_{t+}$. For example the events $\{X_{t+1/n} \neq X_t$ for infinitely many $n\}$ and $\{X_{t+1/n} = X_t$ for infinitely many $n\}$ are both in $\mathcal{B}_{t+}$ but neither is in $\mathcal{B}_t$.

**Theorem 3 (Extended Markov property).**

$$P_a(\theta_t\omega \in \Lambda \mid \mathcal{B}_{t+}) = P_{X_t}(\Lambda) \qquad \textit{a. s. } (P_a) \textit{ for } \Lambda \in \mathcal{B}$$

*and so the corollary above holds for* $F \in \mathbf{B}(\mathcal{B}_{t+})$ *and* $M \in \mathcal{B}_{t+}$.

*Proof.* It is enough to prove (1) for $M \in \mathcal{B}_{t+}$. For this purpose we will just prove

$$\begin{aligned} &E_a(f_1(X_{s_1}(\theta_t\omega)) \cdots f_n(X_{s_n}(\theta_t\omega)), M) \\ &\quad = E_a(E_{X_t}(f_1(X_{s_1}) \cdots f_n(X_{s_n})), M) \end{aligned} \tag{2}$$

for $f_i \in C(S)$, $M \in \mathcal{B}_{t+}$, and $0 \le s_1 < s_2 < \cdots < s_n$. Since $M \in \mathcal{B}_{t+h}$ $(h > 0)$, we have by the Corollary

$$\begin{aligned} &E_a(f_1(X_{s_1}(\theta_{t+h}\omega)) \cdots f_n(X_{s_n}(\theta_{t+h}\omega)), M) \\ &\quad = E_a(E_{X_{t+h}}(f_1(X_{s_1}) \cdots f_n(X_{s_n})), M) \,. \end{aligned} \tag{2'}$$

Noticing

$$E_a(f_1(X_{s_1}) \cdots f_n(X_{s_n})) = H_{s_1}[f_1 \cdots (H_{s_{n-1}-s_{n-2}}(f_{n-1} \cdot H_{s_n - s_{n-1}} f_n)) \cdots )]$$

and $H_s \colon C \longrightarrow C$, we see that $E_a(f_1(X_{s_1}) \cdots f_n(X_{s_n}))$ is continuous in $a$. Since $X_t(\omega)$ is right continuous in $t$,

$$f_i(X_{s_i}(\theta_{t+h}\omega)) = f_i(X_{s_i+t+h}(\omega)) \to f_i(X_{s_i+t}(\omega)) = f_i(X_{s_i}(\theta_t\omega))$$

as $h \downarrow 0$. Taking the limit in (2′) as $h \downarrow 0$, we have (2).

Since the indicator of every open set in $S$ is an increasing limit of continuous functions, we obtain from (2)

$$\begin{aligned} &E_a(X_{s_i}(\theta_t\omega) \in G_1, \ldots, X_{s_n}(\theta_t\omega) \in G_n, M) \\ &\quad = E_a(P_{X_t}(X_{s_1} \in G_1, \ldots, X_{s_n} \in G_n), M) \end{aligned}$$

for $G_i$ open in $S$. Now use Dynkin's theorem.

**Theorem 4.** [17] *For every $\Lambda \in \mathcal{B}_{t+}$ we have $\Lambda_1 \in \mathcal{B}_t$ such that*

$$P_a(\Lambda \vartriangle \Lambda_1) = 0 \,,$$

*where $\Lambda_1$ may depend on $a$.*[18]

*Proof.* It is enough to prove that if $f \in \mathbf{B}(\mathcal{B}_{t+})$, then

$$E_a(f \mid \mathcal{B}_t) = f \qquad \text{a.s. } (P_a) \,. \tag{3}$$

Let $g_1 \in \mathbf{B}(\mathcal{B}_t)$ and $g_2 \in \mathbf{B}(\mathcal{B})$. Then

$$\begin{aligned}
&E_a(E_a(f \mid \mathcal{B}_t) \cdot g_1 \cdot (g_2 \circ \theta_t)) \\
&\quad = E_a(E_a(f \mid \mathcal{B}_t) \cdot g_1 \cdot E_{X_t}(g_2)) \qquad \text{(Markov property)} \\
&\quad = E_a(f \cdot g_1 \cdot E_{X_t}(g_2)) \qquad \text{(definition of conditional expectation)} \\
&\quad = E_a(f \cdot g_1 \cdot (g_2 \circ \theta_t)). \qquad \text{(extended Markov property)}
\end{aligned}$$

Replacing $g_1$ and $g_2$ respectively by the indicators of

$$\{X_{t_1} \in B_1, \ldots, X_{t_n} \in B_n\} \text{ and } \{X_{s_1} \in C_1, \ldots, X_{s_m} \in C_m\}$$

with $t_1 \leq t_2 \leq \cdots \leq t_n \leq t$ and $0 \leq s_1 \leq s_2 \cdots \leq s_m$, we have

$$E_a(E_a(f \mid \mathcal{B}_t), \Lambda) = E_a(f, \Lambda)$$

for

$$\Lambda = \{X_t \in B_1, \ldots, X_{t_n} \in B_n, X_{s_1+t} \in C_1, \ldots, X_{s_m+t} \in C_m\} \,.$$

This is true for general $\Lambda \in \mathcal{B}$ by virtue of Dynkin's theorem.

**Corollary to Theorem 4 (Blumenthal's 0-1 law).** *If $\Lambda \in \mathcal{B}_{0+}$, then $P_a(\Lambda) = 0$ or 1. Namely $\mathcal{B}_{0+}$ is trivial.*

*Proof.* Since $P_a(X_0 = a) = 1$ and $\mathcal{B}_0 = \mathcal{B}[X_0]$, $\mathcal{B}_0$ is trivial and so is $\mathcal{B}_{0+}$ by Theorem 4.

Let

$$e(\omega) = \inf\{t : X_t \neq X_0\} \,.$$

This is called the *waiting time* of the path at the starting point.

**Theorem 5.** *For each $a$, only the following three cases are possible:*

(i) $P_a(e \geq t) = \mathrm{e}^{-\lambda_a t}$ *for some* $\lambda_a \in (0, \infty)$ .

(ii) $P_a(e = 0) = 1$ .

(iii) $P_a(e = \infty) = 1$ .

---

[17] This theorem remains true if we replace $P_a$ by $P_\mu$. This fact will be used in Section 2.6.

[18] $\Lambda \vartriangle \Lambda_1 = (\Lambda \cap \Lambda_1^{\mathrm{C}}) \cup (\Lambda_1 \cap \Lambda^{\mathrm{C}})$, the symmetric difference of $\Lambda$ and $\Lambda_1$.

*Remark.* (ii) is considered as the limiting case of (i) ($\lambda_a = \infty$) and (iii) is also a limiting case ($\lambda_a = 0$).

*Proof.* We have

$$\{e \geq t\} = \bigcap_{s<t} \{X_s = X_0\} = \bigcap_{s<t, s \text{ rational}} \{X_s = X_0\} \in \mathcal{B}_t ,$$

$$\{e > t\} = \bigcap_{n \geq m} \{e \geq t + 1/n\} \in \mathcal{B}_{t+1/m} ,$$

for every $m$, and so $\{e > t\} \in \mathcal{B}_{t+}$ .

Observing the *composition rule*

$$e(\omega) = t + e(\theta_t \omega) \qquad \text{if } e(\omega) > t ,$$

we have

$$\begin{aligned} P_a(e > t + s) &= P_a(e > t + s,\ e > t) \\ &= P_a(e \circ \theta_t > s,\ e > t) \\ &= E_a(P_{X_t}(e > s),\ e > t) \quad \text{(extended Markov property)} \\ &= E_a(P_a(e > s),\ e > t) \\ &= P_a(e > s) P_a(e > t) . \end{aligned}$$

If $P_a(e > t) = 0$ for some $t$, then $P_a(e > u) = 0$ for $u > t$. But $P_a(e > t) = P_a(e > t/n)^n$ and so $P_a(e > t/n) = 0$. Therefore

$$P_a(e > u) = 0 \qquad \text{for every } u > 0 .$$

This is the case (ii). Suppose that $P_a(e > t) > 0$ for every $t$. Then $f(t) = \log P_a(e > t)$ satisfies

$$f(t + s) = f(t) + f(s), \qquad -\infty < f(t) \leq 0 .$$

Therefore[19] $f(t) = -\lambda_a t$ for a constant $\lambda_a \in [0, \infty)$. If $\lambda_a = 0$, then $P_a(e > t) = 1$ for every $t > 0$ and so $P_a(e = \infty) = 1$. This is the case (iii). The remaining case is

$$P_a(e > t) = \mathrm{e}^{-\lambda_a t}, \qquad \lambda_a \in (0, \infty) ,$$

and so

$$P_a(e \geq t) = \lim_n P_a(e > t - 1/n) = \lim_n \mathrm{e}^{-\lambda_a (t - 1/n)} = \mathrm{e}^{-\lambda_a t}.$$

This completes the proof.

[19] Note that $f(t)$ is right continuous.

A point $a$ is called

an *exponential holding point* in case (i),
an *instantaneous point* in case (ii),
a *trap* in case (iii),
a *stable point* in case (i) and (iii).

If $a$ is an isolated point in $S$, $a$ cannot be instantaneous by virtue of right-continuity of the path.

## 2.6 The $\sigma$-Algebras $\overline{\mathcal{B}}$, $\overline{\mathcal{B}}_t$, and $\overline{\mathcal{B}(S)}$

Let $\lambda$ be a probability measure on a measurable space $(E, \mathcal{E})$. A set $A \subset E$ is called $\lambda$-*measurable* if there exist $B_1, B_2 \in \mathcal{E}$ such that

$$B_1 \subset A \subset B_2, \qquad \lambda(B_2 - B_1) = 0 . \tag{1}$$

The family of all $\lambda$-measurable sets is denoted by $\mathcal{E}^\lambda$. This family $\mathcal{E}^\lambda$ is a $\sigma$-algebra and includes $\mathcal{E}$. The measure $\lambda$ can be extended to a probability measure on $(E, \mathcal{E}^\lambda)$, called the *Lebesgue extension* of $\lambda$. We denote the Lebesgue extension by the same symbol $\lambda$. The family of all probability measures on $(E, \mathcal{E})$ is denoted by $\mathfrak{P}(E, \mathcal{E})$ or $\mathfrak{P}(E)$.

Let $X = \{X_t(\omega),\ t \in T,\ \omega \in (\Omega, \mathcal{B}, P_a)\}_{a \in S}$ be a Markov process (Definition 2.5.1). $P_\mu(\cdot)$ denotes $\int_S \mu(da) P_a(\cdot)$ for $\mu \in \mathfrak{P}(S)$.

Now we will extend $\mathcal{B}$, $\mathcal{B}_t$, and $\mathcal{B}(S)$ and define $\overline{\mathcal{B}}$, $\overline{\mathcal{B}}_t$, and $\overline{\mathcal{B}(S)}$. Although we use the bar (overline) in all cases, its meanings are not the same. (See Remark at the end of this section.)

**Definition 1.**

$$\overline{\mathcal{B}} = \bigcap_{\mu \in \mathfrak{P}(S)} \mathcal{B}^{P_\mu}, \tag{2}$$

$$\overline{\mathcal{B}}_t = \{A \in \overline{\mathcal{B}} : \text{for every } \mu \text{ there is } B \in \mathcal{B}_t \text{ such that } P_\mu(A \triangle B) = 0\} . \tag{3}$$

It is easy to see that $\overline{\mathcal{B}}$ is a $\sigma$-algebra on $\Omega$ including $\mathcal{B}$ and that $\overline{\mathcal{B}}_t$ is a $\sigma$-algebra including $\mathcal{B}_t$. Theorem 2.5.4 shows that

$$\mathcal{B}_t \subset \mathcal{B}_{t+} \subset \overline{\mathcal{B}}_t . \tag{4}$$

It is also easy to see

$$\overline{\mathcal{B}}_t \supset \bigcap_\mu \mathcal{B}_t^{P_\mu | \mathcal{B}_t} \quad \text{where } P_\mu | \mathcal{B}_t = \text{ the restriction of } P_\mu \text{ to } \mathcal{B}_t . \tag{5}$$

Now we will further extend the extended Markov property (Theorem 2.5.3).

**Theorem 1.** *$P_\mu(\theta_t^{-1}\Lambda \mid \overline{\mathcal{B}}_t) = P_{X_t}(\Lambda)$ a. s. $(P_\mu)$ for $\Lambda \in \overline{\mathcal{B}}$.*

*Proof.* It is enough to prove that

$$P_\mu(\theta_t^{-1}\Lambda \cap M) = E_\mu(P_{X_t}(\Lambda), M)$$

for $\Lambda \in \overline{\mathcal{B}}$ and $M \in \overline{\mathcal{B}}_t$. By the definition of $\overline{\mathcal{B}}_t$ we can assume with no loss of generality that $M \in \mathcal{B}_t$. Let $\Lambda \in \overline{\mathcal{B}}$ and

$$\nu(B) = P_\mu X_t^{-1}(B)\,, \qquad B \in \mathcal{B}(S)\,.$$

Then $\nu \in \mathfrak{P}(S)$. Therefore we can find $\Lambda_1, \Lambda_2 \in \mathcal{B}$ such that

$$\Lambda_1 \subset \Lambda \subset \Lambda_2 \quad \text{and} \quad P_\nu(\Lambda_2 - \Lambda_1) = 0\,.$$

Then we have

$$\theta_t^{-1}\Lambda_1 \subset \theta_t^{-1}\Lambda \subset \theta_t^{-1}\Lambda_2\,.$$

By the Markov property we obtain

$$\begin{aligned} P_\mu(\theta_t^{-1}\Lambda_2 - \theta_t^{-1}\Lambda_1) &= P_\mu(\theta_t^{-1}(\Lambda_2 - \Lambda_1)) \\ &= E_\mu(P_{X_t}(\Lambda_2 - \Lambda_1)) = \int_S P_a(\Lambda_2 - \Lambda_1) P_\mu X_t^{-1}(\mathrm{d}a) \\ &= \int_S P_a(\Lambda_2 - \Lambda_1)\nu(\mathrm{d}a) = P_\nu(\Lambda_2 - \Lambda_1) = 0\,. \end{aligned}$$

Therefore $\theta_t^{-1}\Lambda \in \mathcal{B}^{P_\mu}$ (and so $\theta_t^{-1}\Lambda \in \overline{\mathcal{B}}$) and

$$\theta_t^{-1}\Lambda = \theta_t^{-1}\Lambda_1 \qquad \text{a.s. } (P_\mu)\,. \tag{6}$$

Since $\Lambda \in \overline{\mathcal{B}}$, $P_a(\Lambda)$ is meaningful for every $a$ and so is $P_{X_t}(\Lambda)$. Since $\Lambda_1 \subset \Lambda \subset \Lambda_2$, we have

$$P_{X_t}(\Lambda_1) \le P_{X_t}(\Lambda) \le P_{X_t}(\Lambda_2)\,.$$

But

$$E_\mu(P_{X_t}(\Lambda_2) - P_{X_t}(\Lambda_1)) = E_\mu(P_{X_t}(\Lambda_2 - \Lambda_1)) = 0$$

as we have proved above. $P_{X_t(\omega)}(\Lambda)$ is therefore measurable $(P_\mu)$ in $\omega$ and

$$P_{X_t}(\Lambda) = P_{X_t}(\Lambda_1) \qquad \text{a.s. } (P_\mu)\,. \tag{7}$$

By the Markov property

$$P_\mu(\theta_t^{-1}\Lambda_1 \cap M) = E_\mu(P_{X_t}(\Lambda_1), M)\,,$$

which combined with (6) and (7) implies

$$P_\nu(\theta_t^{-1}\Lambda \cap M) = E_\mu(P_{X_t}(\Lambda), M)\,.$$

**Theorem 2.** *$\overline{\mathcal{B}}_t$ increases with $t$ and is right continuous i. e.*

$$\overline{\mathcal{B}}_t = \bigcap_n \overline{\mathcal{B}}_{t+1/n} .$$

*Proof.* Increasingness is obvious. Therefore

$$\overline{\mathcal{B}}_t \subset \bigcap_n \overline{\mathcal{B}}_{t+1/n} .$$

Suppose $A$ is a member of the right-hand side. Then

$$P_\mu(A \triangle B_n) = 0 \qquad \text{for some } B_n \in \mathcal{B}_{t+1/n} ,$$

where $B_n$ depends on $\mu$ in general. Set

$$B \equiv \limsup_{n\to\infty} B_n \in \bigcap_n \mathcal{B}_{t+1/n} = \mathcal{B}_{t+} \subset \overline{\mathcal{B}}_t$$

by (4). Then[20]

$$P_\mu(A \triangle B) = 0 .$$

Therefore $A \in \overline{\mathcal{B}}_t$. This completes the proof.

Now we will define $\overline{\mathcal{B}(S)}$. A set $A \subset S$ is called *nearly Borel measurable* if for every $\mu \in \mathfrak{P}(S)$, we have $B_1, B_2 \in \mathcal{B}(S)$ such that[21]

$$B_1 \subset A \subset B_2, \qquad P_\mu(e_{B_1}(X_t) = e_{B_2}(X_t) \text{ for all } t) = 1 , \tag{8}$$

where $e_B$ denotes the indicator of $B$. The family of all nearly Borel measurable sets, $\overline{\mathcal{B}(S)}$ in notation, is a $\sigma$-algebra on $S$ and includes $\mathcal{B}(S)$. It follows from (8) that

$$E_\mu(e_{B_1}(X_0)) = E_\mu(e_{B_2}(X_0)) \qquad \text{i.e. } \mu(B_1) = \mu(B_2) .$$

Therefore

$$\overline{\mathcal{B}(S)} \subset \bigcap_{\mu\in\mathfrak{P}(S)} \mathcal{B}(S)^\mu . \tag{9}$$

It is easy to see that a real-valued function $f$ on $S$ is $\overline{\mathcal{B}(S)}$-measurable (that is, nearly Borel measurable), if and only if for every $\mu \in \mathfrak{P}(S)$ we have $\mathcal{B}(S)$-measurable functions $f_1, f_2$ such that $f_1 \le f \le f_2$ and

$$P_\mu(f_1(X_t) = f_2(X_t) \text{ for all } t) = 1 .$$

[20] Note that $P_\mu(\bigcup_n B_n - \bigcap_n B_n) = 0$.

[21] The event $\{e_{B_1}(X_t) = e_{B_2}(X_t) \text{ for all } t\}$ belongs to $\overline{\mathcal{B}}$, but the proof is not easy. It is given in the book of Blumenthal and Getoor, p. 57–60, mentioned in the Preface to the Original.

*Remark.*

$$\overline{\mathcal{B}} = \bigcap_{\mu \in \mathfrak{P}(S)} \mathcal{B}^{P_\mu} \supset \bigcap_{Q \in \mathfrak{P}(\Omega, \mathcal{B})} \mathcal{B}^Q \qquad \text{(by Definition 1)},$$

$$\overline{\mathcal{B}}_t \supset \bigcap_{\mu \in \mathfrak{P}(S)} \mathcal{B}_t^{P_\mu | \mathcal{B}_t} \supset \bigcap_{Q \in \mathfrak{P}(\Omega, \mathcal{B}_t)} \mathcal{B}^Q \qquad \text{(by (5))},$$

$$\overline{\mathcal{B}(S)} \subset \bigcap_{\mu \in \mathfrak{P}(S)} \mathcal{B}(S)^\mu \qquad \text{(by (9))}.$$

## 2.7 Strong Markov Property

The intuitive meaning of the Markov property is that under the condition that the path is known up to time $t$, the future motion would be as if it started at the point $X_t(\omega) \in S$. What will happen if we replace $t$ by a random time $\sigma(\omega)$? Then the same is true provided $\sigma$ is a stopping time; see Theorem 3 below.

A random time $\sigma\colon \Omega \to [0, \infty]$ is called a *stopping time* with respect to $\{\overline{\mathcal{B}}_t\}$ if

$$\text{(1)} \qquad \{\sigma \le t\} \in \overline{\mathcal{B}}_t \qquad \text{for every } t\,.$$

This condition is equivalent to the following

$$\text{(2)} \qquad \{\sigma < t\} \in \overline{\mathcal{B}}_t \qquad \text{for every } t\,.$$

In fact if (1) holds, then

$$\{\sigma < t\} = \bigcup_n \{\sigma \le t - 1/n\} \in \overline{\mathcal{B}}_t\,,$$

while if (2) holds, then

$$\{\sigma \le t\} = \bigcap_{n \ge m} \{\sigma < t + 1/n\} \in \overline{\mathcal{B}}_{t+1/m}$$

for every $m$ and so

$$\{\sigma \le t\} \in \bigcap_m \overline{\mathcal{B}}_{t+1/m} = \overline{\mathcal{B}}_t$$

by right continuity of $\overline{\mathcal{B}}_t$ (Theorem 2.6.2).

A trivial example is the *deterministic time* $\sigma \equiv t$. An important example is the *hitting time* $\sigma_G$ of an open set $G \subset S$:

$$\sigma_G = \inf\{t > 0\colon X_t \in G\}\,.$$

In fact

$$\{\sigma_G < t\} = \bigcup_{s<t} \{X_s \in G\} = \bigcup_{r<t,\ r \text{ rational}} \{X_r \in G\} \in \mathcal{B}_t \subset \overline{\mathcal{B}}_t$$

by right continuity of the path and the assumption that $G$ is open. We will later prove that the hitting time $\sigma_F$ of a closed set $F$ is also a stopping time (see Section 2.10).

It is easy to see that

**Theorem 1.** *The family of stopping times is closed under the following operations:*

$$\sigma_1 \vee \sigma_2 \quad \sigma_1 \wedge \sigma_2, \quad \sigma_1 + \sigma_2 \quad (\textit{in particular } \sigma + t)$$

*and monotone limits.*[22]

Every stopping time is approximated from above by discrete stopping times. More precisely we have

**Theorem 2.** *Let $\sigma$ be a stopping time. Then*

$$\sigma_n = \frac{[2^n \sigma] + 1}{2^n}, \qquad n = 1, 2, \ldots, \qquad [\cdot] = \textit{integral part},$$

*are stopping times with values $k/2^n$, $k = 1, 2, \ldots, \infty$ and $\sigma_n \downarrow \sigma$.*

*Proof.*

$$\begin{aligned}\{\sigma_n = k/2^n\} &= \{(k-1)/2^n \leq \sigma < k/2^n\} \\ &= \{\sigma < k/2^n\} - \{\sigma < (k-1)/2^n\} \in \overline{\mathcal{B}}_{k/2^n}.\end{aligned}$$

The rest of the proof is easy.

**Definition 1.**

$$\overline{\mathcal{B}}_\sigma = \{\Lambda \in \overline{\mathcal{B}} \colon \Lambda \cap \{\sigma \leq t\} \in \overline{\mathcal{B}}_t\}.$$

By right continuity of $\overline{\mathcal{B}}_t$, this can also be stated as

$$\overline{\mathcal{B}}_\sigma = \{\Lambda \in \overline{\mathcal{B}} \colon \Lambda \cap \{\sigma < t\} \in \overline{\mathcal{B}}_t\}.$$

The family $\overline{\mathcal{B}}_\sigma$ is obviously a sub-$\sigma$-algebra of $\overline{\mathcal{B}}$ and its intuitive meaning is the knowledge of the path up to time $\sigma(\omega)$. If $\sigma \equiv t$, then the $\overline{\mathcal{B}}_\sigma$ in this definition coincides with the $\overline{\mathcal{B}}_t$ introduced in Section 2.6.

[22] See Problem 2.6 for a proof.

**Theorem 3.** *Let $\sigma$ and $\sigma_n$, $n = 1, 2, 3, \ldots$, be stopping times.*

*(i) $\sigma$ is $\overline{\mathcal{B}}_\sigma$-measurable.*

*(ii) If $\sigma_1 \le \sigma_2$, then $\overline{\mathcal{B}}_{\sigma_1} \subset \overline{\mathcal{B}}_{\sigma_2}$.*

*(iii) $\{\sigma_1 < \sigma_2\}$, $\{\sigma_1 = \sigma_2\}$, $\{\sigma_1 \le \sigma_2\} \in \overline{\mathcal{B}}_{\sigma_1} \cap \overline{\mathcal{B}}_{\sigma_2}$.*

*(iv) If $\sigma_n \downarrow \sigma$, then $\overline{\mathcal{B}}_\sigma = \bigcap_n \overline{\mathcal{B}}_{\sigma_n}$.*

*(v) $X_\sigma$ is $\overline{\mathcal{B}}_\sigma$-measurable on $\{\sigma < \infty\}$, namely $X_\sigma$ coincides with a $\overline{\mathcal{B}}_\sigma$-measurable function on $\{\sigma < \infty\}$.*

*Proof.* (i) $\{\sigma \le s\} \cap \{\sigma \le t\} = \{\sigma \le s \wedge t\} \in \overline{\mathcal{B}}_{s\wedge t} \subset \overline{\mathcal{B}}_t$, and so $\{\sigma \le s\} \in \overline{\mathcal{B}}_\sigma$.

(ii) If $\Lambda \in \overline{\mathcal{B}}_{\sigma_1}$, then $\Lambda \cap \{\sigma_1 \le t\} \in \overline{\mathcal{B}}_t$ and

$$\begin{aligned}\Lambda \cap \{\sigma_2 \le t\} &= \Lambda \cap \{\sigma_1 \le t\} \cap \{\sigma_2 \le t\} \quad (\text{by } \sigma_1 \le \sigma_2)\\ &\in \overline{\mathcal{B}}_t\,.\end{aligned}$$

(iii) $\{\sigma_1 < \sigma_2\} \cap \{\sigma_i < t\} = \bigcup_{r<t,\, r \text{ rational}} \{\sigma_1 \le r < \sigma_2\} \cap \{\sigma_i < t\} \in \overline{\mathcal{B}}_t$ since $\{\sigma_1 \le r < \sigma_2\} = \{\sigma_1 \le r\} - \{\sigma_2 \le r\} \in \overline{\mathcal{B}}_r \subset \overline{\mathcal{B}}_t$). Therefore, for $\mathcal{C} \equiv \overline{\mathcal{B}}_{\sigma_1} \wedge \overline{\mathcal{B}}_{\sigma_2}$,

$$\begin{aligned}&\{\sigma_1 < \sigma_2\} \in \mathcal{C}\,,\\ &\{\sigma_1 = \sigma_2\} = \Omega - \{\sigma_1 < \sigma_2\} - \{\sigma_2 < \sigma_1\} \in \mathcal{C}\,,\\ &\{\sigma_1 \le \sigma_2\} = \{\sigma_1 < \sigma_2\} \cup \{\sigma_1 = \sigma_2\} \in \mathcal{C}\,.\end{aligned}$$

(iv) It is obvious that $\overline{\mathcal{B}}_\sigma \subset \bigcap_n \overline{\mathcal{B}}_{\sigma_n}$. If $B \in \bigcap_n \overline{\mathcal{B}}_{\sigma_n}$, then $B \cap \{\sigma_n < t\} \in \overline{\mathcal{B}}_t$, $n = 1, 2, \ldots$, and so

$$B \cap \{\sigma < t\} = \bigcup_n (B \cap \{\sigma_n < t\}) \in \overline{\mathcal{B}}_t\,.$$

(v) Take the $\sigma_n$ in Theorem 2. Then $\sigma_n \downarrow \sigma$ and so

$$\overline{\mathcal{B}}_\sigma = \bigcap_n \overline{\mathcal{B}}_{\sigma_n} \tag{3}$$

by (iv) and

$$X_\sigma = \lim_n X_{\sigma_n} \qquad (\text{by right continuity}), \tag{4}$$

$$\{X_{\sigma_n} \in B\} \cap \{\sigma_n < t\} = \bigcup_{k/2^n < t} \{X_{k/2^n} \in B\} \cap \{\sigma_n = k/2^n\} \in \overline{\mathcal{B}}_t\,.$$

Therefore $\{X_{\sigma_n} \in B\} \in \overline{\mathcal{B}}_{\sigma_n}$. Hence it follows that $X_{\sigma_n}$ is $\overline{\mathcal{B}}_{\sigma_n}$-measurable and so $\overline{\mathcal{B}}_{\sigma_m}$-measurable for $n > m$. Then $X_\sigma$ is $\overline{\mathcal{B}}_{\sigma_m}$-measurable for every $m$ by (4) and so $\overline{\mathcal{B}}_\sigma$-measurable by (3).

**Theorem 4 (Strong Markov property).** [23]

*(i)* $P_\mu(\theta_\sigma^{-1}\Lambda \mid \overline{\mathcal{B}}_\sigma) = P_{X_\sigma}(\Lambda)$ *a. s.* $(P_\mu)$ *on* $\{\sigma < \infty\}$, *where* $\Lambda \in \overline{\mathcal{B}}$.

*(ii)* $E_\mu(F(\theta_\sigma\omega) \mid \overline{\mathcal{B}}_\sigma) = E_{X_\sigma}(F)$ *a. s.* $(P_\mu)$ *on* $\{\sigma < \infty\}$, *where* $F$ *is bounded and* $\overline{\mathcal{B}}$*-measurable.*

*Proof.* (i) follows from (ii). (Set $F = e_\Lambda$). For the proof of (ii) it is enough to show that

$$E_\mu(F(\theta_\sigma\omega), M \cap \{\sigma < \infty\}) = E_\mu(E_{X_\sigma}(F), M \cap \{\sigma < \infty\}) \tag{5}$$

for $M \in \overline{\mathcal{B}}_\sigma$; note that $\{\sigma < \infty\} \in \overline{\mathcal{B}}_\sigma$.

Take the $\sigma_n$ in Theorem 1. Then

$$\begin{aligned} &E_\mu(F(\theta_{\sigma_n}\omega), M \cap \{\sigma < \infty\}) \\ &\quad = \sum_k E_\mu(F(\theta_{k/2^n}\omega), M \cap \{(k-1)/2^n \le \sigma < k/2^n\}) \\ &\quad = \sum_k E_\mu(E_{X_{k/2^n}}(F), M \cap \{(k-1)/2^n \le \sigma < k/2^n\}) \\ &\qquad\qquad\qquad\qquad\qquad\qquad \text{(by Markov property)} \\ &\quad = E_\mu(E_{X_{\sigma_n}}(F), M \cap \{\sigma < \infty\}), \end{aligned}$$

namely

$$E_\mu(F(\theta_{\sigma_n}\omega), M \cap \{\sigma < \infty\}) = E_\mu(E_{X_{\sigma_n}}(F), M \cap \{\sigma < \infty\}). \tag{6}$$

If $F(\omega)$ is of the form $f_1(X_{t_1}) \cdots f_m(X_{t_m})$, $f_i \in C(S)$, then

$$F(\theta_{\sigma_n}\omega) = f_1(X_{t_1+\sigma_n}) \cdots f_m(X_{t_m+\sigma_n})$$

and so

$$F(\theta_{\sigma_n}\omega) \to F(\theta_\sigma\omega) \qquad \text{as } n \to \infty.$$

Since $E_a(F)$ is continuous in $a$ (see the proof of the extended Markov property, Section 2.5), $E_{X_{\sigma_n}}(F) \to E_{X_\sigma}(F)$ as $n \to \infty$. Therefore (5) holds for such $F$ as the limit of (6). Then (5) can be proved for bounded $\mathcal{B}$-measurable $F$ by routine reasoning.

If $F$ is bounded and $\overline{\mathcal{B}}$-measurable, we can find bounded $\mathcal{B}$-measurable $F_1, F_2$ such that

$$F_1 \le F \le F_2, \qquad E_\nu(F_2 - F_1) = 0$$

where $\nu = P_\mu X_\sigma^{-1}$; $F_i$ depends on $\nu$ and so on $\mu$ in general and $\mu$ is an arbitrary probability measure on $(S, \mathcal{B}(S))$. Then

$$F_1(\theta_\sigma\omega) \le F(\theta_\sigma\omega) \le F_2(\theta_\sigma\omega), \tag{7}$$

[23] $\theta_\sigma\omega = \theta_{\sigma(\omega)}\omega$ for each $\omega$ with $\sigma(\omega) < \infty$, and $\theta_\sigma^{-1}\Lambda = \{\omega : \sigma(\omega) < \infty$ and $\theta_\sigma\omega \in \Lambda\}$.

and

$$(8) \qquad E_{X_\sigma}(F_1) \le E_{X_\sigma}(F) \le E_{X_\sigma}(F_2)\,;$$

note that $E_a(F)$ is meaningful by $\overline{\mathcal{B}}$-measurability and boundedness of $F$. Since (5) holds for bounded $\mathcal{B}$-measurable $F$ and since $F_2 - F_1$ is bounded and $\mathcal{B}$-measurable, we have

$$(7') \qquad \begin{aligned} &E_\mu(F_2(\theta_\sigma\omega) - F_1(\theta_\sigma\omega),\ \sigma < \infty) \\ &\quad = E_\mu(E_{X_\sigma}(F_2 - F_1),\ \sigma < \infty) = E_\nu(F_2 - F_1) = 0\,. \end{aligned}$$

It holds also that

$$(8') \qquad \begin{aligned} &E_\mu(E_{X_\sigma}(F_2) - E_{X_\sigma}(F_1),\ \sigma < \infty) \\ &\quad = E_\mu(E_{X_\sigma}(F_2 - F_1),\ \sigma < \infty) = E_\nu(F_2 - F_1) = 0\,. \end{aligned}$$

Thus $F(\theta_\sigma\omega)$ is $P_\mu$-measurable and equals $F_1(\theta_\sigma\omega)$ a. s. $(P_\mu)$ by (7) and (7′) and $E_{X_\sigma}(F)$ is $P_\mu$-measurable and equals $E_{X_\sigma}(F_1)$ a. s. by (8) and (8′). Since $F_1$ is bounded and $\mathcal{B}$-measurable, (5) holds for $F_1$ and so for this $F$.

*Remark.* It should be noted that $F(\theta_\sigma\omega)$ and $E_{X_\sigma}(F)$ are both $\overline{\mathcal{B}}$-measurable because $\mu$ is arbitrary.

The following variant of the strong Markov property is often useful.

**Theorem 5 (Time-dependent strong Markov property).** *Let $F(t,\omega)$ be bounded and $\mathcal{B}[0,\infty] \times \overline{\mathcal{B}}$-measurable. Then*

$$(9) \qquad E_\mu(F(\sigma, \theta_\sigma\omega) \mid \overline{\mathcal{B}}_\sigma) = E_a(F(t,\omega))|_{t=\sigma,\, a=X_\sigma} \qquad a.\ s.\ (P_\mu)\,.$$

*Proof.* Since $\sigma$ is $\overline{\mathcal{B}}_\sigma$-measurable, this is true for

$$F(t,\omega) = f(t)G(\omega)\,.$$

The general case can be derived from this by virtue of additivity of both sides of (9) in $F$.

**Example.** Let $\sigma$ be a stopping time. Then

$$P_\mu(X_{t_0} \in B, \sigma < t_0) = \int_{t=0}^{t_0-} \int_{a \in S} p_{t_0-t}(a,B) P_\mu((\sigma, X_\sigma) \in \mathrm{d}t\,\mathrm{d}a)\,.$$

*Proof.* Set $F(t,\omega) = e_B(X_{t_0-t}(\omega))$. Then

$$\begin{aligned} F(\sigma, \theta_\sigma\omega) &= e_B(X_{t_0-\sigma}(\theta_\sigma\omega)) = e_B(X_{t_0})\,, \\ E_a(F(t,\omega)) &= P_a(X_{t_0-t} \in B) = p_{t_0-t}(a,B)\,. \end{aligned}$$

Observing $\{\sigma < t_0\} \in \overline{\mathcal{B}}_\sigma$, we can use Theorem 5 to get

$$\begin{aligned} P_\mu(X_{t_0} \in B, \sigma < t_0) &= E_\mu(e_B(X_{t_0}), \sigma < t_0) \\ &= E_\mu(F(\sigma, \theta_\sigma\omega), \sigma < t_0) = E_\mu(p_{t_0-\sigma}(X_\sigma, B), \sigma < t_0) \\ &= \int_{t=0}^{t_0-} \int_{a \in S} p_{t_0-t}(a,B) P_\mu((\sigma, X_\sigma) \in \mathrm{d}t\,\mathrm{d}a)\,. \end{aligned}$$

## 2.8 Superposition of Stopping Times

The purpose of this section is to prove

**Theorem 1.** *Let $\sigma_1$ and $\sigma_2$ be stopping times. Then*

$$\sigma(\omega) = \begin{cases} \sigma_1(\omega) + \sigma_2(\theta_{\sigma_1(\omega)}\omega) & \text{if } \sigma_1(\omega) < \infty \\ \infty & \text{if otherwise} \end{cases}$$

*is also a stopping time.*

Before proving this, we will prove a preliminary fact. Recall that $\sigma + t$ is a stopping time (Theorem 2.7.1).

**Lemma 1.** $\theta_\sigma^{-1}\mathcal{B}_t \subset \overline{\mathcal{B}}_{\sigma+t}$ *if* $\sigma < \infty$ .

*Proof.* First we will prove that if

$$F(\omega) = f_1(X_{t_1}(\omega)) \cdots f_m(X_{t_m}(\omega)), \quad f_i \in C(S),\ t_i \leq t\,,$$

then $F(\theta_\sigma\omega)$ is $\overline{\mathcal{B}}_{\sigma+t}$-measurable. We have

$$F(\theta_\sigma\omega) = f_1(X_{t_1+\sigma}) \cdots f_m(X_{t_m+\sigma})\,.$$

Since $X_{t_i+\sigma}$ is $\overline{\mathcal{B}}_{t_i+\sigma}$-measurable (Theorem 2.7.3 (v)) and so $\overline{\mathcal{B}}_{\sigma+t}$-measurable by $t_i \leq t$, $F(\theta_\sigma\omega)$ is $\overline{\mathcal{B}}_{\sigma+t}$-measurable.

Since the indicator of an open subset of $S$ is an increasing limit of continuous functions,

$$\theta_\sigma^{-1}\Lambda \in \overline{\mathcal{B}}_{\sigma+t} \qquad \text{for } \Lambda = \{X_{t_1} \in G_1, \ldots, X_{t_m} \in G_m\}\,,$$

where $t_i \leq t$ and all $G_i$ are open. Then we have

$$\theta_\sigma^{-1}\Lambda \in \overline{\mathcal{B}}_{\sigma+t} \qquad \text{for } \Lambda \in \mathcal{B}_t\,,$$

because $\{\Lambda\colon \theta_\sigma^{-1}\Lambda \in \overline{\mathcal{B}}_{\sigma+t}\}$ is a $\sigma$-algebra on $\Omega$.

Let $\Lambda \in \overline{\mathcal{B}}_t$ . Then $\Lambda \in \overline{\mathcal{B}}$ and so $\theta_\sigma^{-1}\Lambda \in \overline{\mathcal{B}}$ (see Remark at the end of the proof of Theorem 2.7.4). Let $\mu$ be an arbitrary probability measure on $(S, \mathcal{B}(S))$ and set

$$\nu = P_\mu X_\sigma^{-1}\,.$$

Since $\Lambda \in \overline{\mathcal{B}}_t$, we have $\Lambda_1 \in \mathcal{B}_t$ such that

$$P_\nu(\Lambda \bigtriangleup \Lambda_1) = 0\,.$$

Then

$$P_\mu(\theta_\sigma^{-1}\Lambda \bigtriangleup \theta_\sigma^{-1}\Lambda_1) = P_\nu(\Lambda \bigtriangleup \Lambda_1) = 0\,.$$

Since $\theta_\sigma^{-1}\Lambda_1 \in \overline{\mathcal{B}}_{\sigma+t}$ as is proved above and since $\mu$ is arbitrary, $\theta_\sigma^{-1}\Lambda \in \overline{\mathcal{B}}_{\sigma+t}$. This completes the proof of our lemma.

Now we will prove Theorem 1. We have

$$\{\sigma < t\} = \bigcup_{r<t,\, r \text{ rational}} \{\sigma_1 < r,\, \sigma_2(\theta_{\sigma_1}\omega) < t - r\}\,.$$

Set $\tau_r = \sigma_1 \wedge r$. Then

$$\begin{aligned}
&\{\sigma_1 < r,\, \sigma_2(\theta_{\sigma_1}\omega) < t - r\} = \{\sigma_1 < r,\, \sigma_2(\theta_{\tau_r}\omega) < t - r\} \\
&\qquad = \{\sigma_1 < r\} \cap \theta_{\tau_r}^{-1}\{\sigma_2 < t - r\}\,, \\
&\{\sigma_1 < r\} \in \overline{\mathcal{B}}_r \subset \overline{\mathcal{B}}_t \qquad \text{(by } r < t\text{)}, \\
&\theta_{\tau_r}^{-1}\{\sigma_2 < t - r\} \in \overline{\mathcal{B}}_{\tau_r + t - r} \subset \overline{\mathcal{B}}_t \qquad \text{(by Lemma 1).}
\end{aligned}$$

Therefore $\{\sigma < t\} \in \overline{\mathcal{B}}_t$.

**Example.** Consider a Markov process on $S = \{1, 2, 3, \ldots, n\}$ (Example in Section 2.4). All paths are step functions. Take a path $\omega$ and let $\sigma_1(\omega), \sigma_2(\omega), \ldots$ be the successive jump time points along the path $\omega$. Then

$$\sigma_1(\omega) = \inf\{t\colon X_t(\omega) \neq X_0(\omega)\}$$

is a stopping time, because

$$\{\sigma_1 < t\} = \bigcup_{r<t,\, r \text{ rational}} \{X_r(\omega) \neq X_0(\omega)\} \in \mathcal{B}_t \subset \overline{\mathcal{B}}_t$$

by right continuity of $\omega$. Therefore $\sigma_1$ is a stopping time. Then

$$\sigma_2 = \sigma_1 + \sigma_1(\theta_{\sigma_1}\omega), \quad \sigma_3 = \sigma_2 + \sigma_1(\theta_{\sigma_2}\omega), \quad \sigma_4 = \sigma_3 + \sigma_1(\theta_{\sigma_3}\omega), \quad \ldots.$$

Therefore all $\sigma_n$ are stopping times by Theorem 1.

## 2.9 An Inequality of Kolmogorov Type and its Application

We have shown in Lemma 1 in the Remark at the end of Section 2.4 that for every $\varepsilon > 0$,

$$(1) \qquad P_a(d(X_h, X_0) > \varepsilon) \to 0 \text{ uniformly in } a \text{ as } h \downarrow 0.$$

Using the Markov property we can derive immediately from this that

$$(1') \qquad P_\mu(d(X_t, X_s) > \varepsilon) \to 0 \text{ uniformly in } \mu \text{ as } t - s \downarrow 0.$$

Now we want to prove a stronger assertion:

$$(2) \qquad P_\mu\left(\sup_{s \le u, v < t} d(X_u, X_v) > \varepsilon\right) \to 0 \text{ uniformly in } \mu \text{ as } t - s \downarrow 0.$$

For this purpose it is enough to prove

**Theorem 1 (Inequality of Kolmogorov type).** *There exists $\delta = \delta(\varepsilon) > 0$ for $\varepsilon > 0$ such that*

$$(3) \qquad P_\mu\left(\sup_{s\le u,v<t} d(X_u, X_v) > 4\varepsilon\right) \le 4P_\mu(d(X_s, X_t) > \varepsilon)$$

*as far as $0 \le t - s < \delta$.*

*Proof.* By (1) we have $\delta = \delta(\varepsilon) > 0$ for $\varepsilon > 0$ such that

$$\sup_{0\le h<\delta,\, a\in S} P_a(d(X_h, X_0) > \varepsilon) < 1/2$$

i. e.

$$(1'') \qquad \inf_{0\le h<\delta,\, a\in S} P_a(d(X_h, X_0) \le \varepsilon) > 1/2\,.$$

Now[24] we will prove (3) for $0 < t - s < \delta$. For this purpose it is enough to prove that

$$(3') \qquad P_\mu\left(\sup_{s\le u<t} d(X_u, X_s) > 2\varepsilon\right) \le 2P_\mu(d(X_s, X_t) > \varepsilon)$$

for $0 \le t - s < \delta$. Since this can be written as

$$E_\mu\left[P_{X_s}\left(\sup_{0\le u<t-s} d(X_0, X_u) > 2\varepsilon\right)\right] \le 2E_\mu[P_{X_s}(d(X_0, X_{t-s}) > \varepsilon)]$$

by the Markov property, it is enough to prove

$$P_a\left(\sup_{0\le u<t-s} d(X_0, X_u) > 2\varepsilon\right) \le 2P_a(d(X_0, X_{t-s}) > \varepsilon)$$

i. e.

$$(3'') \qquad P_a\left(\sup_{0\le u<t-s} d(a, X_u) > 2\varepsilon\right) \le 2P_a(d(a, X_{t-s}) > \varepsilon)$$

for $0 \le t - s < \delta$.

Let $\sigma$ be the hitting time of the open set $\overline{U_{2\varepsilon}(a)}^{C}$. Then the left-hand side of (3″) is $P_a(\sigma < t - s)$. Using the time dependent strong Markov property, we have, for $0 \le t - s < \delta$,

[24] Notice that, by right continuity of the path, $\sup_{s\le u,v<t} d(X_u(\omega), X_v(\omega))$ is measurable ($\mathcal{B}$) in $\omega$.

$$\begin{aligned}
P_a(d(a, X_{t-s}) > \varepsilon) &\geq P_a(d(a, X_{t-s}) > \varepsilon,\ \sigma < t-s) \\
&\geq P_a(d(a, X_\sigma) \geq 2\varepsilon,\ d(X_\sigma, X_{t-s}) < \varepsilon,\ \sigma < t-s) \\
&= P_a(d(a, X_\sigma) \geq 2\varepsilon,\ \sigma < t-s,\ d(X_0(\theta_\sigma\omega), X_{t-s-\sigma}(\theta_\sigma\omega)) < \varepsilon) \\
&= E_a[P_b(d(X_0, X_{t-s-u}) < \varepsilon)|_{b=X_\sigma,\, u=\sigma},\ d(a, X_\sigma) \geq 2\varepsilon,\ \sigma < t-s] \\
&\geq (1/2)P_a(d(a, X_\sigma) \geq 2\varepsilon,\ \sigma < t-s) \qquad \text{(by (1''))} \\
&= (1/2)P_a(\sigma < t-s) \qquad \text{(by right continuity of the path)},
\end{aligned}$$

which proves (3″). This completes the proof of Theorem 1.

Although $X_t(\omega)$ is right continuous in $t$ and has left limits by the definition, it is not necessarily left continuous. However, if we fix a time point $t_0$, $X_t(\omega)$ is left continuous and so continuous at $t_0$ a. s.; the exceptional $\omega$-set may depend on $t_0$. To prove this, we need only to observe

$$\begin{aligned}
P_\mu(d(X_{t_0-}, X_{t_0}) > \varepsilon) &\leq P_\mu(d(X_{t_0}, X_s) > \varepsilon/2) + P_\mu(d(X_{t_0-}, X_s) > \varepsilon/2) \\
&\to 0 \qquad \text{as } s \uparrow t_0
\end{aligned}$$

by virtue of (1′) and the definition of $X_{t_0-}$.

What will happen if we replace the deterministic time $t_0$ by a random time $\sigma$? The conclusion is not true in general, even if $\sigma$ is assumed to be a stopping time. For example, consider a Markov process with a finite state space (see Example in Section 2.4). The first jump time

$$\sigma(\omega) = \inf\{t\colon X_t(\omega) \neq X_0(\omega)\}$$

is a stopping time and the sample path is a right continuous step function. If the state $i$ is an exponential holding point, then

$$P_i(X_{\sigma-}(\omega) = i,\ X_\sigma(\omega) \neq i) = 1\,.$$

To determine those stopping times at which the sample path is left-continuous a. s., we will introduce a notion "accessible stopping time".

**Definition 1.** A stopping time $\sigma$ is called *accessible* (from the left) a. s. $(P_\mu)$ on a set $A \in \mathcal{B}$ with $P_\mu(A) > 0$ if there exists a sequence of stopping times $\{\sigma_n\}$ with the following properties:

(a) $\sigma_1 \leq \sigma_2 \leq \cdots \to \sigma < \infty$ a. s. $(P_\mu)$ on $A$,
(b) $\sigma_n < \sigma$ for every $n$ a. s. $(P_\mu)$ on $A$.

For example the deterministic time $\sigma \equiv t$ is accessible on $\Omega$; set $\sigma_n = t - 1/n$. If $\sigma$ is an arbitrary stopping time, then $\tau \equiv \sigma + t$ ($t$ positive constant) is accessible on $\Omega$.

If $f \in \mathbf{B}(S)$ and $f > 0$, then

$$\sigma(\omega) = \sup\left\{t\colon \int_0^t f(X_s(\omega))\mathrm{d}s \le 1\right\},$$

is accessible on $\{\sigma < \infty\}$. Indeed if we set

$$\sigma_n(\omega) = \sup\left\{t\colon \int_0^t f(X_s(\omega))\mathrm{d}s \le 1 - \frac{1}{n}\right\},$$

then

$$\sigma_1 < \sigma_2 < \cdots \to \sigma \quad \text{on } \Omega .$$

**Theorem 2 (Quasi-left continuity).** *If $\sigma$ is accessible a. s. ($P_\mu$) on $A$, then*

$$P_\mu(X_\sigma = X_{\sigma-} \mid A) = 1$$

*i. e.*

$$P_\mu(\{X_\sigma \neq X_{\sigma-}\} \cap A) = 0 .$$

*Proof.* Let $\{\sigma_n\}$ be a sequence of stopping times as above. Then

$$\begin{aligned}
&P_\mu(\{d(X_\sigma, X_{\sigma-}) > \varepsilon\} \cap A) \\
&\quad\le P_\mu\left(\sup_{\sigma_m \le u,v < \sigma_m+\delta} d(X_u, X_v) > \varepsilon,\ \sigma_m < \sigma < \sigma_m + \delta < \infty\right) \\
&\qquad + P_\mu(\{\sigma_m + \delta \le \sigma\} \cap A) \\
&\quad\le P_\mu\left(\sup_{\sigma_m \le u,v < \sigma_m+\delta} d(X_u, X_v) > \varepsilon,\ \sigma_m < \infty\right) \\
&\qquad + P_\mu(\{\sigma_m + \delta \le \sigma\} \cap A) \\
&\quad\le E_\mu\left[P_{X_{\sigma_m}}\left(\sup_{0 \le u,v < \delta} d(X_u, X_v) > \varepsilon\right),\ \sigma_m < \infty\right] \\
&\qquad + P_\mu(\{\sigma_m + \delta \le \sigma\} \cap A) \qquad \text{(by strong Markov property)} \\
&\quad\le \sup_{a \in S} P_a\left(\sup_{0 \le u,v < \delta} d(X_u, X_v) > \varepsilon\right) + P_\mu(\{\sigma_m + \delta \le \sigma\} \cap A) .
\end{aligned}$$

The first term is less than $\varepsilon$ for $\delta = \delta(\varepsilon)$ small enough by (2) and the second term is less than $\varepsilon$ by taking $m$ big enough for this $\delta$. Then

$$P_\mu(\{d(X_\sigma, X_{\sigma-}) > \varepsilon\} \cap A) < 2\varepsilon .$$

Now let $\varepsilon \downarrow 0$ to complete the proof.

The path of the Brownian motion is continuous a. s. We will prove a general theorem which covers this fact.

**Theorem 3.** *If for every $\varepsilon > 0$*

$$\frac{1}{t} p_t(a, U_\varepsilon(a)^{\mathrm{C}}) \to 0 \qquad \textit{uniformly in } a \in S$$

*as $t \downarrow 0$, then the path is continuous a. s. for every initial distribution.*

*Proof.* First fix $\varepsilon > 0$ and an integer $n$. By writing $P_\mu^*$ for the outer measure of $P_\mu$, we have

$$\begin{aligned}
&P_\mu^*(d(X_{t-}, X_t) > 4\varepsilon \quad \text{for some } t \le n) \\
&\quad \le P_\mu\left( \sup_{(k-1)/m \le u,v < (k+1)/m} d(X_u, X_v) > 4\varepsilon \quad \text{for some } k \le mn \right) \\
&\quad \le \sum_{k=1}^{mn} P_\mu\left( \sup_{(k-1)/m \le u,v < (k+1)/m} d(X_u, X_v) > 4\varepsilon \right) \\
&\quad \le \sum_{k=1}^{mn} 4P_\mu(d(X_{(k-1)/m}, X_{(k+1)/m} > \varepsilon) \quad \text{(by Theorem 1)} \\
&\quad \le 4mn \sup_{a \in S} p_{2/m}(a, U_\varepsilon(a)^{\mathrm{C}}) \quad \text{(by Markov property)} \\
&\quad = 8n \sup_{a \in S} p_{2/m}(a, U_\varepsilon(a)^{\mathrm{C}})/(2/m) \to 0 \qquad \text{as } m \to \infty \,.
\end{aligned}$$

Therefore $P_\mu^*(d(X_{t-}, X_t) > 4\varepsilon$ for some $t \le n) = 0$. Letting $\varepsilon \downarrow 0$ and then $n \uparrow \infty$, we have

$$P_\mu^*(X_{t-} \ne X_t \quad \text{for some } t) = 0 \,.$$

This proves that $\{X_{t-} \equiv X_t\}$ is $\mathcal{B}^{P_\mu}$-measurable and has $P_\mu$-measure 1.

**Example.** We will use Theorem 3 to prove the continuity of the path of the Brownian motion[25]. A metric in $S = R^1 \cup \{\infty\}$ is

$$\begin{aligned}
d(a,b) &= |\arctan a - \arctan b\,| && a, b, \in R^1 \,, \\
d(a,\infty) &= \begin{cases} |\arctan a - \pi/2\,| & 0 \le a < \infty \\ |\arctan a + \pi/2\,| & 0 \ge a > -\infty \,. \end{cases}
\end{aligned}$$

If $a \in R^1$, then

[25] The Markov process $\{X_t(\omega),\, t \in T,\, \omega \in (\Omega, \mathcal{B}, P_a)\}_{a \in R^1 \cup \{\infty\}}$ constructed from the Brownian transition probabilities of Example 1 of Section 2.1 is called the Brownian motion. Let $\{W_t\}$ be the Wiener process defined in Example 1 of Section 1.4. Then formula (1) in Section 2.4 shows that $\{X_t\}$ under $P_a$ is identical in law with $\{a + W_t\}$. Hence the continuity of the path of the Brownian motion is concluded also from the results of Section 1.4.

$$
\begin{aligned}
p_t(a, U_\varepsilon(a)^{\mathrm{C}}) &= \frac{1}{\sqrt{2\pi t}} \int_{d(b,a)\geq\varepsilon} \mathrm{e}^{-(b-a)^2/(2t)} \mathrm{d}b \\
&\leq \frac{1}{\sqrt{2\pi t}} \int_{|b-a|\geq\varepsilon} \mathrm{e}^{-(b-a)^2/(2t)} \mathrm{d}b \\
&\leq \frac{1}{\sqrt{2\pi}} \int_{|x|\geq\varepsilon/\sqrt{t}} \mathrm{e}^{-x^2/2} \mathrm{d}x \leq \sqrt{\frac{2}{\pi}} \frac{\sqrt{t}}{\varepsilon} \mathrm{e}^{-\varepsilon^2/(2t)}
\end{aligned}
$$

and hence

$$
\frac{1}{t} p_t(a, U_\varepsilon(a)^{\mathrm{C}}) \leq \sqrt{\frac{2}{\pi}} \frac{1}{\varepsilon\sqrt{t}} \mathrm{e}^{-\varepsilon^2/(2t)} \leq \sqrt{\frac{2}{\pi}} \frac{1}{\varepsilon\sqrt{t}} \frac{2t}{\varepsilon^2} \to 0
$$

uniformly in $a \in R^1$ as $t \downarrow 0$. Evidently

$$
\frac{1}{t} p_t(\infty, U_\varepsilon(\infty)^{\mathrm{C}}) \equiv 0 .
$$

Thus the condition in Theorem 3 is verified.

## 2.10 Hitting Times of Closed Sets

The hitting time $\sigma_B$ of $B$ is defined by

$$
\sigma_B(\omega) = \begin{cases} \inf\{t > 0 \colon X_t(\omega) \in B\} & \text{if there is such } t \\ \infty & \text{if otherwise.} \end{cases}
$$

We proved in Section 2.7 that $\sigma_B$ is a stopping time if $B$ is open. Here we will prove

**Theorem 1.** *If $F$ is closed, then the hitting time $\sigma_F$ of $F$ is a stopping time.*

*Proof.* It is enough to prove that

$$
\widetilde{\sigma}_F = \begin{cases} \inf\{t \geq 0 \colon X_t \in F\} & \text{if there is such } t \\ \infty & \text{if otherwise} \end{cases}
$$

is a stopping time, because

$$
\sigma_F = \inf_n (1/n + \widetilde{\sigma}_F(\theta_{1/n}\omega)) .
$$

Let $\{G_F\}$ be a sequence of open sets such that

$$
G_1 \supset \overline{G}_2 \supset G_2 \supset \overline{G}_3 \supset G_3 \supset \cdots \to F ,
$$

and $\sigma_n$ the hitting time of $G_n$. Then $\sigma_1 \leq \sigma_2 \leq \cdots$ . Set $\sigma = \lim_n \sigma_n$ . Then $\sigma$ is a stopping time.

If $\sigma = \infty$, then $\widetilde{\sigma}_F = \infty$.
If $\sigma_n = \sigma_{n+1} = \cdots = \sigma < \infty$ for some $n$, then

$$X_\sigma \in \overline{G}_m \qquad \text{for } m = n, n+1, \ldots$$

by right continuity of the path. Thus we have

$$X_\sigma \in F\,,$$

while for $t < \sigma$ we have $X_t \in G_n^{\mathrm{C}}$ and so $X_t \in F^{\mathrm{C}}$. This proves $\sigma = \widetilde{\sigma}_F$.

If $\sigma_n < \sigma < \infty$ for every $n$ (write this event as $A$), then it is obvious that

$$X_t \in F^{\mathrm{C}} \qquad \text{for } t < \sigma$$

and that

$$X_{\sigma-} \in F\,.$$

Since $\sigma$ is accessible (by $\{\sigma_n\}_n$) on $A$, $X_t$ is left continuous at $\sigma$ a. s. $(P_\mu)$ on $A$ by the quasi-left continuity. Therefore

$$X_\sigma = X_{\sigma-} \in F \qquad \text{a. s. } (P_\mu)$$

on $A$. Thus

$$P_\mu^*(\{\sigma \neq \widetilde{\sigma}_F\} \cap A) = 0\,.$$

This completes the proof.

## 2.11 Dynkin's Formula

**Theorem 1.** *Suppose that $\sigma$ is a stopping time. If $f \in C(S)$ and if $u = R_\alpha f$, then*

$$u(a) = E_a\left(\int_0^\sigma \mathrm{e}^{-\alpha t} f(X_t)\mathrm{d}t\right) + E_a(\mathrm{e}^{-\alpha\sigma} u(X_\sigma),\ \sigma < \infty)\,.$$

*Proof.* We have

$$\begin{aligned} u(a) &= E_a\left(\int_0^\infty \mathrm{e}^{-\alpha t} f(X_t)\mathrm{d}t\right) \\ &= E_a\left(\int_0^\sigma \mathrm{e}^{-\alpha t} f(X_t)\mathrm{d}t\right) + E_a\left(\int_\sigma^\infty \mathrm{e}^{-\alpha t} f(X_t)\mathrm{d}t,\ \sigma < \infty\right)\,. \end{aligned}$$

The second term is

$$= E_a\left(\mathrm{e}^{-\alpha\sigma} \int_0^\infty \mathrm{e}^{-\alpha t} f(X_{t+\sigma})\mathrm{d}t,\ \sigma < \infty\right)$$

$$= E_a\left(e^{-\alpha\sigma}\int_0^\infty e^{-\alpha t} f(X_t(\theta_\sigma\omega))dt,\ \sigma < \infty\right)$$

$$= E_a\left[e^{-\alpha\sigma} E_{X_\sigma}\left(\int_0^\infty e^{-\alpha t} f(X_t)dt\right),\ \sigma < \infty\right]$$

(by strong Markov property)

$$= E_a(e^{-\alpha\sigma} u(X_\sigma),\ \sigma < \infty)\ .$$

**Theorem 2 (Dynkin's formula).** *Suppose that $\sigma$ is a stopping time with $E_a(\sigma) < \infty$. If $u \in \mathfrak{D}(A)$, then*

$$E_a\left(\int_0^\sigma Au(X_t)dt\right) = E_a(u(X_\sigma)) - u(a)\ .$$

*Proof.* Since $E_a(\sigma) < \infty$, $P_a(\sigma < \infty) = 1$. By $u \in \mathfrak{D}(A)$ we have

$$u = R_\alpha f \qquad \text{for some } f \in C(S)$$

and

$$Au = \alpha u - f\ .$$

By Theorem 1 we get

$$u(a) = E_a\left(\int_0^\sigma e^{-\alpha t} f(X_t)dt\right) + E_a(e^{-\alpha\sigma}u(X_\sigma))$$

$$= E_a\left(\int_0^\sigma e^{-\alpha t}(\alpha u(X_t) - Au(X_t))dt\right) + E_a(e^{-\alpha\sigma}u(X_\sigma))$$

$$\to -E_a\left(\int_0^\sigma Au(X_t)dt\right) + E_a(u(X_\sigma))$$

as $\alpha \downarrow 0$ by $E_a(\sigma) < \infty$ and $P_a(\sigma < \infty) = 1$.

Let $\tau_U$ be the exit time from an open set $U$, that is, the hitting time of $U^{\mathrm{C}}$. By Theorem 2.10.1, $\tau_U$ is a stopping time.

**Theorem 3.** *Suppose that $a$ is not a trap. Then*

$$E_a(\tau_U) < \infty$$

*for some neighborhood $U = U(a)$ of $a$.*

*Proof.* Using the assumption that $a$ is not a trap, we will prove that there exists $u \in \mathfrak{D}(A)$ such that $Au(a) \neq 0$. Suppose that

$$Au(a) = 0 \qquad \text{for every } u \in \mathfrak{D}(A)\ .$$

Then

$$\alpha R_\alpha f(a) - f(a) = 0$$

for every $\alpha > 0$ and every $f$. Thus

$$\int_0^\infty e^{-\alpha t} H_t f(a) dt = R_\alpha f(a) = \frac{1}{\alpha} f(a) = \int_0^\infty e^{-\alpha t} f(a) dt .$$

Taking the inverse Laplace transform we have

$$H_t f(a) = f(a) \qquad \text{a. e. in } 0 \le t < \infty .$$

By continuity of $H_t f(a)$ in $t$ we have

$$H_t f(a) = f(a) \qquad \text{for every } t .$$

Since this is true for every $f \in C(S)$, we obtain

$$p_t(a, E) = \delta_a(E)$$

namely

$$P_a(X_t = a) = 1 \qquad \text{for every } t .$$

By right continuity of the path,

$$P_a(X_t = a \text{ for every } t) = 1 ,$$

which means that $a$ is a trap, in contradiction with our assumption.

Therefore we have $Au(a) \neq 0$ for some $u \in \mathfrak{D}(A)$. We can assume $Au(a) > 0$ by replacing $u$ by $-u$ if necessary. By continuity of $Au$ we have $U = U(a)$ such that

$$Au(b) > (1/2)Au(a) > 0 \qquad \text{for } b \in U .$$

Then Dynkin's formula shows that

$$(1/2)Au(a)E_a(\tau_U^n) \le 2\|u\| < \infty \qquad \text{for } \tau_U^n = \tau_U \wedge n ,$$

which implies $E_a(\tau_U) \le \lim_{n\to\infty} E_a(\tau_U^n) \le 4\|u\|/Au(a) < \infty$.

**Theorem 4 (Dynkin's representation of the generator).** *If $a$ is not a trap, then*

$$Au(a) = \lim_{U \downarrow a} \frac{E_a(u(X_{\tau_U})) - u(a)}{E_a(\tau_U)} \qquad \textit{for } u \in \mathfrak{D}(A) .$$

*Proof.* Since $Au$ is continuous, we have $U_0 = U_0(a, \varepsilon)$ such that

$$|Au(b) - Au(a)| < \varepsilon \qquad \text{for } b \in U_0 .$$

By Theorem 3 we can take $U_1 = U_1(a, \varepsilon)$, such that

$$E_a(\tau_{U_1}) < \infty .$$

By Dynkin's formula we have

$$\begin{aligned}|E_a(u(X_{\tau_U})) - u(a) - Au(a)E_a(\tau_U)| &\le E_a\left(\int_0^{\tau_U} |Au(X_t) - Au(a)|\mathrm{d}t\right) \\ &\le \varepsilon E_a(\tau_U)\end{aligned}$$

for every neighborhood $U = U(a) \subset U_0 \cap U_1$. This proves our theorem.

**Example.** Let $X$ be a Markov process with a finite state space $S$. Since $S$ is discrete, $\{i\}$ is a neighborhood of $i$. Therefore if $i$ is not a trap, then

$$Au(i) = \frac{E_i(u(X_e)) - u(i)}{E_i(e)}$$

where $e$ is the waiting time. Rewriting this, we have

$$Au(i) = \frac{1}{E_i(e)}\left\{\sum_{j\neq i} u(j)P(X_e = j) - u(i)\right\}.$$

Comparing this with the result in Example in Section 2.4, we have

$$a(i) = \frac{1}{E_i(e)} \quad \text{and} \quad a(i,j) = \frac{P_i(X_e = j)}{E_i(e)} \quad (j \neq i)$$

or

$$E_i(e) = \frac{1}{a(i)} \quad \text{and} \quad P_i(X_e = j) = \frac{a(i,j)}{a(i)} \quad (j \neq i)\,.$$

*Construction of the path of the process.* First we will prove the following

**Proposition.**

$$\begin{aligned}P_i(e > t, X_e = j) &= P_i(e > t)P_i(X_e = j) \\ &= \mathrm{e}^{-a(i)t}\,\frac{a(i,j)}{a(i)}\end{aligned}$$

where $a(i,i) = 0$.

*Proof.* We have

$$\begin{aligned}P_i(e > t, X_e = j) &= P_i(e > t, X_{t+e(\theta_t\omega)} = j) \\ &= P_i(e > t, X_{e(\theta_t\omega)}(\theta_t\omega) = j) = E_i(P_{X_t}(X_e = j),\ e > t) \\ &= E_i(P_i(X_e = j),\ e > t) = P_i(e > t)\,P_i(X_e = j)\,.\end{aligned}$$

It is proved above that $P_i(X_e = j) = a(i,j)/a(i)$. Since $e$ is exponentially distributed,

$$P_i(e > t) = e^{-\lambda t}$$

where $\lambda = 1/E_i(e) = a(i)$.

Let $e_n$ be the $n$-th waiting time and set

$$Y_n = X_{e_1+\cdots+e_n} \, .$$

It is obvious that

$$e_1 = e, \qquad e_{n+1} = e(\theta_{e_1+\cdots+e_n}\omega) \, .$$

The path $X_t$ is completely determined by the sequence $e_1, Y_1, e_2, Y_2, \ldots$. It

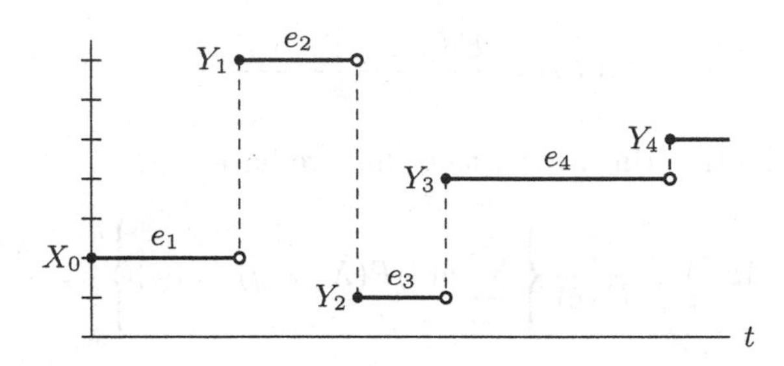


is enough to give the finite joint distribution of these random variables. They are as follows:

$$\begin{aligned}
P_i(e_1 > t,\ Y_1 = i_1) &= \mathrm{e}^{-a(i)t}\frac{a(i,i_1)}{a(i)} \, , \\
P_i(e_1 > t,\ Y_1 = i_1,\ e_2 > s,\ Y_2 = i_2)& \\
&= P_i(e_1 > t,\ Y_{e_1} = i_1,\ e_1(\theta_{e_1}\omega) > s,\ Y_1(\theta_{e_1}\omega) = i_2) \\
&= E_i(P_{X_{e_1}}(e_1 > s,\ Y_1 = i_2),\ e_1 > t,\ X_{e_1} = i_1) \\
&= P_i(e_1 > t,\ X_{e_1} = i_1)\, P_{i_1}(e_1 > s,\ Y_1 = i_2) \\
&= \mathrm{e}^{-a(i)t}\,\frac{a(i,i_1)}{a(i)}\,\mathrm{e}^{-a(i_1)s}\,\frac{a(i_1,i_2)}{a(i_1)} \, ,
\end{aligned}$$

and so on.

**Theorem 5 (Converse of Theorem 4).** *Let $X$ be a Markov process and define the operator $\widetilde{A}$ by*

$$\widetilde{A}u(a) = \begin{cases} \lim\limits_{r\downarrow 0} \dfrac{E_a(u(X_{\tau(r)})) - u(a)}{E_a(\tau(r))} & \textit{if } a \textit{ is not a trap} \\ 0 & \textit{if } a \textit{ is a trap} \end{cases}$$

*with $\tau(r)$ = the exit time from the ball $U(a,r) = \{b\colon d(a,b) < r\}$, where the domain of definition of $\widetilde{A}$ is the set of all $u \in C(S)$ for which $\widetilde{A}u(a)$ is continuous in $a$. Then $\widetilde{A} = A\,(=$ the generator of the process$)$.*

*Proof.* Let us verify conditions (a) – (d) in Theorem 2.3.2. Conditions (a), (b), and (c) are obvious. Since $(\alpha - A)u = f$, $f \in C(S)$, has a solution $u = R_\alpha f \in C(S)$ and $\widetilde{A}$ is an extension of $A$ by Theorem 4, $(\alpha - \widetilde{A})u = f$ has a solution $u = R_\alpha f$. Hence (d) is satisfied. Therefore, by Theorem 2.3.2, $\widetilde{A}$ is the generator of a transition semi-group. Moreover we have $(\alpha - \widetilde{A})^{-1} = (\alpha - A)^{-1}$, and thus $\widetilde{A} = A$.

## 2.12 Markov Processes in Generalized Sense

The class of Markov processes which we have dealt with so far is so restricted that it does not cover many interesting processes. Therefore we will extend the notion in this section.

**1. Definition.** The collection of stochastic processes

$$X = \{X_t(\omega) \equiv \omega(t) \in S,\ t \in T,\ \omega \in (\Omega, \mathcal{B}, P_a)\}_{a \in S}$$

is called a *Markov process*, if the following conditions are satisfied:

1. $S$ (the *state space*) is a complete separable metric space,[26] and $\mathcal{B}(S)$ is the topological $\sigma$-algebra on $S$.

2. $T$ (the *time interval*) $= [0, \infty)$.

3. $\Omega$ (the *space of paths*) is the space of all right continuous functions: $T \longrightarrow S$ and $\mathcal{B}$ is the $\sigma$-algebra $\mathcal{B}[X_t : t \in T]$ on $\Omega$.

4. $P_a(\Lambda)$ (the probability law of the path starting at $a$) is a probability measure on $(\Omega, \mathcal{B})$ for every $a \in S$ satisfying the following conditions.

(a) $P_a(\Lambda)$ is $\mathcal{B}(S)$-measurable in $a$ for every $\Lambda \in \mathcal{B}$.

(b) $P_a(X_0 = a) = 1$.

(c) $P_a(X_{t_1} \in E_1, \ldots, X_{t_n} \in E_n)$

$$= \int \cdots \int_{a_i \in E_i} P_a(X_{t_1} \in da_1) P_{a_1}(X_{t_2 - t_1} \in da_2) \cdots P_{a_{n-1}}(X_{t_n - t_{n-1}} \in da_n)$$

$$(0 < t_1 < t_2 < \cdots < t_n)\,.$$

The main differences between this definition and the previous definition are as follows.

(i) $S$ is not necessarily compact.

(ii) The existence of the left limits of the path is not assumed.

(iii) The transition operator $f \longrightarrow H_t f(\cdot) = E_\cdot(f(X_t))$ does not necessarily carry $C(S)$ into $C(S)$.[27]

---

[26] All results in this section hold also when $S$ is a locally compact separable metric space.

[27] Here $C(S)$ is the space of all real-valued bounded continuous functions on $S$.

**2. Markov property.** We will define $P_\mu, \mathcal{B}_t, \overline{\mathcal{B}}$ and $\overline{\mathcal{B}}_t$ as before. Then we can use the same arguments to prove

$$P_\mu(\Lambda \cap \theta_t^{-1} M) = E_\mu(P_{X_t}(M); \Lambda) \qquad \text{for } \Lambda \in \overline{\mathcal{B}}_t,\ M \in \overline{\mathcal{B}} .$$

**3. The transition semi-group.** The transition probabilities are defined by

$$p_t(a, B) = P_a(X_t \in B) .$$

The transition operators are defined by

$$H_t f(a) = \int_S p_t(a, \mathrm{d}b) f(b) = E_a(f(X_t)), \qquad t \geq 0. .$$

Let $\mathbf{B}(S)$ denote the space of all bounded real $\mathcal{B}(S)$-measurable functions. Then it is obvious that $\{H_t\}$ is a semi-group of linear operators on $\mathbf{B}(S)$ in the sense that

$$\begin{aligned} &H_t \colon \mathbf{B}(S) \longrightarrow \mathbf{B}(S) \text{ linear},\\ &H_0 = I ,\\ &H_{t+s} = H_t H_s . \end{aligned}$$

To prove the last equation, observe

$$\begin{aligned} H_{t+s} f(a) &= E_a(f(X_{t+s}))\\ &= E_a(E_{X_t} f(X_s))) \qquad \text{(Markov property)}\\ &= E_a(H_s f(X_t))\\ &= H_t H_s f(a) . \end{aligned}$$

**4. Resolvent operator.** $R_\alpha f$ is defined by

$$R_\alpha f(a) = \int_0^\infty \mathrm{e}^{-\alpha t} H_t f(a) \mathrm{d}t = E_a\left(\int_0^\infty \mathrm{e}^{-\alpha t} f(X_t) \mathrm{d}t\right), \qquad \alpha > 0 .$$

Notice that $f(X_t(\omega))$ is $\mathcal{B}(T) \times \mathcal{B}$-measurable in $(t, \omega)$ since

$$X_t(\omega) = \lim_{n \to \infty} X\left(\frac{[2^n t] + 1}{2^n}, \omega\right) ,$$

so that Fubini's theorem can be used.

We have

$$\begin{aligned} H_t R_\alpha &= R_\alpha H_t ,\\ R_\alpha - R_\beta + (\alpha - \beta) R_\alpha R_\beta &= 0 . \end{aligned}$$

Indeed,

$$\begin{aligned}
H_t R_\alpha f(a) &= E_a(R_\alpha f(X_t)) \\
&= E_a\left(E_{X_t}\left(\int_0^\infty e^{-\alpha s} f(X_s)ds\right)\right) \\
&= E_a\left(\int_0^\infty e^{-\alpha s} f(X_{s+t})ds\right) = \int_0^\infty e^{-\alpha s} E_a(f(X_{s+t}))ds \\
&= \int_0^\infty e^{-\alpha s} H_{s+t} f(a)ds = \int_0^\infty e^{-\alpha s} H_s H_t f(a)ds \\
&= R_\alpha H_t f(a) \, , \\
R_\alpha R_\beta f(a) &= \int_0^\infty e^{-\alpha t} H_t R_\beta f(a)dt = \int_0^\infty e^{-\alpha t} R_\beta H_t f(a)dt \\
&= \int_0^\infty e^{-\alpha t} \int_0^\infty e^{-\beta s} H_s H_t f(a)ds\,dt \\
&= \int_0^\infty e^{-\alpha t} \int_0^\infty e^{-\beta s} H_{s+t} f(a)ds\,dt \\
&= \int_0^\infty e^{(\beta-\alpha)t} \int_t^\infty e^{-\beta s} H_s f(a)ds\,dt \\
&= \int_0^\infty \left(\int_0^s e^{(\beta-\alpha)t}dt\right) e^{-\beta s} H_s f(a)ds \\
&= \frac{1}{\beta-\alpha}\int_0^\infty (e^{(\beta-\alpha)s} - 1)e^{-\beta s} H_s f(a)ds \\
&= \frac{1}{\beta-\alpha}(R_\alpha f(a) - R_\beta f(a)) \, .
\end{aligned}$$

**5. The subspace $\mathbf{L}$ of $\mathbf{B}(S)$ and the restriction of $\{H_t\}$ to $\mathbf{L}$.** We will introduce the subspace $\mathbf{L}$ following Dynkin:

$$\mathbf{L} = \{f \in \mathbf{B}(S) \colon H_h f(a) \to f(a) \quad (h \downarrow 0) \quad \text{for every } a\} \, .$$

(i) $\mathbf{L}$ is a linear space.

(ii) $\mathbf{L} \supset C(S)$ (= the space of all bounded real continuous functions). Indeed,

$$H_t f(a) = E_a(f(X_t)) \to f(a) \qquad \text{for } f \in C(S)$$

as $t \downarrow 0$. If $P_a(f(X_t) \to f(a)) = 1$, then $f \in \mathbf{L}$ by the same reason.

(iii) $H_t f(a)$ is right continuous in $t$ for $f \in \mathbf{L}$. Indeed,

$$H_{t+h} f(a) = H_t H_h f(a) = E_a(H_h f(X_t)) \to E_a(f(X_t)) = H_t f(a) \, .$$

(iv) $H_t \colon \mathbf{L} \longrightarrow \mathbf{L}$ (linear), since

$$H_h H_t f(a) = H_{t+h} f(a) \to H_t f(a)$$

by (iii).

(v) $R_\alpha \colon \mathbf{L} \longrightarrow \mathbf{L}$ (linear) since, for $f \in \mathbf{L}$,

$$\begin{aligned} H_h R_\alpha f(a) = R_\alpha H_h f(a) &= \int_0^\infty \mathrm{e}^{-\alpha t} H_t H_h f(a) \mathrm{d}t \\ &= \int_0^\infty \mathrm{e}^{-\alpha t} H_{t+h} f(a) \mathrm{d}t \to \int_0^\infty \mathrm{e}^{-\alpha t} H_t f(a) \mathrm{d}t = R_\alpha f(a) \end{aligned}$$

by (iii).

(vi) $\alpha R_\alpha f(a) \to f(a)$ $(\alpha \to \infty)$ for every $a$ if $f \in \mathbf{L}$, because

$$\alpha R_\alpha f(a) = \int_0^\infty \alpha \mathrm{e}^{-\alpha t} H_t f(a) \mathrm{d}t = \int_0^\infty \mathrm{e}^{-t} H_{t/\alpha} f(a) \mathrm{d}t \to f(a) .$$

(vii) $\mathfrak{R} = R_\alpha \mathbf{L}$ is independent of $\alpha$. Since

$$R_\alpha f = R_\beta (f + (\beta - \alpha) R_\alpha f)$$

we have $R_\alpha \mathbf{L} \subset R_\beta \mathbf{L}$. Similarly $R_\beta \mathbf{L} \subset R_\alpha \mathbf{L}$ and so $R_\alpha \mathbf{L} = R_\beta \mathbf{L}$ .

(viii) If $R_\alpha f = 0$ and $f \in \mathbf{L}$, then $f = 0$. Indeed we have

$$R_\beta f = R_\alpha f + (\alpha - \beta) R_\beta R_\alpha f = 0$$

and hence $\beta R_\beta f = 0$. Letting $\beta \to \infty$, we get $f = 0$ by (vi).

It follows from this that $(R_\alpha | \mathbf{L})^{-1}$ is determined uniquely.

**6. The generator** $A$**.** Define

$$Au(a) = \lim_{h \downarrow 0} \frac{H_h u(a) - u(a)}{h}$$

with $\mathfrak{D}(A)$ being the class of $u \in \mathbf{L}$ such that $\frac{1}{h}(H_h u(a) - u(a))$ converges to a function in $\mathbf{L}$ boundedly as $h \downarrow 0$. We will prove that

$$\mathfrak{D}(A) = \mathfrak{R} \quad \textit{and} \quad Au = \alpha u - (R_\alpha | \mathbf{L})^{-1} u .$$

*First step.* If $u \in \mathfrak{D}(A)$, then

$$\begin{aligned} H_t u \in \mathfrak{D}(A), &\quad A H_t u = H_t A u , \\ R_\alpha u \in \mathfrak{D}(A), &\quad A R_\alpha u = R_\alpha A u . \end{aligned}$$

Indeed,

$$\begin{aligned} \frac{1}{h}(H_h H_t u(a) - H_t u(a)) &= \frac{1}{h}(H_{h+t} u(a) - H_t u(a)) = H_t \left[ \frac{1}{h}(H_h - I) \right] u(a) \\ &= E_a \left[ \frac{1}{h}(H_h u(X_t) - u(X_t)) \right] \to E_a(Au(X_t)) = H_t A u(a) \end{aligned}$$

by bounded convergence theorem, and so

$$H_t u \in \mathfrak{D}(A) \quad \text{and} \quad A H_t u = H_t A u .$$

We have

$$\begin{aligned}
\frac{1}{h}(H_h R_\alpha u(a) - R_\alpha u(a)) &= \frac{1}{h}(R_\alpha H_h u(a) - R_\alpha u(a)) \\
&= R_\alpha \left[\frac{1}{h}(H_h u - u)\right](a) = E_a\left(\int_0^\infty \mathrm{e}^{-\alpha t}\frac{1}{h}(H_t u(X_t) - u(X_t))\mathrm{d}t\right) \\
&\to E_a\left(\int_0^\infty \mathrm{e}^{-\alpha t} Au(X_t)\mathrm{d}t\right) = R_\alpha Au(a)\,,
\end{aligned}$$

and so

$$R_\alpha u \in \mathfrak{D}(A) \quad \text{and} \quad AR_\alpha u = R_\alpha Au\,.$$

*Second step.* Let us show that if $u \in \mathfrak{R}$, then $u \in \mathfrak{D}(A)$ and $Au = \alpha u - (R_\alpha|\mathbf{L})^{-1}u$. Set $u = R_\alpha f$, $f \in \mathbf{L}$. Then

$$\begin{aligned}
H_h R_\alpha f(a) &= R_\alpha H_h f(a) = \int_0^\infty \mathrm{e}^{-\alpha t} H_t H_h f(a)\mathrm{d}t = \int_0^\infty \mathrm{e}^{-\alpha t} H_{t+h} f(a)\mathrm{d}t \\
&= \mathrm{e}^{\alpha h}\int_h^\infty \mathrm{e}^{-\alpha t} H_t f(a)\mathrm{d}t = \mathrm{e}^{\alpha h} R_\alpha f(a) - \mathrm{e}^{\alpha h}\int_0^h \mathrm{e}^{-\alpha t} H_t f(a)\mathrm{d}t \\
&= \mathrm{e}^{\alpha h} u(a) - \mathrm{e}^{\alpha h}\int_0^h \mathrm{e}^{-\alpha t} H_t f(a)\mathrm{d}t
\end{aligned}$$

and so

$$\begin{aligned}
\frac{1}{h}[H_h u(a) - u(a)] &= \frac{\mathrm{e}^{\alpha h} - 1}{h} u(a) - \mathrm{e}^{\alpha h}\frac{1}{h}\int_0^h \mathrm{e}^{-\alpha t} H_t f(a)\mathrm{d}t \\
&\to \alpha u(a) - f(a)\ (\in \mathbf{L}) \quad \text{(bounded convergence)}.
\end{aligned}$$

*Third step.* We show that $\mathfrak{D}(A) = \mathfrak{R}$. It is enough to prove that $u \in \mathfrak{D}(A)$ implies $u \in \mathfrak{R}$. Let $v = R_\alpha(\alpha u - Au) \in \mathfrak{R}$. Then

$$\begin{aligned}
&Av = \alpha v - (\alpha u - Au)\,, \\
&\alpha(v - u) = A(v - u)\,, \\
&\alpha R_\alpha(v - u) = R_\alpha A(v - u) = AR_\alpha(v - u) = \alpha R_\alpha(v - u) - (v - u)
\end{aligned}$$

and so $u = v \in \mathfrak{R}$.

**7.** The generator $A$ and the resolvent $R_\alpha$ are related as $(\alpha - A)^{-1} = R_\alpha|\mathbf{L}$. Indeed, $R_\alpha\colon \mathbf{L} \longrightarrow \mathfrak{R} = R_\alpha\mathbf{L}$ (one-to-one, onto) and $(\alpha - A)R_\alpha f = f$ for $f \in \mathbf{L}$ by (f). This proves $(\alpha - A)^{-1} = R_\alpha|\mathbf{L}$.

**8.** Let $X_1$ and $X_2$ be two Markov prosesses with the same generator $A$. Then $X_1 = X_2$. To prove this, note that, since $(\alpha - A)\colon \mathbf{L}^i \longrightarrow \mathfrak{R}^i$ (one-to-one, onto),

$$\mathbf{L}^1 = \mathbf{L}^2, \quad \mathfrak{R}^1 = \mathfrak{R}^2, \quad \text{and} \quad R^1_\alpha = (\alpha - A)^{-1} = R^2_\alpha\,.$$

Let $f \in C(S)$. Then $f \in \mathbf{L}^i$ and $R^1_\alpha f = R^2_\alpha f$. Hence

$$H^1_t f(a) = H^2_t f(a) \qquad \text{for almost every } t\,.$$

By the right continuity of $H^i_t f(a)$, we get $H^1_t f = H^2_t f$, that is,

$$\int_S p^1_t(a, \mathrm{d}b) f(b) = \int_S p^2_t(a, \mathrm{d}b) f(b), \qquad f \in C(S)\,.$$

Since $S$ is a complete separable metric space, it follows that $p^1_t(a, B) = p^2_t(a, B)$ for $B \in \mathcal{B}(S)$. Thus

$$P^1_a = P^2_a\,.$$

**9. Strong Markov property.** In the previous definition the strong Markov property was a consequence of the definition. In our new definition it does not hold automatically. The strong Markov property is a condition we have to impose if we want.

For example, let $S = [0, \infty)$. Define $P_a$, $a > 0$, to be concentrated on a single path $\omega(t) \equiv a + t$ and $P_0$ to be concentrated on the paths of the type

$$\omega(t) = (t - \xi) \vee 0$$

such that

$$P_0(X_t = 0 \text{ for all } t \le s) = \mathrm{e}^{-s}\,.$$

It is not difficult to prove that this is Markov in our new sense and that the strong Markov property fails to hold at the stopping time

$$e = \inf\{t\colon X_t \ne X_0\}\,.$$

The *Dynkin representation* of the generator is as follows:

**Theorem 1.** *If $X$ is strong Markov and if $Au$ is continuous at $a$, then*

$$Au(a) = \lim_{r \downarrow 0} \frac{E(u(X_{\tau(r)}) - u(a)}{E(\tau(r))} \qquad \textit{for } a \textit{ not a trap},$$

*where $\tau(r)$ is the exit time from the closed ball $\overline{U(a,r)} = \{b\colon d(a,b) \le r\}$.*

**Theorem 2.** *If $X$ is strong Markov, then the waiting time $e = \inf\{t\colon X_t \ne X_0\}$ has one of the following distributions:*

*(i) $P_a(e = \infty) = 1$ (a is a trap).*
*(ii) $P_a(e = 0) = 1$ (a is an instantaneous state).*
*(iii) $P_a(e > t) = \mathrm{e}^{-\lambda t}$, $0 < \lambda < \infty$ (a is an exponential holding state).*

**10. Conditions for strong Markov property.** Following Blumenthal and Getoor (Markov Processes and Potential Theory, Academic Press) we will list several useful sufficient (but not necessary) conditions for a Markov process to be strong Markov.

**Proposition 1.** *If*

$$(1) \qquad E_a[f(X_{\sigma+t}),\ \{\sigma<\infty\}\cap\Lambda] = E_a[E_{X(\sigma)}(f(X_t)),\ \{\sigma<\infty\}\cap\Lambda] \qquad a\in S,\ f\in\mathbf{B}(S),\ \Lambda\in\overline{\mathcal{B}}_\sigma,\ t\in T$$

*for every stopping time $\sigma$, then $X$ is strong Markov.*

*Proof.* It follows from the assumption that

$$E_a[f(X_{\sigma+t})G(\omega),\ \{\sigma<\infty\}\cap\Lambda] = E_a[E_{X(\sigma)}[f(X_t)]G(\omega),\ \{\sigma<\infty\}\cap\Lambda]$$

for every bounded and $\overline{\mathcal{B}}_\sigma$-measurable $G(\omega)$. For the proof of the sufficiency of condition (1), it is enough to prove

$$\begin{aligned}&E_a[f_1(X_{\sigma+t_1})\cdots f_n(X_{\sigma+t_n}),\ \{\sigma<\infty\}\cap\Lambda]\\ &\quad = E_a[E_{X(\sigma)}[f_1(X_{t_1})\cdots f_n(X_{t_n})],\ \{\sigma<\infty\}\cap\Lambda]\end{aligned}$$

for $f_i\in\mathbf{B}(S)$, $i=1,2,\ldots,n$, and $0<t_1<t_2<\cdots<t_n$. This identity is obvious for $n=1$ by the assumption. Suppose that it is true for $n=k$. Then it holds for $n=k+1$ and so for every $n$. Indeed, since $\{\sigma<\infty\}=\{\sigma+t_k<\infty\}$ and $\Lambda\in\overline{\mathcal{B}}_\sigma\subset\overline{\mathcal{B}}_{\sigma+t_k}$, we have

$$\begin{aligned}&E_a[f_1(X_{\sigma+t_1})f_2(X_{\sigma+t_2})\cdots f_{k+1}(X_{\sigma+t_{k+1}}),\ \{\sigma<\infty\}\cap\Lambda]\\ &\quad = E_a[f_1(X_{\sigma+t_1})\cdots f_k(X_{\sigma+t_k})E_{X_{\sigma+t_k}}(f_{k+1}(X_{t_{k+1}-t_k})),\ \{\sigma<\infty\}\cap\Lambda]\\ &\quad = E_a[E_{X_\sigma}[f_1(X_{t_1})\cdots f_k(X_{t_k})E_{X_{t_k}}(f_{k+1}(X_{t_{k+1}-t_k}))],\ \{\sigma<\infty\}\cap\Lambda]\\ &\quad = E_a[E_{X_\sigma}[f_1(X_{t_1})\cdots f_k(X_{t_k})f_{k+1}(X_{t_{k+1}})],\ \{\sigma<\infty\}\cap\Lambda]\,.\end{aligned}$$

**Proposition 2.** *Suppose that*

$$E_a[f(X_{\sigma+t}),\ \sigma<\infty] = E_a[E_{X_\sigma}(f(X_t)),\ \sigma<\infty] \qquad a\in S,\ f\in\mathbf{B}(S),\ t\in T$$

*for every stopping time $\sigma$. Then $X$ is strong Markov.*

*Proof.* Let $\Lambda\in\overline{\mathcal{B}}_\sigma$. Then

$$\sigma_\Lambda = \begin{cases}\sigma & \omega\in\Lambda\\ \infty & \omega\in\Lambda^{\mathrm{C}}\end{cases}$$

is also a stopping time. Then

$$E_a[f(X_{\sigma_\Lambda+t}),\ \sigma_\Lambda<\infty] = E_a[E_{X(\sigma_\Lambda)}(f(X_t)),\ \sigma_\Lambda<\infty]\,.$$

Since

$$\{\sigma_\Lambda<\infty\} = \{\sigma<\infty\}\cap\Lambda\subset\{\sigma_\Lambda=\sigma\}\,,$$

we have

$$E_a[f(X_{\sigma+t}),\ \{\sigma<\infty\}\cap\Lambda] = E_a[E_{X(\sigma)}(f(X_t)),\ \{\sigma<\infty\}\cap\Lambda]\,,$$

which verifies the assumption of Proposition 1.

**Proposition 3.** *If the assumption of Proposition 2 holds for $f \in C(S)$, then $X$ is strong Markov.*

*Proof.* Both sides of the equation in Proposition 2 are closed under linear combination and bounded convergence with respect to $f$. Therefore, if it holds for $f \in C(S)$, then it will hold for $f \in \mathbf{B}(S)$, since $S$ is a complete separable metric space.

**Proposition 4.** *Let $\Gamma$ be a linear subspace of $C(S)$ such that every $f \in C(S)$ is the limit of a bounded convergent sequence $\{f_n\} \subset \Gamma$. If the assumption of Proposition 2 holds for every $f \in \Gamma$, then $X$ is strong Markov.*

**Theorem 3 (Blumenthal and Getoor).** *If $R_\alpha f(X_t)$ is right-continuous a. s. $(P_a)$ for $f \in \Gamma$ (in Proposition 4), $\alpha > 0$, and $a \in S$, then $X$ is strong Markov.*

*Proof.* For $\sigma_n = ([2^n\sigma] + 1)/2^n$ we can easily see

$$E_a\left[\int_0^\infty \mathrm{e}^{-\alpha t} f(X_{\sigma_n+t})\mathrm{d}t,\ \sigma_n < \infty\right] = E_a[R_\alpha f(X_{\sigma_n}),\ \sigma_n < \infty]\,,$$

because the strong Markov property at a discrete valued stopping time follows at once from the Markov property (see the proof of Theorem 2.7.4). Since $f$ is continuous, $f(X_t)$ is right-continuous in $t$; $R_\alpha f(X_t)$ is right continuous in $t$ by our assumption. Therefore by letting $n \uparrow \infty$ we have

$$\begin{aligned} E_a\left[\int_0^\infty \mathrm{e}^{-\alpha t} f(X_{\sigma+t})\mathrm{d}t,\ \sigma < \infty\right] &= E_a[R_\alpha f(X_\sigma),\ \sigma < \infty] \\ &= E_a\left[E_{X(\sigma)}\left(\int_0^\infty \mathrm{e}^{-\alpha t} f(X_t)\mathrm{d}t\right),\ \sigma < \infty\right] \end{aligned}$$

and so

$$\int_0^\infty \mathrm{e}^{-\alpha t} E_a[f(X_{\sigma+t}),\ \sigma < \infty]\mathrm{d}t = \int_0^\infty \mathrm{e}^{-\alpha t} E_a[E_{X(\sigma)}(f(X_t)),\ \sigma < \infty]\,\mathrm{d}t\,.$$

By taking the inverse Laplace transform we have

$$E_a[f(X_{\sigma+t}),\ \sigma < \infty] = E_a[E_{X(\sigma)}(f(X_t)),\ \sigma < \infty]$$

for almost every $t$. But both sides are right continuous in $t$. Therefore the identity holds for every $t$. Now use Proposition 4.

**Corollary 1.** *If $H_t\colon \Gamma \longrightarrow C(S)$ for $t \geq 0$, then $X$ is strong Markov. In particular if $H_t\colon C(S) \longrightarrow C(S)$, then $X$ is strong Markov.*

**Corollary 2.** *If $R_\alpha\colon \Gamma \longrightarrow C(S)$ (in particular if $R_\alpha\colon C(S) \longrightarrow C(S)$) for $\alpha > 0$, then $X$ is strong Markov.*

## 2.13 Examples

**1. The deterministic linear motion on $(-\infty, \infty)$.**
Let $S = R^1$ and let $P_a$ be concentrated at a single path:

$$\omega(t) \equiv a + t\,.$$

This is strong Markov and[28]

$$\begin{aligned}
&H_t f(a) = f(a+t)\,,\\
&R_\alpha f(a) = \int_0^\infty \mathrm{e}^{-\alpha t} f(a+t)\mathrm{d}t = \mathrm{e}^{\alpha a}\int_a^\infty \mathrm{e}^{-\alpha s} f(s)\mathrm{d}s\,,\\
&\mathbf{L} = \{f\colon \text{ right continuous and bounded}\}\,,\\
&H_t\colon C(S) \longrightarrow C(S)\,,\\
&Au = u^+\,,\\
&\mathfrak{D}(A) = \mathfrak{R} = R_\alpha \mathbf{L} = \{u\colon u \text{ continuous and bounded,}\\
&\qquad\qquad u^+ \text{ right continuous and bounded}\}\,.
\end{aligned}$$

**2. The deterministic linear motion on $(0, 1)$ with instantaneous return to $1/2$.**
The particle makes a motion with a constant speed 1 until it hits 1. As soon as it hits 1, it returns to 1/2 and makes the same motion as before and repeats the same jump. This is Markov.

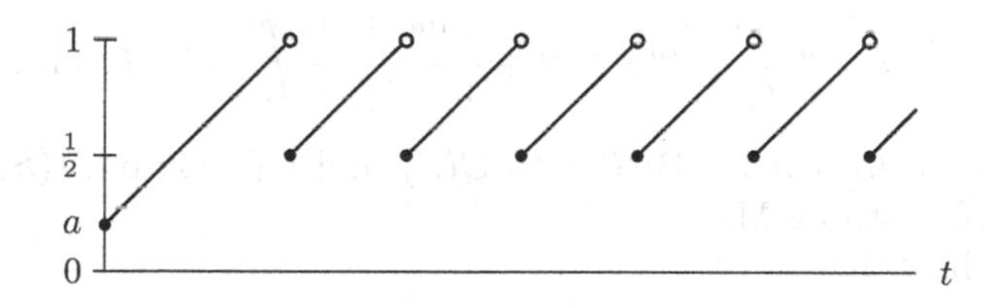


Since $H_t f(a) = f(a+t)$ for small $t$, we have

$$\mathbf{L} = \{f\colon f \text{ right continuous, bounded}\}\,.$$

But $H_t$ does not carry $C(0,1)$ into $C(0,1)$, as you see in the illustration below. We have

---

[28] $u^+(a) = \lim_{h\downarrow 0} \frac{1}{h}(u(a+h) - u(a))$, the right derivative of $u$.

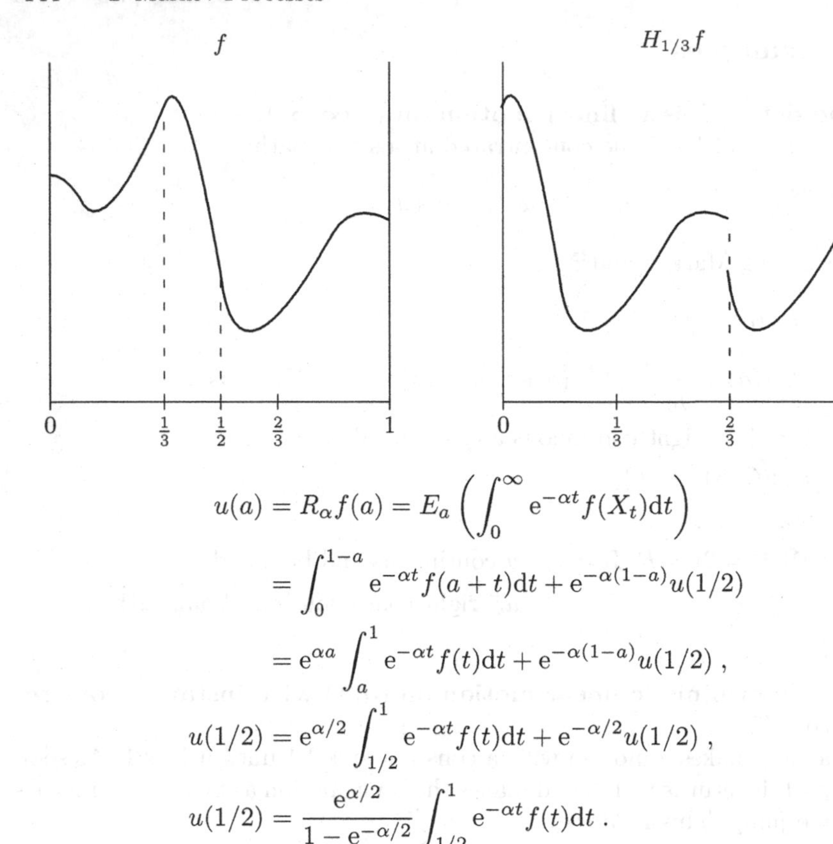


$$\begin{aligned}
u(a) = R_\alpha f(a) &= E_a\left(\int_0^\infty \mathrm{e}^{-\alpha t} f(X_t)\mathrm{d}t\right) \\
&= \int_0^{1-a} \mathrm{e}^{-\alpha t} f(a+t)\mathrm{d}t + \mathrm{e}^{-\alpha(1-a)} u(1/2) \\
&= \mathrm{e}^{\alpha a} \int_a^1 \mathrm{e}^{-\alpha t} f(t)\mathrm{d}t + \mathrm{e}^{-\alpha(1-a)} u(1/2)\ ,
\end{aligned}$$

$$u(1/2) = \mathrm{e}^{\alpha/2} \int_{1/2}^1 \mathrm{e}^{-\alpha t} f(t)\mathrm{d}t + \mathrm{e}^{-\alpha/2} u(1/2)\ ,$$

$$u(1/2) = \frac{\mathrm{e}^{\alpha/2}}{1-\mathrm{e}^{-\alpha/2}} \int_{1/2}^1 \mathrm{e}^{-\alpha t} f(t)\mathrm{d}t\ .$$

Therefore

$$u(a) = \mathrm{e}^{\alpha a} \int_a^1 \mathrm{e}^{-\alpha t} f(t)\mathrm{d}t + \frac{\mathrm{e}^{\alpha(a-1/2)}}{1-\mathrm{e}^{-\alpha/2}} \int_{1/2}^1 \mathrm{e}^{-\alpha t} f(t)\mathrm{d}t\ .$$

It follows that $R_\alpha$ carries $\mathbf{B}(S)$ into $C(S)$ and $C(S)$ into $C^1(S) \subset C(S)$. Therefore $X$ is strong Markov.

Let us show that

(1) $\mathfrak{D}(A) = \{u\colon u$ continuous and bounded, $u^+$ right continuous and bounded, $u(1/2) = u(1-)\}$ ,

(2) $Au = u^+$ .

To prove this, write $\widetilde{\mathfrak{D}}$ for the right-hand side of (1). It is obvious that $\mathfrak{D}(A) = R_\alpha \mathbf{L} \subset \widetilde{\mathfrak{D}}$. If $u \in \mathfrak{D}(A)$, then $u = R_\alpha f$, $f \in \mathbf{L}$, $\alpha > 0$. By computation we get

$$u^+ = \alpha u - f = Au\ .$$

To prove $\widetilde{\mathfrak{D}} = \mathfrak{D}(A)$, take an arbitrary $v \in \widetilde{\mathfrak{D}}$ and set

$$u = R_\alpha(\alpha v - v^+) \in \mathfrak{D}(A)\ .$$

It is to be noted that $\alpha v - v^+ \in \mathbf{L}$. Then $\alpha u - u^+ = \alpha v - v^+$ . Set $w = v - u$. Then

$$\alpha w = w^+$$

and so

$$w = C e^{\alpha a} .$$

Since $w = v - u \in \widetilde{\mathfrak{D}}$, we have $w(1/2) = w(1-)$. Therefore $C = 0$. Thus we have $w = 0$ i. e. $v = u \in \mathfrak{D}(A)$.

**3. The linear motion on $(0, 1)$ with a fixed return distribution.** Let $\nu$ be a distribution on $(0, 1)$. The particle makes a motion with speed 1 and as soon as it hits 1, it returns in $(0, 1)$ with the probability law $\nu$ and continues the same motion as before. We assume that the sizes of the successive returning jumps are independent. This process is Markov. Notice

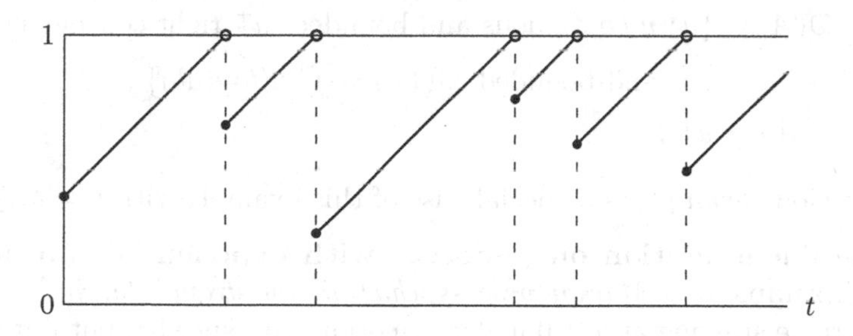


that the successive jump times tend to $\infty$ by the law of large numbers. We have

$$\mathbf{L} = \{f : f \text{ right continuous, bounded}\} ,$$

and $H_t$ does not carry $C(0, 1)$ into $C(0, 1)$. Let $u = R_\alpha f$. Then

$$\begin{aligned} u(a) &= E_a \left( \int_0^\infty e^{-\alpha t} f(X_t) dt \right) \\ &= E_a \int_0^{1-a} e^{-\alpha t} f(a + t) dt + E_a(e^{-\alpha(1-a)} u(X_{1-a})) \\ &= \int_0^{1-a} e^{-\alpha t} f(a + t) dt + e^{-\alpha(1-a)} \int_0^1 u(b) \nu(db) \\ &= e^{\alpha a} \int_a^1 e^{-\alpha t} f(t) dt + e^{-\alpha(1-a)} \int_0^1 u(b) \nu(db) . \end{aligned}$$

Therefore

$$\int_0^1 u(b) \nu(db) = \int_0^1 \nu(db) \int_0^{1-b} e^{-\alpha t} f(b + t) dt$$

$$+ \left( \int_0^1 \mathrm{e}^{-\alpha(1-b)} \nu(\mathrm{d}b) \right) \left( \int_0^1 u(b) \nu(\mathrm{d}b) \right),$$

$$\int_0^1 u(b)\nu(\mathrm{d}b) = \frac{\int_0^1 \nu(\mathrm{d}b) \int_0^{1-b} \mathrm{e}^{-\alpha t} f(b+t) \mathrm{d}t}{1 - \int_0^1 \mathrm{e}^{-\alpha(1-b)} \nu(\mathrm{d}b)},$$

$$u(a) = \mathrm{e}^{\alpha a} \int_a^1 \mathrm{e}^{-\alpha t} f(t) \mathrm{d}t + \frac{\mathrm{e}^{-\alpha(1-a)} \int_0^1 \nu(\mathrm{d}b) \int_0^{1-b} \mathrm{e}^{-\alpha t} f(b+t) \mathrm{d}t}{1 - \int_0^1 \mathrm{e}^{-\alpha(1-b)} \nu(\mathrm{d}b)}.$$

It follows that

$$R_\alpha \colon \mathbf{B}(0,1) \longrightarrow C(0,1).$$

Therefore $X$ is strong Markov,

$$\mathfrak{D}(A) = \{u \colon u \text{ continuous and bounded, } u^+ \text{ right continuous and bounded, } u(1-) = \textstyle\int_0^1 u(b)\nu(\mathrm{d}b)\},$$

$$Au = u^+ .$$

The previous example is a special case of this example with $\nu = \delta_{1/2}$.

**4. The linear motion on $(-\infty, \infty)$ with exponential waiting time at 0.** (Example of a *Markov process which is not strong Markov.*)
The particle starting at $a > 0$ makes a motion with speed 1, but if it starts at $a \leq 0$, it goes with speed 1 until it comes to 0. It must wait for an exponential holding time at 0 until it begins to start a motion with speed 1. We have

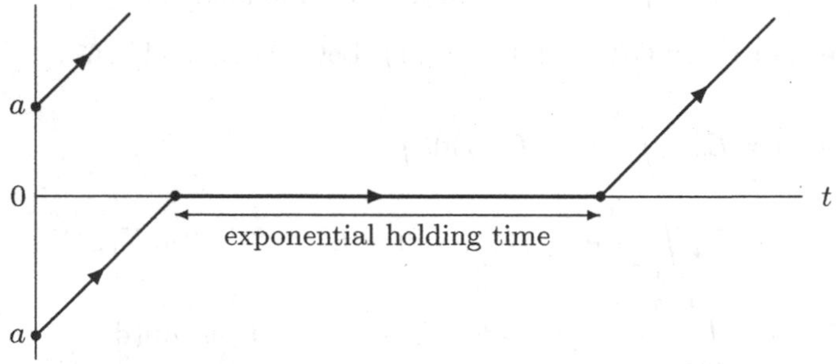

a
0
a
t
exponential holding time

$$H_t f(a) = f(a+t) \qquad \text{for } a \neq 0,$$
$$H_t f(0) = f(0) P_0(t < \sigma) + E_0(f(X_t),\ \sigma \leq t) \to f(0), \quad t \downarrow 0$$

where $\sigma$ is the waiting time under $P_0$ :

$$P_0(\sigma > t) = \mathrm{e}^{-\lambda t} \qquad \text{with } 0 < \lambda < \infty .$$

Therefore

$$\mathbf{L} = \{f\colon \text{right continuous at } a \neq 0 \text{ and bounded}\} .$$

Let $u = R_\alpha f$, $f \in \mathbf{L}$ . If $a > 0$, then

$$u(a) = \int_0^\infty \mathrm{e}^{-\alpha t} f(a+t)\mathrm{d}t = \mathrm{e}^{\alpha a} \int_a^\infty \mathrm{e}^{-\alpha t} f(t)\mathrm{d}t .$$

If $a = 0$, then

$$\begin{aligned} u(a) = u(0) &= E_0\left(\int_0^\infty \mathrm{e}^{-\alpha t} f(X_t)\mathrm{d}t\right) \\ &= E_0\left(\int_0^\sigma \mathrm{e}^{-\alpha t} f(0)\mathrm{d}t\right) + E_0\left(\mathrm{e}^{-\alpha\sigma}\int_0^\infty \mathrm{e}^{-\alpha t} f(t)\mathrm{d}t\right) \\ &= E_0\left(\frac{1-\mathrm{e}^{-\alpha\sigma}}{\alpha}\right) f(0) + E_0(\mathrm{e}^{-\alpha\sigma})\, u(0+) \\ &= \frac{1}{\alpha+\lambda} f(0) + \frac{\lambda}{\alpha+\lambda}\, u(0+) . \end{aligned}$$

Thus

$$\begin{aligned} f(0) &= (\alpha+\lambda)u(0) - \lambda u(0+) , \\ \alpha u(0) - f(0) &= \lambda(u(0+) - u(0)) . \end{aligned}$$

If $a < 0$, then

$$\begin{aligned} u(a) &= \int_0^{|a|} \mathrm{e}^{-\alpha t} f(a+t)\mathrm{d}t + \mathrm{e}^{-\alpha|a|} E_0\left(\frac{1-\mathrm{e}^{-\alpha\sigma}}{\alpha}\right) f(0) \\ &\quad + \mathrm{e}^{-\alpha|a|} E_0(\mathrm{e}^{-\alpha\sigma})\, u(0+) \\ &= \int_0^{|a|} \mathrm{e}^{-\alpha t} f(a+t)\mathrm{d}t + \mathrm{e}^{-\alpha|a|} u(0) . \end{aligned}$$

Hence $u(0-) = u(0)$. We can show that[29]

$$\begin{aligned} \mathfrak{D}(A) &= R_\alpha \mathbf{L} \\ &= \{u\colon u \text{ continuous at } a \neq 0 \text{ and bounded}, \\ &\qquad u^+(a) \text{ exists on } \{a\colon a \neq 0\}, \text{ right continuous and bounded}, \\ &\qquad u(0-), u(0+) \text{ exist}, u(0-) = u(0)\} \end{aligned}$$

(notice that the existence of $u(0-)$ and $u(0+)$ follows automatically from the other conditions since $u(-\varepsilon) = u(-1) + \int_{-1}^{-\varepsilon} u^+(a)\mathrm{d}a$, $u(\varepsilon) = u(1) - \int_1^\varepsilon u^+(a)\mathrm{d}a$), and that

[29] A proof can be given along the same line as in Example 2.

$$Au(a) = \begin{cases} u^+(a) & a \neq 0 \\ \lambda(u(0+) - u(0)) & a = 0\,. \end{cases}$$

Now we will prove that this is not strong Markov. Let $\sigma$ be the hitting time of the open half line $(0, \infty)$. Then it is obvious that

$$\sigma(\theta_\sigma \omega) = 0, \quad X_\sigma = 0 \quad \text{a.s.}\ (P_0)\,.$$

Therefore, if the strong Markov property holds, then

$$P_0(\sigma(\theta_\sigma \omega) = 0) = 1\,,$$

while the left-hand side will be

$$E_0(P_{X_\sigma}(\sigma = 0)) = P_0(\sigma = 0) = 0\,,$$

which is a contradiction.

## 2.14 Markov Processes with a Countable State Space

Let $S$ be a countable space with discrete topology and let $X$ be a Markov process with state space $S$.

**1.** Since $\mathbf{B}(S) = C(S)$, it is trivially true that $H_t\colon C(S) \longrightarrow C(S)$. Therefore $X$ is strong Markov.

**2.** Each state $a \in S$ is either a trap or an exponential holding state. For a trap $a$ we have

$$H_t f(a) = f(a)$$

and for an exponential holding state we have ($e$ = waiting time)

$$\begin{aligned} H_t f(a) &= f(a) P_a(t < e) + O(P_a(t \geq e)) \\ &= f(a) \mathrm{e}^{-\lambda(a)t} + O(1 - \mathrm{e}^{-\lambda(a)t}) \\ &\to f(a), \qquad t \downarrow 0\,. \end{aligned}$$

Thus we have

$$\mathbf{B}(S) = \mathbf{L} = C(S)\,.$$

**3.** Let $u \in \mathfrak{D}$. If $a$ is a trap, then

$$Au(a) = 0\,.$$

If $a$ is an exponential holding state ($P_a(e > t) = \mathrm{e}^{-\lambda(a)t}$), then

$$
\begin{aligned}
Au(a) &= \frac{E_a(u(X_e)) - u(a)}{E_a(e)} \\
&= \lambda(a)\left(\sum_b \pi(a,b)u(b) - u(a)\right) \\
&= \lambda(a)\sum_b \pi(a,b)(u(b) - u(a)) ,
\end{aligned}
$$

where $\pi(a,b) = P_a(X_e = b)$.

**4.** Let $a_0$ be a fixed point in $S$. We introduce the following stopping times

$e = \inf\{t\colon X_t \neq X_0\}$ waiting time (as before),

$\sigma = \inf\{t\colon X_t = a_0\}$ hitting time of $a_0$,

$$\tau = \begin{cases} e + \sigma(\theta_e\omega) & \text{if } e < \infty \\ \infty & \text{if } e = \infty \end{cases}$$ hitting time of $a_0$ after waiting.

Let $e_n$ be the $n$-th waiting time at $a_0$ and $\sigma_n$ the $n$-th hitting time of $a_0$. Let

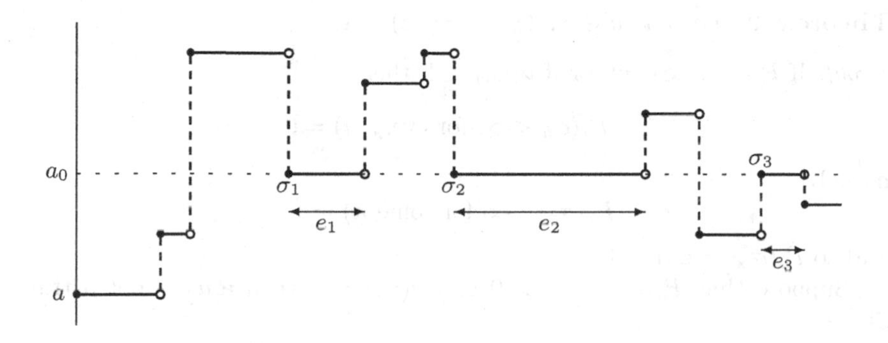


$\nu(a_0)$ be the return probability to $a_0$:

$$\nu(a_0) = P_{a_0}(\tau < \infty) .$$

**Theorem 1.**

$$P_{a_0}(\sigma_n < \infty \textit{ for every } n) = \begin{cases} 1 & \textit{if } \nu(a_0) = 1 \\ 0 & \textit{if } \nu(a_0) < 1 . \end{cases}$$

*Proof.* Since $\sigma_1 \leq \sigma_2 \leq \cdots$, $\sigma_1 = \sigma$, and $\sigma_n = \sigma_{n-1} + \tau(\theta_{\sigma_{n-1}}\omega)$, $n \geq 2$, we have

$$\begin{aligned}
P_{a_0}(\sigma_n < \infty \text{ for every } n) &= \lim_n P_{a_0}(\sigma_n < \infty) \\
&= \lim_n P_{a_0}(\sigma_{n-1} < \infty,\ \tau(\theta_{\sigma_{n-1}}\omega) < \infty) \\
&= \lim_n E_{a_0}(P_{X_{\sigma_{n-1}}}(\tau < \infty),\ \sigma_{n-1} < \infty) \\
&= \lim_n P_{a_0}(\tau < \infty)\, P_{a_0}(\sigma_{n-1} < \infty) \\
&= \lim_n \nu(a_0)\, P_{a_0}(\sigma_{n-1} < \infty) \\
&= \lim_n \nu(a_0)^n ,
\end{aligned}$$

which proves our theorem.

**Corollary 1.** *For any $a \in S$*

$$P_a(\sigma_n < \infty \textit{ for every } n) = \begin{cases} P_a(\sigma < \infty) & \textit{if } \nu(a_0) = 1 \\ 0 & \textit{if } \nu(a_0) < 1 . \end{cases}$$

**5.** Fix $a_0$ as above. Let $\sigma_\infty = \lim_n \sigma_n$. Then

**Theorem 2.** *For any $a \in S$, $P_a(\sigma_\infty = \infty) = 1$.*

*Proof.* If $P_a(\sigma < \infty) = 0$ or if $\nu(a_0) < 1$, then

$$P_a(\sigma_n < \infty \text{ for every } n) = 0$$

namely

$$P_a(\sigma_n = \infty \text{ for some } n) = 1 ,$$

and so $P_a(\sigma_\infty = \infty) = 1$.

Suppose that $P_a(\sigma < \infty) > 0$ and $\nu(a_0) = 1$ (hence $a_0$ is not a trap). Since

$$e_1 + e_2 + \cdots + e_{n-1} \le \sigma_n, \qquad e_n = e(\theta_{\sigma_n}\omega)$$

we have

$$\begin{aligned}
&E_a(\mathrm{e}^{-\alpha\sigma_\infty},\ \sigma_\infty < \infty) \le E_a(\mathrm{e}^{-\alpha\sigma_n},\ \sigma_n < \infty) \\
&\quad \le E_a(\mathrm{e}^{-\alpha(e_1+\cdots+e_{n-1})},\ e_1 + \cdots + e_{n-1} < \infty) \\
&\quad = E_a(\mathrm{e}^{-\alpha(e_1+\cdots+e_{n-2})}\mathrm{e}^{-\alpha e_{n-1}},\ e_1 + \cdots + e_{n-2} < \infty,\ e_{n-1} < \infty) \\
&\quad = E_a(\mathrm{e}^{-\alpha(e_1+\cdots+e_{n-2})}\mathrm{e}^{-\alpha e(\theta_{\sigma_{n-1}}\omega)},\ e_1 + \cdots + e_{n-2} < \infty,\ e(\theta_{\sigma_{n-1}}\omega) < \infty) \\
&\quad = E_a(\mathrm{e}^{-\alpha(e_1+\cdots+e_{n-2})},\ e_1 + \cdots + e_{n-2} < \infty) E_{a_0}(\mathrm{e}^{-\alpha e},\ e < \infty) \\
&\qquad\qquad\qquad\qquad\qquad\qquad\qquad\qquad (\text{since } P_a(X_{\sigma_{n-1}} = a_0) = 1) \\
&\quad = (E_{a_0}(\mathrm{e}^{-\alpha e},\ e < \infty))^{n-1} = \left(\frac{\lambda}{\alpha + \lambda}\right)^{n-1} \to 0, \quad n \to \infty
\end{aligned}$$

for $\alpha > 0$. It follows that $P_a(\sigma_\infty < \infty) = 0$.

**6.** Let $j_n$ be the $n$-th jump time and set

$$j_\infty = \lim_{n\to\infty} j_n \, .$$

Then $j_\infty$ is also a stopping time.

**Theorem 3.** *(i) $X_t$, $t < j_\infty$, is a step function a. s.*

*(ii) $X_t$ tends to the point at infinity (in the one point compactification of $S$) as $t \uparrow j_\infty$ a. s. on $\{j_\infty < \infty\}$.*

*Proof.* (i) Obvious.

(ii) It is enough to prove that

$$\begin{aligned} &P_a(\{\text{for any } K \text{ compact } \subset S \text{ there is } s < j_\infty \text{ such that} \\ &\quad X_t \in K^{\mathrm{C}} \text{ for all } t \in (s, j_\infty)\} \cap \{j_\infty < \infty\}) \\ &\quad = P_a(j_\infty < \infty) \end{aligned}$$

i. e.

$$\begin{aligned} &P_a(\{\text{there is } K \text{ compact } \subset S \text{ such that, for every } s < j_\infty, \\ &\quad \text{there is } t \in (s, j_\infty) \text{ satisfying } X_t \in K\} \cap \{j_\infty < \infty\}) \\ &\quad = 0 \, . \end{aligned}$$

Since every compact subset of $S$ is a finite set, the class of all compact subsets of $S$ is countable. Therefore it is enough to show that

$$\begin{aligned} &P_a(\{\text{for every } s < j_\infty \text{ there is } t \in (s, j_\infty) \text{ satisfying } X_t \in K\} \cap \{j_\infty < \infty\}) \\ &\quad = 0 \end{aligned}$$

for every compact $K$. Since $K$ is finite, this probability is

$$\begin{aligned} \le \sum_{b \in K} &P_a(\{\text{for every } s < j_\infty \text{ there is } t \in (s, j_\infty) \text{ satisfying } X_t = b\} \\ &\cap \{j_\infty < \infty\}) \, . \end{aligned}$$

But each term is 0 by Theorem 2.

**7.** The *construction of the path* in $t < j_\infty$ is similar to the finite state case (Section 2.11, Example). If

$$P_a(j_\infty = \infty) = 1 \, ,$$

then the construction is completed; it is easy to see by the argument above that this is the case for $S$ finite, though we did not mention this in Section 2.11.

Suppose that

$$P_a(j_\infty < \infty) > 0 .$$

Then we have to construct $X_t$ for $t \geq j_\infty$ on $\{j_\infty = \infty\}$. There are many possibilities for the further development of the process. We will consider the simplest case that $X(j_\infty)$ is independent of the past and has a distribution $\nu$, called the (instantaneous) *return probability distribution.* Then $X(j_\infty)$ is determined. Once $X(j_\infty)$ is determined, we can define $X(t)$ for $t < j_\infty +$

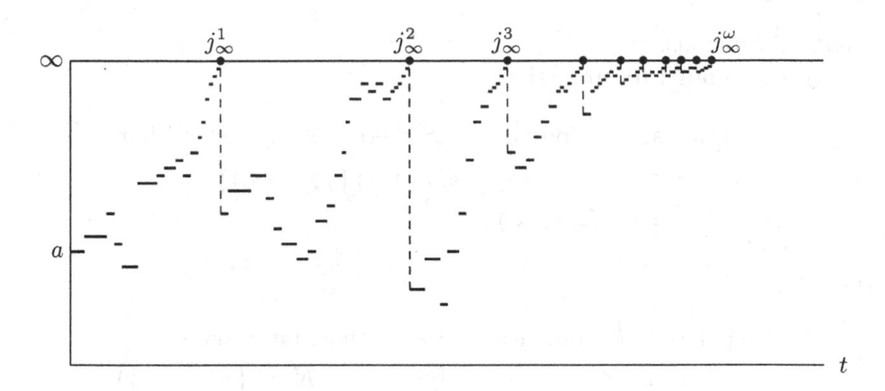


$j_\infty(\theta_{j_\infty}\omega) \equiv j_\infty^2$ in the same way as above. Let $j_\infty^1 \equiv j_\infty$ and $j_\infty^n \equiv j_\infty^{n-1} + j_\infty(\theta_{j_\infty^{n-1}}\omega)$, $n = 2, 3, \ldots$. Then, continuing this we can construct the path for $t < j_\infty^n$ recursively and so for

$$t < \lim_{n\to\infty} j_\infty^n = j_\infty^\omega ,$$

$\omega$ here being the first infinite ordinal.

**8.** We have proved that if $u \in \mathfrak{D}(A)$, then

$$Au(a) = \begin{cases} 0 & \text{if } a \text{ is a trap} \\ \lambda(a) \sum_b \pi(a,b)(u(b) - u(a)) & \text{if } a \text{ is not a trap.} \end{cases}$$

For the complete description of $A$ we have to determine $\mathfrak{D}(A)$. This was initiated by W. Feller and is a very interesting subject. Since it is beyond the scope of these lectures, we will only present a very simple example which illustrates some important aspects.

**Example.** Let $S = \{0, \pm 1, \pm 2, \ldots\}$,

$$\begin{aligned} &\lambda(a) = \mathrm{e}^a , \\ &\pi(a, a+1) = 1, \qquad \pi(a,b) = 0 \quad \text{for } b \neq a+1 , \\ &\nu(\,\cdot\,) = \text{ arbitrary probability distribution on } S . \end{aligned}$$

It is easy to see that

$$E_a(j_\infty) = \sum_{k=0}^{\infty} \mathrm{e}^{-(a+k)} < \infty$$

and so

$$P_a(j_\infty < \infty) = 1\ .$$

Here

$$j_\infty^2 - j_\infty^1 = j_\infty(\theta_{j_\infty^1}\omega), \quad j_\infty^3 - j_\infty^2 = j_\infty(\theta_{j_\infty^2}\omega), \quad \ldots$$

are independent and have the identical distribution

$$P(E) = \sum_b \nu(b) P_b(j_\infty \in E)\ ,$$

which clearly has positive mean value. Therefore

$$j_\infty^n \geq \sum_{k=2}^{n} (j_\infty^k - j_\infty^{k-1}) \to \infty \qquad \text{a. s.},$$

namely

$$j_\infty^\omega = \infty \quad \text{a. s.}$$

It is obvious that

$$Au(a) = \mathrm{e}^a(u(a+1) - u(a)) \qquad \text{for } u \in \mathfrak{D}(A)\ .$$

We will prove that

$$\mathfrak{D}(A) = \left\{ u \in \mathbf{B}(S)\colon \lim_{a \to +\infty} u(a) = \sum_b \nu(b) u(b) \right\}.$$

It should be noted here that $+\infty$ is not the point at infinity in the one-point compactification but the right-hand one of the points at infinity (i.e. $\pm\infty$) in the two point compactification of the set of integers. As in the method used before, we need only prove that

(1) For $f \in \mathbf{B}(S)$, $u = R_\alpha f$ satisfies $u(+\infty) = \sum_b \nu(b) u(b)$.

(2) If $\alpha w = \lambda(a)(w(a+1) - w(a))$ and $w(+\infty) = \sum_b \nu(b) w(b)$, then $w \equiv 0$.

Consider

$$u(a) = R_\alpha f(a) = E_a \left( \int_0^\infty \mathrm{e}^{-\alpha t} f(X_t) \mathrm{d}t \right)\ .$$

By our construction

$$u(a) = E_a\left(\int_0^{j_\infty} \mathrm{e}^{-\alpha t} f(X_t)\mathrm{d}t\right) + E_a(\mathrm{e}^{-\alpha j_\infty})\sum_b \nu(b)u(b) .$$

Now observing

$$E_a\left(\int_0^{j_\infty} \mathrm{e}^{-\alpha t} f(X_t)\mathrm{d}t\right) \le \|f\| E_a(j_\infty) = \|f\| \sum_{k=0}^{\infty} \mathrm{e}^{-(a+k)} \to 0, \quad a \to +\infty ,$$

$$E_a(\mathrm{e}^{-\alpha j_\infty}) > 0 \qquad \text{by } P_a(j_\infty < \infty) = 1 ,$$

and

$$E_a(\mathrm{e}^{-\alpha j_\infty}) = E_a(\mathrm{e}^{-\alpha j_n}) E_{a+n}(\mathrm{e}^{-\alpha j_\infty}) ,$$

and letting $n \to \infty$, we have

$$\lim_{a\to+\infty} E_a(\mathrm{e}^{-\alpha j_\infty}) = 1 .$$

Thus

$$u(+\infty) = \sum_b \nu(b)u(b) .$$

Suppose that $\alpha w = Aw$ and $w(+\infty) = \sum_b \nu(b)w(b)$. Then

$$\alpha w(a) = \lambda(a)(w(a+1) - w(a)) .$$

Since

$$w(a)(\alpha + \lambda(a))/\lambda(a) = w(a+1) ,$$

$w(a) > 0$ for every $a$ or $w(a) \equiv 0$ or $w(a) < 0$ for every $a$. In the first case $w(a)$ is strictly increasing. Therefore

$$w(+\infty) > w(b) \qquad \text{for every} \quad b$$

and so

$$w(+\infty) > \sum \nu(b)w(b)$$

in contradiction with our assumption. Thus the first case is excluded. Similarly the third case is also excluded. Thus we have $w \equiv 0$. This completes the proof of (1) and (2).

## 2.15 Fine Topology

Let $X = \{X_t(\omega),\ t \in T,\ \omega \in (\Omega, \mathcal{B}, P_a)\}_{a\in S}$ be a strong Markov process in generalized sense. For simplicity we will assume that the state space $S$ is a locally compact separable metric space.

We will define $P_\mu, \overline{\mathcal{B}}, \overline{\mathcal{B}}_t$, etc. as in Section 2.6. Then we have

**Theorem 1.** *If $B \in \mathcal{B}(S)$, then the hitting time $\sigma_B$ of $B$,*

$$\sigma_B(\omega) = \inf\{t > 0 \colon X_t(\omega) \in B\} ,$$

*is $\overline{\mathcal{B}}$-measurable, and is a stopping time with respect to $\{\overline{\mathcal{B}}_t\}$. For every $\mu$ there exists an increasing sequence of compact sets $K_n \subset B$, $n = 1, 2, \ldots$, such that*

$$\sigma_B = \lim_n \sigma_{K_n} \qquad a.\ s.\ (P_\mu).$$

We will admit this important fact without proof; see Blumenthal and Getoor, Markov Processes and Potential Theory, for the proof.[30]

Now we will introduce a new topology in the state space.

**Definition 1.** A subset $V$ of $S$ is called *finely open*, if for every $a \in V$ there exists a Borel subset $B = B(a) \subset V$ such that

$$P_a(\sigma_{B^{\complement}} > 0) = 1 .$$

The collection of all finely open subsets determines a Hausdorff topology in $S$, called the *fine topology* in $S$.

Because of the right continuity of paths the fine topology is stronger than the original topology so that every continuous (in the original topology) function is finely continuous.

**Theorem 2.** *Suppose that $F(\omega)$ is a bounded $\mathcal{B}$-measurable function on $\Omega$ and that $G$ is a finely open set $\in \mathcal{B}(S)$. If*

$$P_a\left(\lim_{h \to 0} F(\theta_h \omega) = F(\omega)\right) = 1, \qquad a \in G ,$$

*then*

$$f(a) = E_a(F)$$

*is finely continuous in $G$.*

*Proof.* Since $-f(a) = E_a(-F)$, it is enough to prove that $f$ is finely upper semi-continuous[31] in $G$. Fix $a \in G$ and $\varepsilon > 0$ and set

$$B = \{b \in G \colon f(b) < f(a) + \varepsilon\} .$$

---

[30] The proof of this theorem in the book of Blumenthal and Getoor requires more assumptions, namely, the existence of left limits of paths and quasi-left continuity. However one can prove this theorem using results in Meyer's book mentioned in the Preface to the Original (private communication with Shinzo Watanabe).

[31] A function $f$ is called finely upper semi-continuous at $a$ if, for every $\varepsilon > 0$, there is a finely open set $E$ containing $a$ such that $f(b) < f(a) + \varepsilon$ for $b \in E$.

It is obvious that $B$ is a Borel set. It is therefore enough to prove that $B$ is finely open. Let $a_1 \in B$. Then

$$f(a_1) + \varepsilon_1 < f(a) + \varepsilon$$

for some $\varepsilon_1 > 0$. Set

$$B_1 = \{b \in G \colon f(b) < f(a_1) + \varepsilon_1\} .$$

Obviously $B_1$ is a Borel subset of $B$ and $a_1 \in B_1$.

For the completion of the proof it is enough to prove that

$$P_{a_1}(\sigma_{B_1^C} > 0) = 1 .$$

Observing that

$$\begin{aligned} P_{a_1}(\sigma_{B_1^C} = 0) &= P_{a_1}(\sigma_{B_1^C} = 0,\ \sigma_{B_1^C}(\theta_{\sigma_{B_1^C}}\omega) = 0) \\ &= E_{a_1}(P_{X(\sigma_{B_1^C})}(\sigma_{B_1^C} = 0),\ \sigma_{B_1^C} = 0) \\ &= P_{a_1}(\sigma_{B_1^C} = 0)^2 , \end{aligned}$$

we see that

$$P_{a_1}(\sigma_{B_1^C} = 0) = 1 \quad \text{or} \quad P_{a_1}(\sigma_{B_1^C} > 0) = 1 .$$

We need only to prove that the first case is impossible. Suppose it occurs. Then

$$P_{a_1}(\sigma_{G-B_1} = 0) = 1$$

since $G$ is finely open and $a_1 \in G$. Then there exists an increasing sequence of compact sets $K_n \subset G - B_1$ such that

$$\sigma_{K_n} \to \sigma_{G-B_1} = 0 \qquad \text{a.s. } (P_{a_1}) .$$

Hence it follows from our assumption that

$$P_{a_1}\left(\lim_{n\to\infty} F(\theta_{\sigma_{K_n}}\omega) = F(\omega)\right) = 1 .$$

Therefore

$$E_{a_1}(F(\theta_{\sigma_{K_n}}\omega),\ \sigma_{K_n} < \infty) \to E_{a_1}(F(\omega)) = f(a_1), \qquad n \to \infty ,$$

while

$$\begin{aligned} E_{a_1}(F(\theta_{\sigma_{K_n}}\omega),\ \sigma_{K_n} < \infty) &= E_{a_1}(E_{X(\sigma_{K_n})}(F),\ \sigma_{K_n} < \infty) \\ &= E_{a_1}(f(X_{\sigma_{K_n}}),\ \sigma_{K_n} < \infty) . \end{aligned}$$

Since $K_n$ is compact, we have

$$X_{\sigma_{K_n}} \in K_n \subset G - B_1 \qquad \text{on } \{\sigma_{K_n} < \infty\}$$

which implies

$$E_{a_1}(f(X_{\sigma_{K_n}}),\ \sigma_{K_n} < \infty) \geq (f(a_1) + \varepsilon_1) P_{a_1}(\sigma_{K_n} < \infty) .$$

Thus we have

$$f(a_1) \geq f(a_1) + \varepsilon_1 ,$$

which is a contradiction.

**Theorem 3.** *Let $f$ be a Borel measurable function defined on a finely open set $G \in \mathcal{B}(S)$. Then $f$ is finely continuous in $G$ if and only if*

$$P_a\left(\lim_{t \downarrow 0} f(X_t) = f(a)\right) = 1 \qquad \textit{for every } a \in G .$$

*Proof.* If this condition is satisfied, then

$$P_a\left(\lim_{t \downarrow 0} f(X_0(\theta_t \omega)) = f(X_0)\right) = 1$$

and so

$$f(a) = E_a(f(X_0))$$

is finely continuous by Theorem 2. Conversely, let $f$ be finely continuous in $G$. Let $a \in G$. Then

$$B = \{b \in G \colon |f(b) - f(a)| < \varepsilon\}$$

is finely open, because $G$ is finely open. Since $B$ is Borel, we have

$$P_a(\sigma_{B^C} > 0) = 1 .$$

For the $\omega$ for which $\sigma_{B^C}(\omega) > 0$ we have

$$|f(X_t(\omega)) - f(a)| < \varepsilon \qquad \text{for } 0 < t < \sigma_{B^C}(\omega)$$

and so

$$\limsup_{t \downarrow 0} |f(X_t(\omega) - f(a))| \leq \varepsilon .$$

Thus

$$P_a\left(\limsup_{t \downarrow 0} |f(X_t) - f(a)| \leq \varepsilon\right) = 1 .$$

Let $\varepsilon \downarrow 0$ to complete the proof.

Let $C_f(S)$ be the space of all finely continuous functions $\in \mathbf{B}(S)$.

**Theorem 4.** $\quad C(S) \subset C_f(S) \subset \mathbf{L}$ .

*Proof.* If $f \in C_f(S)$, then

$$P_a\left(\lim_{t\downarrow 0} f(X_t) = f(X_0)\right) = 1$$

by Theorem 3. Therefore

$$H_t f(a) = E_a(f(X_t)) \to E_a(f(X_0)) = f(a)\ ,$$

which proves $C_f(S) \subset \mathbf{L}$. Obviously $C(S) \subset C_f(S)$ by Theorem 3.

**Theorem 5.** $H_t\colon C_f(S) \longrightarrow C_f(S)$.

*Proof.* Let $f \in C_f(S)$. Then

$$P_a\left(\lim_{h\downarrow 0} f(X_{t+h}) = f(X_t)\right) = E_a\left(P_{X_t}\left(\lim_{h\downarrow 0} f(X_h) = f(X_0)\right)\right) = 1$$

by Theorem 3. Then $H_t f(a) \equiv E_a(f(X_t))$ is in $C_f(S)$ by Theorem 2.

**Theorem 6.** $R_\alpha\colon \mathbf{B}(S) \longrightarrow C_f(S)$, *a fortiori* $R_\alpha\colon C_f(S) \longrightarrow C_f(S)$.

*Proof.* Let $f \in \mathbf{B}(S)$. Set

$$F(\omega) = \int_0^\infty \mathrm{e}^{-\alpha s} f(X_s)\mathrm{d}s\ .$$

Then

$$F(\theta_h\omega) = \int_0^\infty \mathrm{e}^{-\alpha s} f(X_{s+h})\mathrm{d}s = \mathrm{e}^{\alpha h}\int_h^\infty \mathrm{e}^{-\alpha s} f(X_s)\mathrm{d}s \to F(\omega)$$

as $h \downarrow 0$. Now use Theorem 2.

The restriction of $H_t$ and $R_\alpha$ to $C_f(S)$ and the definition of $A$ in $C_f(S)$ can be discussed in the same way as we did for $\mathbf{L}$ (Section 2.12). In the present case we have more advantage in that the Dynkin representation of $A$ holds analogously to the case in Section 2.11. In fact, if $a$ is a trap, $Au(a) = 0$, while if $a$ is not a trap, we have $E_a(\tau_U) < \infty$ $(\tau_U = \sigma_{U^{\mathrm{c}}})$ for some finely open neighborhood of $a$ and

$$Au(a) = \lim_{U\downarrow a} \frac{E_a(u(X_{\tau_U})) - u(a)}{E(\tau_U)}, \qquad u \in \mathfrak{D}(A)$$

where $U$ is an arbitrary finely open Borel set containing $a$.

Now we shall define the characteristic generator $\mathcal{A}_a$ at $a$ as follows:

$$\mathcal{A}_a u = \lim_{U\downarrow a} \frac{E_a(u(X_{\tau_U})) - u(a)}{E(\tau_U)}$$

where $U$ is an arbitrary finely open Borel set containing $a$ and $u$ is finely continuous at $a$. The domain $\mathfrak{D}(\mathcal{A}_a)$ of $\mathcal{A}_a$ is understood to be the family of all functions $\in \mathbf{B}(S)$ finely continuous at $a$ such that the above limit exists and is finite. It is usual to omit $a$ in $\mathcal{A}_a$ and to write $\mathcal{A}u(a)$ and $\mathfrak{D}(\mathcal{A}, a)$ respectively for $\mathcal{A}_a u$ and $\mathfrak{D}(\mathcal{A}_a)$. Let $U$ be a finely open set $\in \mathcal{B}(S)$. Let $\mathfrak{D}(\mathcal{A}, U)$ denote the family of all functions in $\bigcap_{a\in U} \mathfrak{D}(\mathcal{A}, a)$ such that $\mathcal{A}u(a)$ is finely continuous in $U$. Both $\mathfrak{D}(\mathcal{A}, a)$ and $\mathfrak{D}(\mathcal{A}, U)$ are linear.

## 2.16 Generator in Generalized Sense

Let $X$ be a Markov process in generalized sense. We have proved

$$(1) \qquad H_t\colon \mathbf{B}(S) \longrightarrow \mathbf{B}(S) \quad \text{and} \quad \mathbf{L} \longrightarrow \mathbf{L}\,,$$
$$(2) \qquad R_\alpha\colon \mathbf{B}(S) \longrightarrow \mathbf{L} \subset \mathbf{B}(S) \quad \text{and} \quad \mathbf{L} \longrightarrow \mathbf{L}\,.$$

The generator $A$ defined in Section 2.12, 6 is characterized by

$$(3) \quad \mathfrak{D}(A) = R_\alpha \mathbf{L} \text{ (this is independent of } \alpha), \quad Au = \alpha u - (R_\alpha|\mathbf{L})^{-1}u\,,$$

in view of the fact that

$$(4) \qquad f \in \mathbf{L}, \quad R_\alpha f = 0 \ \text{ for some } \alpha \quad \Longrightarrow \quad f = 0$$

(see Section 2.12, 5 (vi) and the resolvent equation).

Keeping this characterization in mind we will extend the definition of $A$ by using $\mathbf{B}(S)$ instead of $\mathbf{L}$ as follows. Define

$$\mathfrak{R}_\mathbf{B} = R_\alpha \mathbf{B}(S) \quad \text{and} \quad \mathfrak{N} = \{f \in \mathbf{B}(S)\colon R_\alpha f = 0\}\,.$$

By virtue of the resolvent equation we can easily see that $\mathfrak{R}_\mathbf{B}$ and $\mathfrak{N}$ are both independent of $\alpha$. Thus $R_\alpha$ can be regarded as a one-to-one map from the quotient space $\mathbf{B}(S)/\mathfrak{N}$ onto $\mathfrak{R}_\mathbf{B}$. Since the correspondence

$$\mathbf{L} \ni f \longrightarrow f + \mathfrak{N} \in \mathbf{B}(S)/\mathfrak{N}$$

is one-to-one (into) by (4), $\mathbf{L}$ is regarded as a subspace of $\mathbf{B}(S)/\mathfrak{N}$. Let us define the generator $A$ by

$$(3') \qquad \mathfrak{D}(A) = \mathfrak{R}_\mathbf{B}, \qquad Au = \alpha u - R_\alpha^{-1}u \pmod{\mathfrak{N}}\,.$$

For $A$ to be well-defined we have to prove that this definition is independent of $\alpha$. Let us write $Au$ in (3′) as $A_\alpha u$ for the moment. By the resolvent equation we have

$$R_\alpha f = R_\beta(f + (\beta - \alpha)R_\alpha f) \in \mathfrak{R}_\mathbf{B}$$

and so

$$\begin{aligned} A_\beta R_\alpha f &= \beta R_\alpha f - (f + (\beta - \alpha)R_\alpha f) \\ &= \alpha R_\alpha f - f \\ &= A_\alpha R_\alpha f\,. \end{aligned}$$

Since $\mathfrak{R}_\mathbf{B} = R_\alpha \mathbf{B}(S)$, this proves $A_\beta = A_\alpha$.

The definition (3′) is a generalization of (3) in view of the fact that $\mathbf{L} \subset \mathbf{B}(S)/\mathfrak{N}$.

As an example we will take the $k$-dimensional Brownian motion.

**Theorem 1.** *Let $X$ be the $k$-dimensional Brownian motion.*[32] *Then*

$$\mathfrak{N} = \{f \in \mathbf{B}(R^k) \colon f = 0 \text{ a. e.}\},$$

$$\mathfrak{D}(A) = \{u \in \mathbf{B}(R^k) \colon \Delta u \text{ is a measurable function bounded a. e.}\},$$

*and*

$$Au = \frac{1}{2}\Delta u,$$

*where the Laplacian $\Delta$ is to be understood in the Schwartz distribution sense.*

*Proof.* Let $L^p$ denote the space of all functions $\in \mathbf{B}(R^k)$ with finite $L^p$-norm. We will use the following inner products:

$$\langle g, f\rangle = \int_{R^k} g(a)f(a)\mathrm{d}a, \qquad g \in L^\infty,\ f \in L^1,$$

$$\langle g, \varphi\rangle = g(\varphi), \qquad g \in \mathfrak{D}',\ \varphi \in C_K^\infty,$$

where $C_K^\infty$ is the space of all $C^\infty$ functions[33] with compact support and $\mathfrak{D}'$ is the space of all Schwartz distributions. It is obvious that $L^\infty = \mathbf{B}(R^k)$. Then,

$$(5) \qquad \begin{aligned} &\varphi \in C_K^\infty, \quad g \in L^\infty \\ &\implies \quad H_t\varphi,\ R_\alpha\varphi \in L^1, \quad H_tg,\ R_\alpha g \in L^\infty, \\ &\qquad\quad \langle g, H_t\varphi\rangle = \langle H_tg, \varphi\rangle, \quad \langle g, R_\alpha\varphi\rangle = \langle R_\alpha g, \varphi\rangle, \end{aligned}$$

because the transition probability $p_t(a, \mathrm{d}b)$ has the Gaussian density $N_t(b-a)$ which is symmetric in $a$ and $b$.

To avoid confusion we will write $A_{\mathbf{L}}$ for the generator in (3). First we will prove that

$$(6) \qquad \varphi \in C_K^\infty \quad \implies \quad \varphi \in \mathfrak{D}(A_{\mathbf{L}}) \text{ and } A_{\mathbf{L}}\varphi = \frac{1}{2}\Delta\varphi.$$

For this purpose it is enough to verify

$$(7) \qquad \frac{1}{t}(H_t\varphi(a) - \varphi(a)) \to \frac{1}{2}\Delta\varphi(a) \quad \text{boundedly as } t \downarrow 0.$$

---

[32] The $k$-dimensional Brownian motion is defined on $S = R^k$ by

$$p_t(a, \mathrm{d}b) = (2\pi t)^{-k/2}\mathrm{e}^{-|b-a|^2/(2t)}\mathrm{d}b = N_t(b-a)\mathrm{d}b,$$

where $|b-a|$ is the norm of $b-a$ in $R^k$.

[33] $C^\infty$ is the class of functions differentiable infinitely many times.

(Recall the definition in Section 2.12, 6). We have

$$\frac{1}{t}(H_t\varphi(a) - \varphi(a)) = \frac{1}{t}\int_{R^k} N_t(b)(\varphi(a+b) - \varphi(a))\mathrm{d}b$$
$$= \frac{1}{t}\int_{|b|\le 1} + \frac{1}{t}\int_{|b|>1} = I_1 + I_2\,.$$

Using the Taylor expansion[34]

$$\varphi(a+b) = \varphi(a) + \sum_i \partial_i\varphi(a)b_i + \frac{1}{2}\sum_{i,j}\partial_i\partial_j\varphi(a)b_ib_j + O(|b|^3)$$

and the rotation invariance of $N_t(b)$, we have

$$I_1 = \frac{1}{t}\int_{|b|\le 1} N_t(b)\left(\frac{1}{2}\sum_i \partial_i^2\varphi(a)b_i^2 + O(|b|^3)\right)\mathrm{d}b$$
$$= \int_{|b|\le 1/\sqrt{t}} N_1(b)\left(\frac{1}{2}\sum_i \partial_i^2\varphi(a)b_i^2 + O(\sqrt{t}\,|b|^3)\right)\mathrm{d}b$$
$$\underset{\text{boundedly}}{\rightarrow} \int_{R^k} N_1(b)\left(\frac{1}{2}\sum_i \partial_i^2\varphi(a)b_i^2\right)\mathrm{d}b = \frac{1}{2}\Delta\varphi(a), \qquad t\downarrow 0\,.$$

It is easy to see that $I_2 \to 0$ boundedly as $t \downarrow 0$.

Let $\varphi \in C_K^\infty$ and $f \in L^\infty = \mathbf{B}(S)$. Then

$$\left\langle \frac{1}{2}\Delta R_\alpha f, \varphi\right\rangle \qquad (\Delta\text{: Schwartz distribution sense})$$
$$= \left\langle R_\alpha f, \frac{1}{2}\Delta\varphi\right\rangle \qquad (\Delta\text{: ordinary sense})$$
$$= \langle R_\alpha f, A_{\mathbf{L}}\varphi\rangle = \langle f, R_\alpha A_{\mathbf{L}}\varphi\rangle = \langle f, A_{\mathbf{L}}R_\alpha\varphi\rangle$$
$$= \langle f, \alpha R_\alpha\varphi - \varphi\rangle = \langle \alpha R_\alpha f - f, \varphi\rangle\,.$$

Therefore we have

$$\text{(8)} \qquad \frac{1}{2}\Delta R_\alpha f \underset{\text{a. e.}}{=} \alpha R_\alpha f - f \underset{\text{mod } \mathfrak{N}}{=} AR_\alpha f$$

and so

$$\mathfrak{N} = \{f \in \mathbf{B}(R^k)\colon f \equiv 0 \text{ a. e.}\}\,,$$
$$\mathfrak{D}(A) \subset \widetilde{\mathfrak{D}} \equiv \{u \in \mathbf{B}(R^k)\colon \Delta u \text{ is a measurable function bounded a. e.}\}$$

---

[34] $\partial_i\varphi$ is the partial derivative of $\varphi$ with respect to the $i$-th coordinate. The last term in the right-hand side is bounded by $C|b|^3$ in modulus, where $C$ is a constant independent of $a$ and $b$.

and

$$Au = \frac{1}{2}\Delta u \qquad \text{for } u \in \mathfrak{D}(A) \,.$$

It remains now only to prove $\widetilde{\mathfrak{D}} \subset \mathfrak{D}(A)$. Let $v \in \widetilde{\mathfrak{D}}$ and set

$$f = \left(\alpha - \frac{1}{2}\Delta\right)v \in \mathbf{B}(S) = L^\infty \quad \text{and} \quad u = R_\alpha f \in \mathfrak{D}(A) \,.$$

Then

$$\left(\alpha - \frac{1}{2}\Delta\right)v = f = \left(\alpha - \frac{1}{2}\Delta\right)u$$

and so $w = v - u$ satisfies

$$\left(\alpha - \frac{1}{2}\Delta\right)w = 0 \,.$$

By the proposition below,[35] for every $\varphi \in C_K^\infty$ we have $\varphi_n \in C_K^\infty$ such that[36]

$$\left\| \left(\alpha - \frac{1}{2}\Delta\right)R_\alpha\varphi - \left(\alpha - \frac{1}{2}\Delta\right)\varphi_n \right\|_1 \to 0 \,.$$

Therefore we get

$$\begin{aligned} \langle w, \varphi \rangle &= \left\langle w, \left(\alpha - \frac{1}{2}\Delta\right)R_\alpha\varphi \right\rangle \\ &= \lim_{n\to\infty} \left\langle w, \left(\alpha - \frac{1}{2}\Delta\right)\varphi_n \right\rangle \qquad (\text{since } \|w\|_\infty < \infty) \\ &= \lim_{n\to\infty} \left\langle \left(\alpha - \frac{1}{2}\Delta\right)w, \varphi_n \right\rangle \\ &= 0 \,, \end{aligned}$$

i. e. $w = 0$. Hence $v = u = R_\alpha f \in \mathfrak{D}(A)$.

**Proposition.** *If $\varphi$, $\partial_i\varphi$, $\partial_i\partial_j\varphi \in C^\infty \cap L^1$, then there exists $\{\varphi_n\} \subset C_K^\infty$ such that*

$$\begin{aligned} \|\varphi_n - \varphi\|_1 &\to 0 \,, \\ \|\partial_i\varphi_n - \partial_i\varphi\|_1 &\to 0 \,, \\ \|\partial_i\partial_j\varphi_n - \partial_i\partial_j\varphi\|_1 &\to 0 \,, \end{aligned}$$

*and so*

$$\left\| \left(\alpha - \frac{1}{2}\Delta\right)\varphi_n - \left(\alpha - \frac{1}{2}\Delta\right)\varphi \right\|_1 \to 0 \,.$$

---

[35] Since

$$R_\alpha\varphi(a) = \int_0^\infty \mathrm{e}^{-\alpha t}\mathrm{d}t \int_{R^k} (2\pi t)^{-k/2}\mathrm{e}^{-|b|^2/(2t)}\varphi(a+b)\mathrm{d}b \,,$$

we can check that $R_\alpha\varphi \in C^\infty$ and $R_\alpha\varphi, \partial_i R_\alpha\varphi, \partial_i\partial_j R_\alpha\varphi \in L^1$.

[36] The $L^1$-norm and the $L^\infty$-norm are denoted by $\|\cdot\|_1$ and $\|\cdot\|_\infty$ respectively.

*Proof.* Consider the $C_K^\infty$ function

$$d(x) = \begin{cases} C\mathrm{e}^{-1/(1-|x|^2)} & |x| < 1 \\ 0 & |x| \geq 1 \,, \end{cases}$$

where $C$ is determined by

$$\int_{R^k} d(x)\mathrm{d}x = 1 \,.$$

Now set[37]

$$d_n = 1_n * d$$

where $1_n$ is the indicator of the set $\{|x| \leq n+1\}$. Then $d_n$ is a $C_K^\infty$ function and $0 \leq d_n \leq 1$. Observe the following relations:

$$\begin{aligned} \partial_i d_n &= 1_n * \partial_i d \,, \\ \partial_i \partial_j d_n &= 1_n * \partial_i \partial_j d \,, \\ d_n = 1, \quad \partial_i d_n = 0, \quad \partial_i \partial_j d_n &= 0 \qquad \text{in } \{|x| < n\} \,, \\ \|\partial_i d_n\|_\infty &\leq \|1_n\|_\infty \|\partial_i d\|_1 = \|\partial_i d\|_1 \,, \\ \|\partial_i \partial_j d_n\|_\infty &\leq \|\partial_i \partial_j d\|_1 \,. \end{aligned}$$

Set $\varphi_n = \varphi \cdot d_n$. Then

$$\|\varphi_n - \varphi\|_1 = \int_{R^k} |\varphi|\, |1 - d_n| \mathrm{d}x \leq \int_{|x| \geq n} |\varphi| \mathrm{d}x \to 0 \,.$$

$$\begin{aligned} \|\partial_i \varphi_n - \partial_i \varphi\|_1 &= \|\partial_i d_n \cdot \varphi + d_n \cdot \partial_i \varphi - \partial_i \varphi\|_1 \\ &\leq \|\partial_i d_n \cdot \varphi\|_1 + \|(d_n - 1)\partial_i \varphi\|_1 \\ &= \|\partial_i d_n\|_\infty \int_{|x| \geq n} |\varphi| \mathrm{d}x + \int_{|x| \geq n} |\varphi| \mathrm{d}x \\ &\leq (\|\partial_i d\|_1 + 1) \int_{|x| \geq n} |\varphi| \mathrm{d}x \\ &\to 0, \qquad n \to \infty \,. \end{aligned}$$

Similarly

$$\|\partial_i \partial_j \varphi_n - \partial_i \partial_j \varphi\|_1 \to 0$$

and so

$$\left\| \left(\alpha - \frac{1}{2}\Delta\right)\varphi_n - \left(\alpha - \frac{1}{2}\Delta\right)\varphi \right\|_1 \to 0 \,.$$

[37] The convolution $u * v$ of $u \in L^1$ and $v \in L^\infty$ is defined by

$$(u * v)(x) = \int_{R^k} u(x - y)\, v(y) \mathrm{d}y \,.$$

As another example we will mention a *linear* (deterministic) *motion*. Let $b \in R^k$ be an arbitrary fixed element $\neq 0$. The linear motion with velocity $b$ is defined to be a Markov process whose probability law $P_a$ is concentrated on a single path

$$\omega(t) = a + tb .$$

Then

$$\mathfrak{N} = \{f \in \mathbf{B}(R^k) \colon f = 0 \text{ a.e.}\} ,$$

$$\mathfrak{D}(A) = \{u \in \mathbf{B}(R^k) \colon (b, \operatorname{grad} u) \text{ is a measurable function bounded a.e.}\} ,$$

$$Au = (b, \operatorname{grad} u) ,$$

where grad $u$ is to be understood in the Schwartz distribution sense.

**Theorem 2 (Dynkin's formula for generators).** *Let $X$ be a strong Markov process. Then*

$$E_a\left(\int_0^\tau Au(X_t)\mathrm{d}t\right) = u(a) - E_a(u(X_\tau)), \qquad u \in \mathfrak{D}(A)$$

*for every stopping time $\tau$ with $E_a(\tau) < \infty$.*

The proof is essentially the same as in Theorem 2.11.3. One point we should observe is that $Au(\in \mathbf{B}(S))$ is determined mod $\mathfrak{N}$ and that this formula holds for every version of $Au$.

## 2.17 The Kac Semi-Group and its Application to the Arcsine Law

Let $X$ be a Markov process in the generalized sense. We use the notation in Section 2.12. Let $k \in \mathbf{B}(S)$. We will write $k^+$ for the positive part of $k$, i.e. $k^+ = \max\{k, 0\}$. Let $\|k\|$ denote the supremum of $|k(a)|$ for $a \in S$. Let us consider the operators:

$$H_t^k f(a) = E_a\left(\mathrm{e}^{\int_0^t k(X_s)\mathrm{d}s} f(X_t)\right) ,$$

$$R_\alpha^k f(a) = \int_0^\infty \mathrm{e}^{-\alpha t} H_t^k f(a)\mathrm{d}t = E_a\left(\int_0^\infty \mathrm{e}^{-\alpha t}\mathrm{e}^{\int_0^t k(X_s)\mathrm{d}s} f(X_t)\mathrm{d}t\right)$$

for $f \in \mathbf{B}(S)$. The latter operator is well-defined for $\alpha > \|k^+\| \equiv \sup_a k^+(a)$.

**Theorem 1.**

$$H_t^k \colon \mathbf{B}(S) \longrightarrow \mathbf{B}(S) ,$$

$$H_{t+s}^k = H_t^k H_s^k, \quad H_s^k = I .$$

**Definition.** The semi-group $\{H_t^k\}_{t\in T}$ is called the *Kac semi-group* of $X$ with the *rate function* $k$.

*Proof of Theorem 1.* We will prove only $H_{t+s}^k = H_t^k H_s^k$, since the rest is obvious. Consider a function of $\omega$ depending on $t$:

$$K_t(\omega) = \int_0^t k(X_u)\mathrm{d}u .$$

Then it is easy to see the following:
(a) $K_t(\omega)$ is measurable $(\mathcal{B}_t)$ in $\omega$ for every $t$,
(b) $K_{t+s}(\omega) = K_t(\omega) + K_s(\theta_t\omega)$ .
(Such a function is called an *additive functional*). Now observe

$$\begin{aligned} H_{t+s}^k f(a) &= E_a(\mathrm{e}^{K_{t+s}} f(X_{t+s})) \\ &= E_a(\mathrm{e}^{K_t}\mathrm{e}^{K_s\circ\theta_t} f(X_s\circ\theta_t)) \qquad ((g\circ\theta_t)(\omega) = g(\theta_t(\omega))) \\ &= E_a(\mathrm{e}^{K_t} E_{X(t)}(\mathrm{e}^{K_s} f(X_s))) \qquad \text{(Markov property)} \\ &= E_a(\mathrm{e}^{K_t} H_s^k f(X_t)) \\ &= H_t^k(H_s^k f)(a) . \end{aligned}$$

**Theorem 2.** *For $\alpha > \|k^+\|$, we have*
*(i)* $R_\alpha^k f = R_\alpha f + R_\alpha^k(k\cdot R_\alpha f)$ ,
*(ii)* $R_\alpha^k f = R_\alpha f + R_\alpha(k\cdot R_\alpha^k f)$ .

*Proof.* Set $v = R_\alpha^k f$ and $u = R_\alpha f$.
(i) We have

$$\begin{aligned} v(a) - u(a) &= E_a\left(\int_0^\infty \mathrm{e}^{-\alpha t} f(X_t)(\mathrm{e}^{K_t} - 1)\mathrm{d}t\right) \\ &= E_a\left(\int_0^\infty \mathrm{e}^{-\alpha t} f(X_t)\int_0^t \mathrm{e}^{K_s} k(X_s)\mathrm{d}s\,\mathrm{d}t\right) \\ &= E_a\left(\int_0^\infty \mathrm{e}^{K_s} k(X_s)\mathrm{d}s\int_s^\infty \mathrm{e}^{-\alpha t} f(X_t)\mathrm{d}t\right) \\ &= E_a\left(\int_0^\infty \mathrm{e}^{K_s} k(X_s)\mathrm{e}^{-\alpha s}\mathrm{d}s\int_0^\infty \mathrm{e}^{-\alpha t} f(X_t\circ\theta_s)\mathrm{d}t\right) \\ &= \int_0^\infty E_a\left(\mathrm{e}^{K_s} k(X_s)\mathrm{e}^{-\alpha s}\int_0^\infty \mathrm{e}^{-\alpha t} f(X_t\circ\theta_s)\mathrm{d}t\right)\mathrm{d}s \\ &= \int_0^\infty E_a\left(\mathrm{e}^{K_s} k(X_s)\mathrm{e}^{-\alpha s} E_{X_s}\left(\int_0^\infty \mathrm{e}^{-\alpha t} f(X_t)\mathrm{d}t\right)\right)\mathrm{d}s \\ &= E_a\left(\int_0^\infty \mathrm{e}^{K_s} k(X_s)\mathrm{e}^{-\alpha s} u(X_s)\mathrm{d}s\right) \\ &= R_\alpha^k(k\cdot u)(a) . \end{aligned}$$

The exchange of integrals above is justified by

$$\begin{aligned}
E_a\left(\int_0^\infty \int_0^t |\mathrm{e}^{-\alpha t} f(X_t)\mathrm{e}^{K_s} k(X_s)|\mathrm{d}s\,\mathrm{d}t\right)
&\le \|f\|\,\|k\| E_a\left(\int_0^\infty \int_0^t \mathrm{e}^{-\alpha t}\mathrm{e}^{\|k^+\|s}\mathrm{d}s\,\mathrm{d}t\right)\\
&= \|f\|\,\|k\| E_a\left(\int_0^\infty \mathrm{e}^{\|k^+\|s}\int_s^\infty \mathrm{e}^{-\alpha t}\mathrm{d}t\,\mathrm{d}s\right)\\
&= \|f\|\,\|k\| E_a\left(\frac{1}{\alpha}\int_0^\infty \mathrm{e}^{\|k^+\|s}\mathrm{e}^{-\alpha s}\mathrm{d}s\right)\\
&= \frac{1}{\alpha}\|f\|\,\|k\| E_a\left(\int_0^\infty \mathrm{e}^{-(\alpha-\|k^+\|)s}\mathrm{d}s\right)\\
&< \infty \qquad\qquad \text{for } \alpha > \|k^+\|\,.
\end{aligned}$$

(ii) We have

$$\begin{aligned}
v(a)-u(a) &= E_a\left(\int_0^\infty \mathrm{e}^{-\alpha t} f(X_t)(\mathrm{e}^{K_t}-1)\mathrm{d}t\right)\\
&= E_a\left(\int_0^\infty \mathrm{e}^{-\alpha t} f(X_t)\int_0^t \mathrm{e}^{K_t-K_s}k(X_s)\mathrm{d}s\mathrm{d}t\right)\\
&\qquad\qquad \left(\mathrm{e}^{K_t}-1 = -\mathrm{e}^{K_t-K_s}\Big|_{s=0}^{t}\right)\\
&= E_a\left(\int_0^\infty k(X_s)\mathrm{d}s\int_s^\infty \mathrm{e}^{-\alpha t} f(X_t)\mathrm{e}^{K_t-K_s}\mathrm{d}t\right)\\
&= E_a\left(\int_0^\infty \mathrm{e}^{-\alpha s}k(X_s)\mathrm{d}s\int_0^\infty \mathrm{e}^{-\alpha t} f(X_{t+s})\mathrm{e}^{K_{t+s}-K_s}\mathrm{d}t\right)\\
&= E_a\left(\int_0^\infty \mathrm{e}^{-\alpha s}k(X_s)\mathrm{d}s\int_0^\infty \mathrm{e}^{-\alpha t} f(X_t\circ\theta_s)\mathrm{e}^{K_t\circ\theta_s}\mathrm{d}t\right)\\
&= E_a\left(\int_0^\infty \mathrm{e}^{-\alpha s}k(X_s)v(X_s)\mathrm{d}s\right) \qquad \text{as in (i)}\\
&= R_\alpha(k\cdot v)(a)\,.
\end{aligned}$$

This completes the proof.

Since the resolvent equation

$$R_\alpha^k - R_\beta^k + (\alpha-\beta)R_\alpha^k R_\beta^k = 0, \qquad \alpha,\beta > \|k^+\|,$$

follows from the group property $H_{t+s}^k = H_t^k H_s^k$, we can see that

$$\mathfrak{R}_{\mathbf{B}}^k \equiv R_\alpha^k \mathbf{B}(S) \quad\text{and}\quad \mathfrak{N}^k \equiv \{f\in\mathbf{B}(S)\colon R_\alpha^k f = 0\}$$

are both independent of $\alpha > \|k^+\|$. The generator $A^k$ of the Kac semi-group $\{H_t^k\}_t$ is therefore defined in the same way as in Section 2.16, namely

$$\mathfrak{D}(A^k) = \mathfrak{R}_{\mathbf{B}}^k, \qquad A^k u = \alpha u - (R_\alpha^k)^{-1} u \pmod{\mathfrak{N}^k}.$$

Then we have

**Theorem 3.** $\mathfrak{D}(A^k) = \mathfrak{D}(A)$, $\mathfrak{N}^k = \mathfrak{N}$, *and* $A^k = A+k$. *(Here* $A^k = A+k$ *means* $A^k u = Au + ku$*).*

*Proof.* By Theorem 2 we have

$$\begin{aligned} R_\alpha f &= R_\alpha^k (f - kR_\alpha f) \in \mathfrak{D}(A^k) \qquad &\text{by (i)},\\ R_\alpha^k f &= R_\alpha (f + kR_\alpha^k f) \in \mathfrak{D}(A) \qquad &\text{by (ii)}, \end{aligned}$$

for $\alpha > \|k^+\|$ and so

$$\mathfrak{D}(A^k) = \mathfrak{D}(A)\,.$$

The equality $\mathfrak{N}^k = \mathfrak{N}$ also follows from Theorem 1.

Suppose $u \in \mathfrak{D}(A)$ and set $u = R_\alpha f$. Then

$$u = R_\alpha^k (f - ku) \in \mathfrak{D}(A^k)$$

and so

$$\begin{aligned} A^k u &= \alpha u - (f - ku)\\ &= \alpha u - f + ku\\ &= Au + ku\,. \end{aligned}$$

The proof is complete.

As an application of Kac semi-groups we will prove the *arcsine law* for the Brownian motion.

**Theorem 4 (P. Lévy).** *Let* $X_t$ *be a Brownian motion in one dimension. Then the Lebesgue measure* $\xi_t$ *of the set* $\{s \in [0,t] : X_s > 0\}$ *has the following probability law (arcsine law):*

$$P_0(\xi_t \in \mathrm{d}s) = \frac{\mathrm{d}s}{\pi\sqrt{s(t-s)}}$$

*namely*

$$P_0(\xi_t \le s) = \frac{2}{\pi} \arcsin\sqrt{\frac{s}{t}}\,,$$

*where* $s \in [0,t]$.

*Proof.* Let us consider the Brownian motion $X_t$ as a Markov process and set

$$k(a) = k_\beta(a) = \begin{cases} -\beta \quad (\beta > 0) & \text{if } a > 0 \\ 0 & \text{if } a \leq 0 \,. \end{cases}$$

Then

$$K_t \equiv \int_0^t k(X_s)\mathrm{d}s = -\beta\xi_t \,,$$

and

$$\begin{aligned} R_\alpha^k \cdot 1(0) &= E_0\left(\int_0^\infty \mathrm{e}^{-\alpha t}\mathrm{e}^{K_t}\mathrm{d}t\right) \\ &= \int_0^\infty \mathrm{e}^{-\alpha t} E_0(\mathrm{e}^{-\beta\xi_t})\mathrm{d}t \\ &= \int_0^\infty \mathrm{e}^{-\alpha t} \int_0^\infty \mathrm{e}^{-\beta s} P_0(\xi_t \in \mathrm{d}s)\mathrm{d}t \,. \end{aligned}$$

If we compute $u(a) = u_{\alpha,\beta}(a) = R_\alpha^k \cdot 1(0)$ with $k = k_\beta$ at $a = 0$, then we can obtain $P_0(\xi_t \in \mathrm{d}s)$ by inverse Laplace transformation. By Theorem 3 we have

$$\frac{1}{2}\partial^2 u + ku = A^k u = \alpha u - 1 \,,$$

where $\partial^2$ is the second order derivative in the Schwartz distribution sense. The equation implies the following:

(i) $u \in C^1$ and $u'$ is absolutely continuous.

(ii) $u''$ (= the Radon–Nikodym derivative of $u'$) has a version which satisfies

$$\frac{1}{2}u'' + ku = \alpha u - 1, \quad \text{i.e.} \quad (\alpha - k)u - \frac{1}{2}u'' = 1 \,,$$

and so

$$\begin{aligned} (\alpha + \beta)u - \frac{1}{2}u'' &= 1 \qquad \text{in } (0, \infty) \,, \\ \alpha u - \frac{1}{2}u'' &= 1 \qquad \text{in } (-\infty, 0) \,. \end{aligned}$$

Since $u$ is continuous, $u$ is $C^2$ on each of the two half lines $(0, \infty)$ and $(-\infty, 0)$. The general solution of these equations is

$$u(a) = \begin{cases} \frac{1}{\alpha+\beta} + A_1\mathrm{e}^{-\sqrt{2(\alpha+\beta)}a} + A_2\mathrm{e}^{\sqrt{2(\alpha+\beta)}a} & a > 0 \\ \frac{1}{\alpha} + B_1\mathrm{e}^{-\sqrt{2\alpha}a} + B_2\mathrm{e}^{\sqrt{2\alpha}a} & a < 0 \,. \end{cases}$$

Since $u \in C^1$, we have

$$u(0+) = u(0-), \qquad u'(0+) = u'(0-)$$

namely

$$\frac{1}{\alpha+\beta} + A_1 + A_2 = \frac{1}{\alpha} + B_1 + B_2$$
$$-\sqrt{\alpha+\beta}A_1 + \sqrt{\alpha+\beta}A_2 = -\sqrt{\alpha}B_1 + \sqrt{\alpha}B_2 \,.$$

Since $u$ is bounded,

$$A_2 = 0, \qquad B_1 = 0 \,.$$

Solving these linear equations, we have

$$u(0) = \frac{1}{\alpha} + B_1 + B_2 = \frac{1}{\sqrt{\alpha(\alpha+\beta)}} \,,$$

namely

$$\int_0^\infty \mathrm{e}^{-\alpha t} \int_0^\infty \mathrm{e}^{-\beta s} P_0(\xi_t \in \mathrm{d}s)\mathrm{d}t = \frac{1}{\sqrt{\alpha(\alpha+\beta)}} \,.$$

By elementary computation we find

$$\begin{aligned}
&\int_0^\infty \mathrm{e}^{-\alpha t} \int_0^\infty \frac{1}{\sqrt{\pi(t-s)}} \frac{1}{\sqrt{\pi s}} 1_{[0,t]}(s) \mathrm{e}^{-\beta s} \mathrm{d}s\, \mathrm{d}t \\
&= \int_0^\infty \mathrm{e}^{-\beta s} \frac{\mathrm{d}s}{\sqrt{\pi s}} \int_s^\infty \mathrm{e}^{-\alpha t} \frac{1}{\sqrt{\pi(t-s)}} \mathrm{d}t \\
&= \int_0^\infty \mathrm{e}^{-\beta s} \frac{\mathrm{d}s}{\sqrt{\pi s}} \int_0^\infty \mathrm{e}^{-\alpha(t+s)} \frac{1}{\sqrt{\pi t}} \mathrm{d}t \\
&= \int_0^\infty \mathrm{e}^{-(\alpha+\beta)s} \frac{\mathrm{d}s}{\sqrt{\pi s}} \int_0^\infty \mathrm{e}^{-\alpha t} \frac{1}{\sqrt{\pi t}} \mathrm{d}t \\
&= \frac{1}{\sqrt{\alpha+\beta}} \frac{1}{\sqrt{\alpha}} \\
&= \int_0^\infty \mathrm{e}^{-\alpha t} \int_0^\infty \mathrm{e}^{-\beta s} P_0(\xi_t \in \mathrm{d}s) \mathrm{d}t \,.
\end{aligned}$$

By the uniqueness of the inverse Laplace transformation we have

$$P_0(\xi_t \in \mathrm{d}s) = \frac{1}{\pi\sqrt{s(t-s)}} \mathrm{d}s, \qquad s \in [0,t] \,,$$

namely

$$P_0(\xi_t \le s) = \int_0^s \frac{\mathrm{d}u}{\pi\sqrt{u(t-u)}} = \frac{2}{\pi} \arcsin\sqrt{\frac{s}{t}} \,.$$

## 2.18 Markov Processes and Potential Theory

Let $X$ be a Markov process in the generalized sense. Let $f \in \mathbf{B}(S)$ and suppose that

$$(1) \qquad E_a\left(\int_0^\infty |f(X_t)|\mathrm{d}t\right) \qquad \text{is bounded in } a.$$

Then

$$(2) \qquad Uf(a) = E_a\left(\int_0^\infty f(X_t)\mathrm{d}t\right)$$

is well-defined, and is a bounded $\mathcal{B}(S)$-measurable function of $a \in S$. This $Uf$ is called the *potential* of $f$ with respect to $X$. Since $Uf = \lim_{\alpha\downarrow 0} R_\alpha f$, it is reasonable to write $R_0$ for $U$. In view of this fact $R_\alpha f$ is called the potential of order $\alpha$ of $f$. Notice that $R_\alpha f \in \mathbf{B}(S)$ for $\alpha > 0$ and general $f \in \mathbf{B}(S)$ while $R_0 f (= Uf) \in \mathbf{B}(S)$ under the condition (1).

The name "potential" is justified by the following fact on the 3-dimensional Brownian motion.

**Theorem 1.** *Let $X$ be the* 3*-dimensional Brownian motion. If $f \in \mathbf{B}(S)$ has compact support, then $f$ satisfies (1) and*

$$Uf(a) = \frac{1}{2\pi}\int_{R^3} \frac{f(b)\mathrm{d}b}{|b-a|} = \frac{1}{2\pi} \times \textit{Newtonian potential of } f\,.$$

*Proof.* First we will prove that $f$ satisfies (1). We have

$$\begin{aligned}
E_a\left(\int_0^\infty |f(X_t)|\mathrm{d}t\right) &= \int_0^\infty E_a(|f(X_t)|)\mathrm{d}t \\
&= \int_0^\infty \int_{R^3} \frac{1}{(2\pi t)^{3/2}} \mathrm{e}^{-|b-a|^2/(2t)} |f(b)|\mathrm{d}b\,\mathrm{d}t \\
&= \int_{R^3} \int_0^\infty \frac{1}{(2\pi t)^{3/2}} \mathrm{e}^{-|b-a|^2/(2t)} \mathrm{d}t\,|f(b)|\mathrm{d}b \\
&= \int_{R^3} \frac{1}{2|b-a|\pi^{3/2}} \int_0^\infty \mathrm{e}^{-s} s^{-1/2}\mathrm{d}s\,|f(b)|\mathrm{d}b \qquad (\,t = |b-a|^2/(2s)\,) \\
&= \int_{R^3} \frac{1}{2|b-a|\pi^{3/2}} \Gamma\left(\frac{1}{2}\right) |f(b)|\mathrm{d}b \\
&= \int_{R^3} \frac{|f(b)|\mathrm{d}b}{2\pi|b-a|} = \frac{1}{2\pi}(I_1 + I_2)\,,
\end{aligned}$$

where

$$I_1 = \int_{|b-a|\le 1} \frac{|f(b)|db}{|b-a|} \le \|f\| \int_{|b-a|\le 1} \frac{db}{|b-a|}$$

$$= \|f\| 4\pi \int_0^1 \frac{r^2 dr}{r} = 2\pi \|f\| < \infty ,$$

$$I_2 = \int_{|b-a|>1} \frac{|f(b)|db}{|b-a|} \le \|f\| \int_{\text{supp } f} db < \infty .$$

The proof of the identity can be carried out in the same way. Notice that Fubini's theorem can be used by the integrability condition verified above.

**Theorem 2.** *Let $f$ satisfy (1) and let $u = R_0 f$. Then $u \in \mathfrak{D}(A)$ and*

(3) $$Au = -f \quad (\textit{Poisson equation}).$$

*Remark 1.* We have seen that $u_\alpha = R_\alpha f$ satisfies

$$(\alpha - A)u_\alpha = f$$

and our theorem is the limiting case $\alpha = 0$. Notice that (3) holds under the condition (1).

*Remark 2.* Let $X$ be the 3-dimensional Brownian motion. Then

$$u = \frac{1}{2\pi} \int_{R^3} \frac{f(b)}{|b-a|} db ,$$

$$Au = \frac{1}{2} \Delta u \qquad \text{(Schwartz distribution sense)}.$$

Therefore (3) is equivalent to the Poisson equation

$$\Delta v = -4\pi f ,$$

if $v$ is the Newtonian potential of $f$, i. e. if

$$v(a) = \int_{R^3} \frac{f(b)}{|b-a|} db .$$

This is why we call (3) the Poisson equation.

*Proof of Theorem 2.* By the resolvent equation we have

$$R_\alpha f - R_\beta f + (\alpha - \beta) R_\alpha R_\beta f = 0 .$$

Since $R_\beta f \to R_0 f \equiv u$ boundedly as $\beta \downarrow 0$, we have

$$R_\alpha f - u + \alpha R_\alpha u = 0$$

and so

$$u = R_\alpha(\alpha u + f) \in \mathfrak{D}(A)$$

and

$$Au = \alpha u - (\alpha u + f) = -f .$$

## 2.19 Brownian Motion and the Dirichlet Problem

Let $D$ be a bounded domain (= connected open set) in $R^k$, $k \geq 1$. A function $u$ is called *harmonic* in $D$ if $u$ is $C^\infty$ in $D$ and if $\Delta u = 0$. Let $f$ be a continuous function defined on the boundary $\partial D$. The Dirichlet problem $(D, f)$ is to find a continuous function $u = u_{D,f}$ on the closure $\overline{D} \equiv D \cup \partial D$ such that $u$ is harmonic in $D$ and $u = f$ on $\partial D$.

Let $X$ be the $k$-dimensional Brownian motion. If there exists a solution $u$ for the Dirichlet problem $(D, f)$, then

$$(1) \qquad u(a) = E_a(f(X_\tau)), \qquad \tau \equiv \tau_D = \text{ exit time from } D,$$ [38]

as we will prove in Theorem 2. If $\partial D$ is smooth, then the Dirichlet problem $(D, f)$ has a solution (Theorem 4).

**Theorem 1.** *If $D$ is a bounded domain, then*

$$P_a(\tau_D < \infty) = 1 \quad \textit{and} \quad E_a(\tau_D) < \infty,$$

*where $\tau_D$ is the exit time from $D$.*

*Proof.* The first identity follows from the second one at once. Because of $p_t(a, \mathrm{d}b) = N_t(b - a)\mathrm{d}b$, $\{X_t\}_{t \in T}$ under $P_a$ is law-equivalent to $\{a + X_t\}_{t \in T}$ under $P_0$, and so $E_a(\tau_D) = E_0(\tau_{D-a})$. Thus it is enough to consider the case $a = 0$.

If $0 \notin \overline{D}$, then $P_0(\tau_D = 0) \equiv 1$ and $E_0(\tau_D) = 0$ and so our theorem holds trivially. Therefore we consider the case $0 \in \overline{D}$.

Take a sufficiently large $r > 0$ such that $D \subset U \equiv U(0, r)$. Then $P_0(\tau_D \leq \tau_U) = 1$. Therefore $E_0(\tau_D) \leq E_0(\tau_U)$. For $u \in L^\infty \cap C^\infty$ in $R^k$ with $\Delta u \in L^\infty$, we have $u \in \mathfrak{D}(A)$ and

$$Au = \frac{1}{2}\Delta u\,.$$

By Dynkin's formula it holds that

$$(2) \qquad E_0\left(\int_0^{\tau \wedge n} \frac{1}{2}\Delta u(X_t)\mathrm{d}t\right) = E_0(u(X_{\tau \wedge n})) - u(0) \qquad (\tau = \tau_U).$$

Setting $u(a) = -\exp\{-|a|^2/(4r^2)\}$ in (2), observing

$$\frac{1}{2}\Delta u = \frac{1}{2}\left[\frac{k}{2r^2} - \frac{|a|^2}{4r^4}\right] \mathrm{e}^{-|a|^2/(4r^2)}$$

and noticing that the minimum of the right-hand side on $\{|a| \leq r\}$ is positive, say $m$, we have

$$mE_0(\tau \wedge n) \leq 2\,.$$

Letting $n \uparrow \infty$, we obtain $E_0(\tau) \leq 2/m < \infty$.

[38] That is, $\tau_D = \inf\{t > 0\colon X_t \notin D\}$, the hitting time of $D^C$.

**Theorem 2.** *If $D$ is a bounded domain and $u$ is a solution of the Dirichlet problem $(D, f)$, then*

$$u(a) = E_a(f(X_\tau)), \qquad a \in D,\ \tau = \tau_D .$$

*Proof.* Let $D_n$ denote the domain $\left\{b \in D: \overline{U(b, 1/n)} \subset D\right\}$. Then $U(D_n, 1/n) \subset D$. Consider a $C^\infty$ function

$$\varphi_n(a) = \begin{cases} C_n \exp\left\{-\dfrac{1}{1 - n^2|a|^2}\right\} & |a| < 1/n \\ 0 & \text{elsewhere,} \end{cases}$$

where the constant factor $C_n$ is determined by

$$\int_{R^k} \varphi_n(a)\mathrm{d}a = 1 .$$

Then

$$e_n = \varphi_{2n} * 1_{\overline{D}_{2n}} \qquad (1_B = \text{indicator of } B)$$

is a $C^\infty$ function with the support $\subset D$ and

$$e_n = 1 \qquad \text{in } \overline{D}_n .$$

Now set

$$u_n = \begin{cases} u \cdot e_n & \text{in } D \\ 0 & \text{elsewhere.} \end{cases}$$

Then $u_n \in C^\infty \cap L^\infty$ and $\Delta u \in L^\infty$. Therefore $u_n \in \mathfrak{D}(A)$ and

$$Au_n - \frac{1}{2}\Delta u_n .$$

Noticing that $E_a(\tau_{D_n}) < \infty$ by Theorem 1, we have by Dynkin's formula

$$\text{(3)} \qquad u_n(a) - E_a(u_n(X_{\tau_n})) - E_a\left(\int_0^{\tau_n} \frac{1}{2}\Delta u_n(X_t)\mathrm{d}t\right), \qquad \tau_n = \tau_{D_n} .$$

Observing $u_n = u$ on $\overline{D}_n$ and

$$P_a(X_t \in D_n \text{ for } t < \tau_n) = 1, \qquad a \in D_n ,$$
$$P_a(X_{\tau_n} \in \partial D_n) = 1, \qquad a \in D_n ,$$

we can write (3) as

$$\text{(4)} \qquad \begin{aligned} u(a) &= E_a(u(X_{\tau_n})) - E_a\left(\int_0^{\tau_n} \frac{1}{2}\Delta u(X_t)\mathrm{d}t\right) \\ &= E_a(u(X_{\tau_n})), \qquad a \in D_n , \end{aligned}$$

because $\Delta u = 0$ in $D_n \subset D$.

By the continuity of the path, $P_a(\tau_n \to \tau) = 1$ for $a \in D$. Since $u$ is continuous on $\overline{D}$, we have, for $a \in D$,

$$u(a) = \lim_{n\to\infty} E_a(u(X_{\tau_n})) = E_a(u(X_\tau)) = E_a(f(X_\tau)) .$$

**Theorem 3.** *Let $D$ be a bounded domain and let $f$ be a bounded Borel measurable function on $\partial D$. Then*

$$u(a) \equiv E_a(f(X_\tau)), \qquad \tau = \tau_D \tag{5}$$

*is harmonic in $D$*

*Proof.* First we will verify the mean value property:

$$u(a) = \frac{1}{S(a,r)} \int_{\partial U(a,r)} u(b) \mathrm{d}S(b) \qquad \text{if } \overline{U(a,r)} \subset D , \tag{6}$$

where $\mathrm{d}S(b)$ is the surface element on the sphere $\partial U(a,r)$ and $S(a,r)$ is its total area.

Let $\sigma = \sigma_r$ be the exit time from the open ball $U(a,r)$. Then

$$P_a(\sigma < \tau) = 1 \quad \text{and so} \quad P_a(\tau = \sigma + \tau \circ \theta_\sigma) = 1 .$$

By the strong Markov property

$$\begin{aligned} u(a) &= E_a(f(X_{\sigma+\tau\circ\theta_\sigma})) = E_a(f((X_\tau) \circ \theta_\sigma)) \\ &= E_a(E_{X_\sigma}(f(X_\tau))) = E_a(u(X_\sigma)) \\ &= \int_{\partial U(a,r)} u(b) P_a(X_\sigma \in \mathrm{d}b) . \end{aligned}$$

For the proof of (6) it is enough to show

$$P_a(X_\sigma \in E) = \frac{1}{S(a,r)} \int_E \mathrm{d}S(b) \tag{7}$$

for all Borel subsets $E$ of the sphere. Let $O$ be a rotation around $a \in R^3$. Since $N_t(b)$ depends only on $|b|$ and $t$, we can easily see that $P_a$ is $O$-invariant,

$$P_a(X_\sigma \in E) = P_a(X_\sigma \in OE) .$$

This proves that $P_a(X_\sigma \in E)$ is the uniform distribution on the sphere. This implies (7).

Next we will prove that $u \in C^\infty$ in $D$. We will use $\varphi_n$, $D_n$, $u_n$ etc. in the proof of Theorem 2. Extend $u$ onto the whole space $R^k$ by putting $u = 0$ outside of $D$. Observing the rotation invariance of $\varphi_n$ and the mean value property (6) of $u$ we have

$$u * \varphi_n = u \qquad \text{in } D_n .$$

But the left side is $C^\infty$ in $R^k$ because $\varphi_n$ is $C^\infty$ and has compact support. Thus $u$ is $C^\infty$ in $D_n$ and so $u$ is $C^\infty$ in $D$ because $n$ is arbitrary. Then $u_n$ is $C^\infty$ in $R^k$ and has compact support. Therefore $u_n \in \mathfrak{D}(A)$ and $Au_n = \frac{1}{2}\Delta u_n$. By Dynkin's formula we have

$$E_a\left(\int_0^{\sigma_r} \frac{1}{2}\Delta u(X_t)\mathrm{d}t\right) = E_a(u(X_{\sigma_r})) - u(a) = 0 \tag{8}$$

as long as $U(a,r) \subset D_n$, since $u_n = u$ on $\overline{D}_n$ . As we can choose $n$ arbitrarily large, (8) holds as long as $\overline{U(a,r)} \subset D$.

Suppose $\Delta u(a) > 0$ at some point $a \in D$. Take $r > 0$ such that $\overline{U(a,r)} \subset D$ and

$$\Delta u(b) > \frac{1}{2}\Delta u(a) \qquad \text{for } b \in U(a,r)\,.$$

Then

$$0 < \frac{1}{2}\Delta u(a) E_a(\sigma_r) \le E_a\left(\int_0^{\sigma_r} \frac{1}{2}\Delta u(X_t)\mathrm{d}t\right) = 0\,,$$

which is a contradiction. Therefore $\Delta u \le 0$ in $D$. Similarly $\Delta u \ge 0$ in $D$ and so $\Delta u = 0$ in $D$. This completes the proof.

**Theorem 4.** *If $\partial D$ is smooth, namely if $\partial D$ has a unique tangent plane at each point $\xi$ of $\partial D$ and the outward unit normal of the tangent plane at $\xi$ moves continuously with $\xi$, then*

$$u(a) = E_a(f(X_\tau)), \qquad \tau = \tau_D = \text{ exit time from } D,$$

*is the solution of the Dirichlet problem $(D, f)$.*

*Proof.* It is enough by Theorem 3 to prove that for every $\xi \subset \partial D$

$$\lim_{a \in D,\ a \to \xi} u(a) = f(\xi) = u(\xi)\,. \tag{9}$$

Let $e_1, e_2, \ldots, e_k$ be an orthonormal base in $R^k$ such that $e_1$ is the direction of the outward normal of $\partial D$ at $\xi$. Then there exist two sequences $\{\varepsilon_n\}$ and $\{\delta_n\}$ such that

$$\varepsilon_n \downarrow 0, \quad \delta_n \downarrow 0, \quad 1/2 > \delta_n/\varepsilon_n \downarrow 0 \tag{10}$$

and that

(11) the $(k-1)$-dimensional square $(x - \xi, e_1) = \delta_n,\ |(x - \xi, e_i)| \le \varepsilon_n$ $(i = 2, 3, \ldots, k)$ does not meet $\partial D$.

Consider the $2k$ hyperplanes

$$\begin{aligned} &\alpha_1 \colon (x-\xi, e_1) = \delta_n \\ &\beta_1 \colon (x-\xi, e_1) = -\varepsilon_n \\ &\alpha_i \colon (x-\xi, e_i) = \varepsilon_n \qquad (i = 2, 3, \ldots, k) \\ &\beta_i \colon (x-\xi, e_i) = -\varepsilon_n \qquad (i = 2, 3, \ldots, k) \end{aligned}$$

and let $U_n$ be the rectangular domain bounded by these hyperplanes.

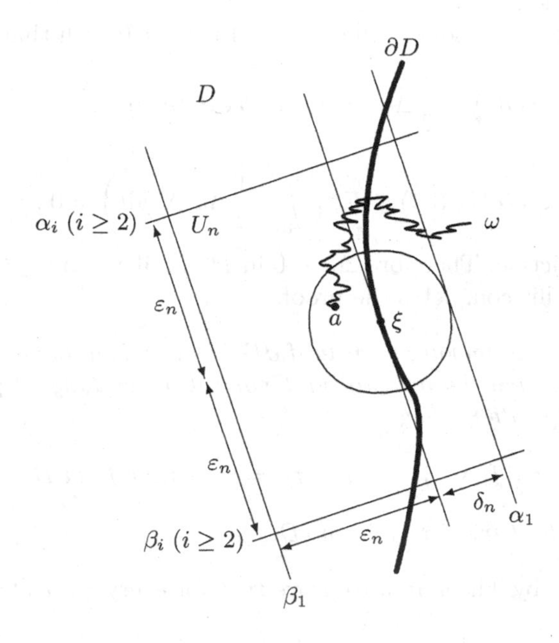


Let $\omega$ be an arbitrary path starting at $\omega(0) = a \in U_n$ and consider the hitting times $\sigma_1$, $\tau_1$, $\sigma_2$, $\tau_2$, $\ldots$, $\sigma_k$, $\tau_k$ of $\alpha_1$, $\beta_1$, $\alpha_2$, $\beta_2$, $\ldots$, $\alpha_k$, $\beta_k$ respectively. Observe the following conditions on such a path $\omega$ :

$$X(\tau, \omega) \in U_n \cap \partial D \,, \tag{12}$$

$$\sigma_1(\omega) < \min\{\tau_1(\omega), \sigma_2(\omega), \tau_2(\omega), \ldots, \sigma_k(\omega), \tau_k(\omega)\} \,, \tag{13}$$

$$\begin{cases} \max_{0 \le s \le t_n} (X_s(\omega) - a, \ e_1) > 2\delta_n \,, \\ \max_{0 \le s \le t_n} (X_s(\omega) - a, -e_1) \le \varepsilon_n/2 \,, \\ \max_{0 \le s \le t_n} (X_s(\omega) - a, \ e_i) \le \varepsilon_n/2 & (i = 2, 3, \ldots, k), \\ \max_{0 \le s \le t_n} (X_s(\omega) - a, -e_i) \le \varepsilon_n/2 & (i = 2, 3, \ldots, k), \end{cases} \tag{14}$$

where $t_n = \varepsilon_n \delta_n$. Then

$$(14) \Longrightarrow (13) \Longrightarrow (12) \qquad \text{if } a \in U(\xi, \delta_n) \cap D \,.$$

Thus for $a \in U(\xi, \delta_n) \cap D$ we have

$$\begin{aligned}
&P_a(X_\tau \in U_n) \geq P_a(\text{the event } (14)) \\
&\geq P_a\Big(\max_{0\leq s\leq t_n}(X_s - a, e_1) > 2\delta_n\Big) - P_a\Big(\max_{0\leq s\leq t_n}(X_s - a, e_1) > \frac{\varepsilon_n}{2}\Big) \\
&- \sum_{i=2}^{k} P_a\Big(\max_{0\leq s\leq t_n}(X_s - a, e_i) > \frac{\varepsilon_n}{2}\Big) - \sum_{i=2}^{k} P_a\Big(\max_{0\leq s\leq t_n}(X_s - a, e_i) > \frac{\varepsilon_n}{2}\Big).
\end{aligned}$$

By observing the joint distributions, we can easily see that $(X_t - a, e_i)$ and $(X_t - a, -e_i)$, $i = 1, 2, \ldots, k$, are all one-dimensional Brownian motions (starting at 0). Therefore if $a \in U(\xi, \delta_n) \cap D$, then[39]

$$\begin{aligned}
&P_a(X_\tau \in U_n) \\
&\geq \frac{2}{\sqrt{2\pi t_n}} \int_{2\delta_n}^{\infty} \mathrm{e}^{-b^2/(2t_n)} \mathrm{d}b - (2k-1)\frac{2}{\sqrt{2\pi t_n}} \int_{\varepsilon_n/2}^{\infty} \mathrm{e}^{-b^2/(2t_n)} \mathrm{d}b \\
&= \frac{2}{\sqrt{2\pi}} \int_{2\sqrt{\delta_n/\varepsilon_n}}^{\infty} \mathrm{e}^{-b^2/2} \mathrm{d}b - (2k-1)\frac{2}{\sqrt{2\pi}} \int_{\sqrt{\varepsilon_n/\delta_n}/2}^{\infty} \mathrm{e}^{-b^2/2} \mathrm{d}b \\
&\to 1, \qquad n \to \infty.
\end{aligned}$$

By taking $n_0 = n_0(\varepsilon)$ large enough we have

$$P_a(X_\tau \in U_n) > 1 - \varepsilon \qquad \text{for } a \in U(\xi, \delta_n) \cap D \text{ and } n \geq n_0 \tag{15}$$

and

$$|f(\eta) - f(\xi)| < \varepsilon \qquad \text{for } \eta \in U_n \cap \partial D\,. \tag{16}$$

Therefore, if $a \in U(\xi, \delta_n) \cap D$ and $n \geq n_0$, then

$$\begin{aligned}
|u(a) - f(\xi)| &\leq E_a[|f(X_\tau) - f(\xi)|, X_\tau \in U_n] + E_a[|f(X_\tau) - f(\xi)|, X_\tau \notin U_n] \\
&\leq \varepsilon + 2M P_a(X_\tau \notin U_n) \\
&\leq \varepsilon + 2M\varepsilon
\end{aligned}$$

where $M \equiv \sup_{\xi \in \partial D} |f(\xi)|$. Thus $u(a) \to f(\xi)$ as $a (\in D) \to \xi$. The proof is complete.[40]

As an application of Theorem 4, we will obtain

[39] Use the result of Problem 2.7 in Exercises.

[40] Since the event (14) is included by the event $\{\tau < t_n\}$, the proof of (15) shows that $P_a(\tau < t_n) > 1 - \varepsilon$ for $a \in U(\xi, \delta_n) \cap D$ and $n \geq n_0$. The same argument shows that $P_\xi(\tau < t_n) > 1 - \varepsilon$ for $n \geq n_0$. Therefore $P_\xi(\tau = 0) = 1$ and we have $u(\xi) = f(\xi)$.

**Theorem 5.** *Let $D$ be a ring domain bounded by*

$$\Gamma_1\colon |x-a_0| = r_1 \quad \textit{and} \quad \Gamma_2\colon |x-a_0| = r_2$$

*with $r_1 < r_2$. Let $\sigma_i$ be the hitting time of $\Gamma_i$ for $i = 1, 2$. Then for $a \in D$ we have*

$$(17) \qquad P_a(\sigma_1 < \sigma_2) = \begin{cases} \dfrac{|a-a_0|^{-(k-2)} - {r_2}^{-(k-2)}}{{r_1}^{-(k-2)} - {r_2}^{-(k-2)}} & k \geq 3 \\ \dfrac{\log(1/|a-a_0|) - \log(1/r_2)}{\log(1/r_1) - \log(1/r_2)} & k = 2 \\ \dfrac{r_2 - |a-a_0|}{r_2 - r_1} & k = 1\,. \end{cases}$$

*Proof.* Set

$$f = \begin{cases} 1 & \text{on } \Gamma_1 \\ 0 & \text{on } \Gamma_2\,. \end{cases}$$

Then $f$ is continuous on $\partial D (= \Gamma_1 \cup \Gamma_2)$. Then

$$u(a) = P_a(\sigma_1 < \sigma_2) = E_a(f(X_{\tau_D}))\,,$$

because $\tau_D$ ($=$ exit time from $D$) $= \sigma_1 \wedge \sigma_2$. Therefore $u(a)$ is the solution of the Dirichlet problem

$$\Delta u = 0 \quad \text{in } D, \qquad \lim_{a \in D,\, a \to \xi} u(a) = \begin{cases} 1 & \xi \in \Gamma_1 \\ 0 & \xi \in \Gamma_2\,. \end{cases}$$

It is easy to verify that the right-hand side of (17) satisfies these conditions.

**Theorem 6.** *Let $\sigma$ be the hitting time of the ball $U = U(a_0, r) \subset R^k$. Then*

$$P_a(\sigma < \infty) \equiv 1 \quad \textit{for } a \in R^k \quad \textit{if } k \leq 2\,,$$

$$P_a(\sigma < \infty) = \begin{cases} 1 & \textit{for } a \in U \\ (r/|a-a_0|)^{k-2} & \textit{for } a \in U^{\mathrm{C}} \end{cases} \quad \textit{if } k \geq 3\,.$$

*Proof.* Let $r_2 \to \infty$ in Theorem 5.

*Remark.* Theorem 6 is often stated as follows: the $k$-dimensional Brownian motion is *transient* or *recurrent* according as $k \geq 3$ or $k \leq 2$.

# Exercises

## E.0 Chapter 0

**0.1.** Let $\varphi(\xi)$ be a characteristic function in one dimension and $\mu$ the corresponding probability distribution.

(i) Prove that $\overline{\varphi(\xi)}$, $|\varphi(\xi)|^2$, $\varphi(\xi)^n$ $(n = 1, 2, \ldots)$ and $e^{ib\xi}\varphi(a\xi)$ are characteristic functions and find the corresponding distributions.

(ii) Prove that if $g(t)$ is the generating function of a distribution on $\{0, 1, 2, 3, \ldots\}$, then $g(\varphi(\xi))$ is also a characteristic function. Find the corresponding distribution.

(iii) Prove the inequality

$$1 - |\varphi(2\xi)|^2 \leq 4(1 - |\varphi(\xi)|^2) \, .$$

**0.2.** Let $X_1, X_2, \ldots$ be independent random variables on a probability space $(\Omega, \mathcal{B}, P)$. Show that

$$X_1, \; X_2 + X_3, \; X_4 + X_5 + X_6, \; \ldots$$

are independent.

**0.3.** Let $\mathcal{C}$ be a multiplicative class, $\mathcal{B}[\mathcal{C}]$ the $\sigma$-algebra generated by $\mathcal{C}$ and $\mathcal{D}[\mathcal{C}]$ the minimal Dynkin class including $\mathcal{C}$. Show that $\mathcal{B}[\mathcal{C}] = \mathcal{D}[\mathcal{C}]$.

**0.4.** Let $X_n$, $n = 1, 2, 3, \ldots$, be independent random variables. Show that there exists $c \in [-\infty, +\infty]$ such that

$$\limsup_{n \to \infty} \frac{X_1 + \cdots + X_n}{n} = c \qquad \text{a. s.}$$

**0.5.** Use Theorem 0.1.2 to prove that if

$$P(X \leq a, \; Y \leq b) = P(X \leq a)\, P(Y \leq b), \qquad a, b \in R^1 \, ,$$

then $X$ and $Y$ are independent.

**0.6.** Prove that if two $n$-dimensional probability measures are equal for all half-spaces, then they are equal.

Hint. Use characteristic functions.

**0.7.** Let $\mathfrak{D}$ be the set of all one-dimensional distribution functions. Verify that

$$\rho(F,G) = \inf\{\delta > 0\colon F(x-\delta) - \delta \le G(x) \le F(x+\delta) + \delta \text{ for all } x\}$$

is a metric on $\mathfrak{D}$ and that $\rho(F_n, F) \to 0$ if and only if $F_n(x) \to F(x)$ at all continuity points of $F$. (This $\rho$ is called the Lévy metric).

**0.8.** Let $\mathcal{L}^2$ be the space of all square integrable real random variables defined on a fixed probability space. If we restrict ourselves to this space, the theory of centralized sums will be rather simple. The centralized sum of $X_1, X_2, \ldots, X_n \in \mathcal{L}^2$ is defined to be $\sum_{1\le i\le n} X_i - m\left(\sum_{1\le i\le n} X_i\right)$. This is denoted by $\sum^{\Delta}_{1\le i\le n} X_i$. It is clear that

$$\begin{aligned}\sum_{1\le i\le n}^{\Delta} X_i &= \sum_{1\le i\le n}^{\circ} X_i + \gamma\left(\sum_{1\le i\le n} X_i\right) - m\left(\sum_{1\le i\le n} X_i\right)\\ &= \sum_{1\le i\le n}^{\circ} X_i + \text{const}\,,\end{aligned}$$

where $\sum^{\circ}_{1\le i\le n} X_i$ is the centralized sum introduced in Section 0.3.

(i) Define the infinite centralized sum $\sum^{\Delta}_{\alpha\in A} X_\alpha$ by the same idea as $\sum^{\circ}_{\alpha\in A} X_\alpha$ was defined.

Hint. Use $\sigma$ instead of $\delta$. Note that $m(X+Y) = m(X) + m(Y)$.

(ii) Prove that $\sum^{\Delta}_{\alpha\in A} X_\alpha = \sum^{\circ}_{\alpha\in A} X_\alpha + \text{const}$ provided the left side is meaningful.

**0.9.** Prove Theorem 0.3.4 and also similar facts for $\sum^{\Delta}$.

**0.10.** Let $X_1, X_2, \ldots$ be independent and identically distributed $(F)$ and let $N(t)$ be a Poisson process with parameter $\lambda > 0$. Suppose that $\mathcal{B}[X_1, X_2, \ldots]$ and $\mathcal{B}[N(t)\colon t \ge 0]$ are independent. Define $Y(t)$ to be $\sum_{i=1}^{N(t)} X_i$.[1]

(i) Find the characteristic function of the random variable $Y(t)$ for $t$ fixed.

(ii) Prove that, for $0 \le t_0 < t_1 < t_2 < \cdots < t_n$, $\{Y(t_i) - Y(t_{i-1})\}_{i=1}^{n}$ is independent.

**0.11.** Let $X_1, X_2, \ldots$ be independent and uniformly distributed on $[0,1]$ and $Y_n(t)$ be the empirical distribution of $X_1, X_2, \ldots, X_n$ i.e.

$$Y_n(t) = \frac{\text{the number of } X_1, \ldots, X_n \le t}{n}, \qquad 0 \le t \le 1.$$

Set $Z_n(t) = \sqrt{n}(Y_n(t) - t)$. Prove that $Z_n(t) \to B(t) - tB(1)$ in joint distribution, where $B(t)$ is a Brownian motion.[2]

---

[1] This process $Y(t)$ is called a compound Poisson process.

[2] The process $\{B(t) - tB(1)\}_{t\in[0,1]}$ is called a pinned Brownian motion.

Hint. Use the central limit theorem in several dimensions.

**0.12.** Let $X_n$, $n = 1, 2, 3, \ldots$, and $X$ be random variables on a probability space $(\Omega, \mathcal{B}, P)$. Prove that $X_n \to X$ i. p. if and only if

$$E\left(\frac{|X_n - X|}{1 + |X_n - X|}\right) \to 0\,.$$

**0.13.** Give a sequence of random variables $\{X_n\}_{n=1}^{\infty}$ such that

$$\limsup_{n\to\infty} X_n = 1, \quad \liminf_{n\to\infty} X_n = -1 \quad \text{a. s.}$$

but $X_n \to 0$ i. p. as $n \to \infty$.

**0.14.** Let $C$ be the space of complex-valued continuous functions defined on $R^1 \equiv (-\infty, \infty)$. Define $\|f\|_n$ and $d(f, g)$ for $f, g \in C$ by

$$\|f\|_n = \max_{|t|\leq n} |f(t)|\,, \quad d(f,g) = \sum_{n=1}^{\infty} 2^{-n} \frac{\|f - g\|_n}{1 + \|f - g\|_n}\,.$$

Prove the following.

(i) $d(f_m, f) \to 0$ if and only if $f_m$ converges to $f$ uniformly on every compact $t$-set.

(ii) This $d$ is a metric in $C$ and, with this metric, $C$ is a complete separable metric space.

(iii) (Ascoli–Arzelà theorem) $A$ subset $M$ of $C$ is conditionally compact if and only if the following two conditions are satisfied.

(a) $M$ is equi-bounded on each bounded $t$-interval, i. e.

$$\sup_{|t|\leq n, f\in M} |f(t)| < \infty \quad \text{for } n = 1, 2, \ldots$$

(b) $M$ is equi-continuous on every bounded $t$-interval, i. e.

$$\sup\{|f(t) - f(s)|\colon |t|, |s| \leq n,\ |t - s| < \delta,\ f \in M\} \to 0, \qquad \delta \to 0$$

for $n = 1, 2, \ldots$.

**0.15.** Let $\mathfrak{P}$ be the space of all one-dimensional distributions and $\widetilde{\mathfrak{P}}$ the space of all one-dimensional characteristic functions. Then $\mathfrak{P}$ is a metric space with the Lévy metric in Problem 0.7 (with identification of $\mathfrak{P}$ and $\mathfrak{D}$); $\widetilde{\mathfrak{P}}$ is a metric space with the metric $d$ in Problem 0.14 (restricted to $\widetilde{\mathfrak{P}}$). Prove that the map $\Phi$ that transforms $\mu \in \mathfrak{P}$ to its characteristic function $(\Phi\mu)(z) = \int_{-\infty}^{\infty} \mathrm{e}^{\mathrm{i}zt} \mu(\mathrm{d}t)$ is a homeomorphism of $\mathfrak{P}$ onto $\widetilde{\mathfrak{P}}$.

**0.16.** Use Problem 0.14(iii) to prove that a subset $\mathfrak{M}$ of $\widetilde{\mathfrak{P}}$ is conditionally compact if and only if $\mathfrak{M}$ is equicontinuous on some neighborhood of 0.

Hint. Observe the properties of characteristic functions: $|\varphi(z)| \leq 1$ and $|\varphi(z + h) - \varphi(z)| \leq \sqrt{2|\varphi(h) - 1|}$.

**0.17.** Use Problems 0.15 and 0.16 to prove that $\{\mu^{n*}\}$ is conditionally compact if and only if $\mu$ is the $\delta$-distribution concentrated at 0.

**0.18.** Prove that each of the following families of one-dimensional distributions is conditionally compact if and only if the parameters involved are bounded.

(i) A family of Gauss distributions.
(ii) A family of Cauchy distributions.
(iii) A family of Poisson distributions.

**0.19.** Let $X_n$ be an independent sequence of real random variables and $S_n = \sum_{k=1}^{n} X_k$. Then the following conditions are equivalent (Lévy's theorem).
(a) $S_n$ is convergent in law i.e. the probability law of $S_n$ is convergent.
(b) $S_n$ is convergent in probability.
(c) $S_n$ is convergent a. s.
Prove this fact using Theorem 0.3.2.

**0.20.** If $\mu$ is a probability distribution on $[0, \infty)$ such that $\mu\{0\} = 0$, then

$$\sum_{n=1}^{\infty} \alpha^n \mu^{n*}[0, t] < \infty$$

for every $t > 0$ and every $\alpha > 0$. Prove this.

Hint. Take $\beta$ such that $\alpha \int_0^\infty \mathrm{e}^{-\beta x} \mu(\mathrm{d}x) < 1$. Then consider the integral

$$I(\alpha, \beta) = \int_0^\infty \mathrm{e}^{-\beta t} \sum_{k=1}^{\infty} \alpha^k \mu^{k*}[0, t] \mathrm{d}t$$

and show that $I(\alpha, \beta)$ is finite for fixed $\alpha$ and $\beta$.

**0.21.** Prove that the Laplace transform of an infinitely divisible distribution with support $\subset [0, \infty)$ is of the form

$$\exp\left\{-m\alpha - \int_0^\infty (1 - \mathrm{e}^{-\alpha u})\, n(\mathrm{d}u)\right\} ,$$

where $\alpha$ is the parameter of the Laplace transform, $m \geq 0$ and $n$ is a measure on $(0, \infty)$ such that

$$\int_0^\infty \frac{u}{1+u} n(\mathrm{d}u) < \infty .$$

Show[3] that $n$ coincides with the Lévy measure in Lévy's formula (Section 0.4) of the characteristic function of $\mu$ and that $m$ coincides with the $m$ that appears in Theorem 1.11.3.

Hint. Use the same idea as in the proof of Lévy's formula. This problem is easier, because everything can be discussed in the field of real numbers.

[3] Added by the Editors.

**0.22.** Let $\varphi(z)$ be the characteristic function of a distribution $\mu$ on $R^1$. Prove that

$$\lim_{z\to 0} \frac{2-\varphi(z)-\varphi(-z)}{z^2} = \int_{-\infty}^{\infty} x^2 \mu(\mathrm{d}x) ,$$

no matter whether the right side is finite or not.

**0.23.** Use Problem 0.22 to prove the equivalence of the following three conditions.
(a) $\mu$ has finite second order moment.
(b) $\varphi$ is continuously twice differentiable.
(c) $\varphi$ is twice differentiable at 0.

**0.24.** Prove that $\mu * \nu$ has finite second order moment if and only if both $\mu$ and $\nu$ have the same property.

**0.25.** Prove that if the characteristic function of $\mu$ is of the form

$$\varphi(z) = \exp\left\{\int_{|u|\le 1} (\mathrm{e}^{\mathrm{i}zu} - 1 - \mathrm{i}zu)\, n(\mathrm{d}u)\right\} ,$$

where $n(\mathrm{d}u) \ge 0$ and $\int_{|u|\le 1} u^2 n(\mathrm{d}u) < \infty$, then $\mu$ has finite second order moment.

**0.26.** Prove that if the characteristic function of $\mu$ is of the form

$$\varphi(z) = \exp\left\{\int_{|u|>1} (\mathrm{e}^{\mathrm{i}zu} - 1)\, n(\mathrm{d}u)\right\} ,$$

where $n(\mathrm{d}u) \ge 0$ and $\int_{|u|>1} n(\mathrm{d}u) < \infty$, then a necessary and sufficient condition for $\mu$ to have finite second order moment is

$$\int_{|u|>1} u^2 n(\mathrm{d}u) < \infty .$$

**0.27.** Let $\mu$ be an infinitely divisible distribution with characteristic function

$$\varphi(z) = \exp\left\{\mathrm{i}mz - \frac{v}{2}z^2 + \int_{-\infty}^{\infty}\left(\mathrm{e}^{\mathrm{i}zu} - 1 - \frac{\mathrm{i}zu}{1+u^2}\right) n(\mathrm{d}u)\right\} .$$

Prove that a necessary and sufficient condition for $\mu$ to have finite second order moment is[4]

$$\int_{|u|>1} u^2 n(\mathrm{d}u) < \infty . \tag{1}$$

[4] For any positive real number $\lambda$ it is known that $\int_{R^1} |x|^\lambda \mu(\mathrm{d}x) < \infty$ if and only if $\int_{|x|>1} |x|^\lambda n(\mathrm{d}x) < \infty$. See Sato's book mentioned in the Foreword for a proof.

**0.28.** Under the condition (1) in Problem 0.27 or, more generally, under the condition that $\int_{|u|>1} |u| n(\mathrm{d}u) < \infty$, $\varphi(z)$ can be written as[5]

$$\varphi(z) = \exp\left\{ \mathrm{i} m' z - \frac{v}{2} z^2 + \int_{-\infty}^{\infty} (\mathrm{e}^{\mathrm{i}zu} - 1 - \mathrm{i}zu) n(\mathrm{d}u) \right\}$$

with some $m'$. Prove this.

**0.29.** If $\mu$ is an infinitely divisible distribution symmetric with respect to 0, then its characteristic function can be written as

$$\varphi(z) = \exp\left\{ -\frac{v}{2} z^2 + \int_{0+}^{\infty} (\cos zu - 1) \overline{n}(\mathrm{d}u) \right\} ,$$

where $v \geq 0$, $\overline{n}(\mathrm{d}u) \geq 0$ and $\int_{0+}^{\infty} u^2/(1+u^2) \overline{n}(\mathrm{d}u) < \infty$. Prove this.

Hint. First prove that $m = 0$ and $n$ is symmetric in Lévy's formula, using the uniqueness of the formula.

**0.30.** Prove that the following distributions are all infinitely divisible and find their canonical forms.

(i) Negative binomial distribution

$$\mu(\{n\}) = \binom{\lambda + n - 1}{n} \theta^n (1-\theta)^{\lambda}, \qquad n = 0, 1, 2, \ldots \quad (\lambda > 0,\ 0 < \theta < 1).$$

(ii) $\Gamma$-distribution

$$\mu(\mathrm{d}x) = \frac{1}{\Gamma(\lambda)} \mathrm{e}^{-x} x^{\lambda - 1} \mathrm{d}x, \qquad x \geq 0 \quad (\lambda > 0).$$

(iii) Cauchy distribution

$$\mu(\mathrm{d}x) = \frac{c}{\pi} \frac{\mathrm{d}x}{(x-m)^2 + c^2}, \qquad -\infty < x < +\infty \quad (c > 0,\ m \text{ real}).$$

Hint. In (i) first prove that the generating function $g(t) = (1-\theta)^{\lambda}(1 - \theta t)^{-\lambda}$. Then consider $\log L(\alpha)$, where $L$ denotes the Laplace transform of $\mu$ (the canonical form is given in Problem 0.21). For (ii) and (iii) prove that

$$\int_{0+}^{\infty} (\mathrm{e}^{-\alpha u} - 1) \frac{\mathrm{e}^{-u}}{u} \mathrm{d}u = -\log(1 + \alpha) \qquad (\alpha > 0)$$

and

$$\int_{0+}^{\infty} (\cos zu - 1) \frac{\mathrm{d}u}{u^2} = -\frac{\pi |z|}{2} \qquad (z \text{ real})$$

and use these facts to find Lévy's canonical forms.

---

[5] Historically, for all infinitely divisible distributions with finite second order moment, this form was derived earlier than Lévy by A. N. Kolmogorov in his papers in 1932.

**0.31.** Prove that $\varphi(z) = e^{-|z|^\alpha}$ $(2 \geq \alpha > 0)$ is the characteristic function of an infinitely divisible distribution.

Hint. First prove that for $2 > \alpha > 0$

$$|z|^\alpha = c \int_{0+}^{\infty} \frac{1 - \cos zu}{u^{\alpha+1}} du \qquad (c = \text{constant depending on } \alpha).$$

**0.32.** Let $\mathcal{C}$, $\mathcal{D}$ and $\mathcal{E}$ be $\sigma$-algebras and suppose that $\mathcal{D} \vee \mathcal{C}$ is independent of $\mathcal{E}$. If $X$ is measurable $(\mathcal{C})$ and if $E|X| < \infty$, then

$$E(X|\mathcal{D} \vee \mathcal{E}) = E(X|\mathcal{D}) \qquad \text{a. s.}$$

Prove this.

**0.33.** Let $\{X_n\}_n$ be a sequence of independent random variables with $E|X_n|^p < \infty$, $n = 1, 2, \ldots$, where $p$ is a constant $\geq 1$. Prove that

$$f(n) = E\left( \left| \sum_{i=1}^{n} (X_i - EX_i) \right|^p \right)$$

is increasing in $n$.

**0.34.** Let $\{X_n\}$ be a martingale $\{\mathcal{B}_n\}$. Prove that

$$X_S = E(X_m|\mathcal{B}_S)$$

if $S$ is a stopping time $\{\mathcal{B}_n\}$ such that $S \leq m$.

**0.35.** Let $\{X_n\}_{n\geq 1}$ be a submartingale $\{\mathcal{B}_n\}_{n\geq 1}$. Prove the existence and the uniqueness of the following decomposition (*Doob's decomposition*):

$$X_n = M_n + A_n \,,$$

where

(a) $\{M_n\}$ is a martingale $\{\mathcal{B}_n\}$,
(b) $A_n$ is measurable $(\mathcal{B}_{n-1})$ for $n = 2, 3, \ldots$,
(c) $0 = A_1 \leq A_2 \leq A_3 \leq \cdots$.

Hint. If there is such a decomposition, then

$$E(X_n - X_{n-1} \mid \mathcal{B}_{n-1}) = A_n - A_{n-1} \,.$$

This will give a hint on how to define $A_n$ and $M_n$.

**0.36.** Let $\{X_n\}_{n\geq 1}$ be a submartingale $\{\mathcal{B}_n\}$ and let $S$ and $T$ be two stopping times such that $0 \leq S \leq T \leq m$, where $m$ is a fixed finite number. Prove that

$$E(X_T \mid \mathcal{B}_S) \geq X_S \,.$$

If $\{X_n\}_{n\geq 1}$ is a martingale $\{\mathcal{B}_n\}$, the identity holds.

Hint. Use Problem 0.34 if $\{X_n\}$ is a martingale and use Problem 0.35 if $\{X_n\}$ is a submartingale.

**0.37.** Give a new proof of

$$cP\left(\sup_n X_n > c\right) \le \sup_n EX_n^+$$

for a submartingale $\{X_n\}$ and $c > 0$ by the following idea. Let $S = \inf\{n\colon X_n > c\}$ ($S = \infty$ if there is no such $n$). Then $S$ is a stopping time $\{\mathcal{B}_n \equiv \mathcal{B}[X_1, \ldots, X_n]\}_n$. Since $X_n^+$ is also a submartingale and since $S \wedge m$ is also a stopping time $\{\mathcal{B}_n\}$, we have

$$EX_m^+ \ge EX_{S\wedge m}^+ \ge E(X_S^+,\, S \le m)\,.$$

**0.38.** A family of random variables $\{X_\lambda\colon \lambda \in \Lambda\}$ is called *uniformly integrable* if

$$\lim_{a\to\infty} \sup_\lambda E(|X_\lambda|,\, |X_\lambda| > a) = 0\,.$$

Prove the following

(i) Let $p > 1$. If $\{E|X_\lambda|^p\colon \lambda \in \Lambda\}$ is bounded, then $\{X_\lambda\colon \lambda \in \Lambda\}$ is uniformly integrable. There is an example to show that this is not true for $p = 1$.

(ii) $\{X_n\}_n$ is convergent in $L^1$-norm if and only if we have
(a) $\{X_n\}$ is uniformly integrable,
(b) $\{X_n\}$ is convergent in probability.

(iii) If $X_t$, $0 \le t \le 1$, is a martingale then $X_t$, $0 \le t \le 1$, is uniformly integrable.

**0.39.** Let $X_n$, $n = 1, 2, \ldots$, be a martingale $\{\mathcal{B}_n\}$ and suppose that $\{E|X_n|\}_n$ is bounded. Prove the following.

(i) $X_n$, $n \to \infty$, is convergent a. s. to a random variable $X_\infty$ satisfying $E(|X_\infty|) < \infty$.

(ii) Suppose further that $\{X_n\}$ is uniformly integrable. Then $E|X_n - X_\infty| \to 0$. Moreover $\{X_n\colon n = 1, 2, \ldots, \infty\}$ is a martingale $\{\mathcal{B}_n\colon n = 1, 2, \ldots, \infty\}$, where $\mathcal{B}_\infty = \bigvee_n \mathcal{B}_n$.

Hint. Use the estimation of the expected upcrossing number to prove (i).

**0.40.** Let $\{B(t)\colon t \in [0, \infty)\}$ be a Brownian motion with $B(0) = 0$. Prove that $\{aB(t) + b\}_t$ is a martingale, where $a$ and $b$ are constants.

**0.41.** Let $B_i(t)$, $i = 1, 2$, be two independent Brownian motions. Prove the following.

(i) $(B_1(t) + \mathrm{i}B_2(t))^n$ is a complex-valued martingale.

(ii) If $u(x, y)$ is a polynomial in $x, y$ with real coefficients such that

$$\frac{\partial^2 u}{\partial x^2} + \frac{\partial^2 u}{\partial y^2} = 0\,,$$

then $u(B_1(t), B_2(t))$ is a martingale.

Hint. (i) Observe

$$\int_{-\infty}^{\infty}\int_{-\infty}^{\infty}(x+\mathrm{i}y)^n\frac{1}{2\pi t}\mathrm{e}^{-(x^2+y^2)/(2t)}\mathrm{d}x\mathrm{d}y=0\,,$$

using polar coordinates. (ii) Take a polynomial $v(x,y)$ with real coefficients such that $u(x,y)+\mathrm{i}v(x,y)$ is expressed as a polynomial of $x+\mathrm{i}y$ with complex coefficients. Such $v$ can be obtained as a solution of the Cauchy–Riemann equations

$$\frac{\partial v}{\partial x}=-\frac{\partial u}{\partial y}$$
$$\frac{\partial v}{\partial y}=\frac{\partial u}{\partial x}\,.$$

**0.42.**[6] Prove the assertion (c) of Theorem 0.4.1 that the family of all infinitely divisible distributions is closed under convergence.

Hint. Use Theorem 0.3.5.

**0.43.**[7] Let $C(R^1)$ be the Banach space of bounded continuous real functions $f$ on $R^1$ with norm $\|f\|=\sup_x|f(x)|$. The space $M$ of bounded countably additive set functions $\mu$ on the Borel sets with the norm of total variation is the conjugate space of $C(R^1)$. The weak* topology of $M$ is, by definition, the weakest topology (fewest open sets) that makes $\int f(x)\mu(\mathrm{d}x)$ continuous with respect to $\mu\in M$ for each $f\in C(R^1)$. The set $\mathfrak{P}$ of 1-dimensional probability measures is a subset of $M$. The topology induced in $\mathfrak{P}$ by the weak* topology of $M$ is called the weak* topology of $\mathfrak{P}$. Considering the distribution function of $\mu\in\mathfrak{P}$, the set $\mathfrak{P}$ is identified with the set $\mathfrak{D}$ of Problem 0.7 and thus the Lévy metric in $\mathfrak{D}$ can be considered as a metric in $\mathfrak{P}$. Show that convergence of a sequence $\{\mu_n\}$ in the Lévy metric is identical with convergence in the weak* topology and that the Lévy metric is complete.[8]

**0.44.** A subset $\mathfrak{M}$ of the set $\mathfrak{P}$ of 1-dimensional probability distributions is called *tight* if

$$\lim_{K\to\infty}\inf_{\mu\in\mathfrak{M}}\mu(-K,K)=1\,.$$

Show that $\mathfrak{M}$ is conditionally compact if and only if it is tight.

**0.45.** A set $E$ is said to be *partially ordered* if there is a relation for certain pairs $(x,y)$ of elements of $E$, expressed by $x\prec y$, satisfying the following: $x\prec x$; if $x\prec y$ and $y\prec x$, then $x=y$; if $x\prec y$ and $y\prec z$, then $x\prec z$. A partially ordered set $E$ is called a *lattice* if, for every pair $(x,y)$ of elements of $E$, the least upper bound $x\vee y$ and the greatest lower bound $x\wedge y$ of $x$

[6] Moved from problems for Chapter 1.

[7] Problems 0.43–0.45 are added by the Editors.

[8] In probability theory, convergence of a sequence $\{\mu_n\}$ in $\mathfrak{P}$ in the weak* topology is usually called weak convergence.

and $y$ with respect to the relation $\prec$ exist in $E$. A lattice $E$ is said to be *complete* if, for every subset $F$ of $E$, the least upper bound $\bigvee_{x \in F} x$ and the greatest lower bound $\bigwedge_{x \in F} x$ of $F$ with respect to the relation $\prec$ exist in $E$. Show that the family $\mathfrak{B}$ of all sub-$\sigma$-algebras of a $\sigma$-algebra $\mathcal{B}$ is a complete lattice with respect to the set-theoretic inclusion.

## E.1 Chapter 1

**1.1.** Suppose that $X_{k,n} \to X_k$ i. p. as $n \to \infty$ for $k = 1, 2, \ldots, m$. Show that if $\{X_{k,n}\}_{k=1}^m$ is independent for each $n$, then $\{X_k\}_{k=1}^m$ is also independent.

**1.2.** Let $\{\mu_n\}_n$ be a sequence of Cauchy distributions. Show that if $\mu_n \to \mu$, then $\mu$ is Cauchy distributed and the parameters of $\mu_n$ converge respectively to those of $\mu$. Prove also similar facts for Poisson distributions.

**1.3.** Let $X(t, \omega)$ be a Lévy process. Prove that if $f \in L^1[0, 1]$, then $\int_0^t f(s) X(s, \omega) ds$ is measurable in $\omega$, i. e. a random variable.

**1.4.** Suppose that $X_n \to X$ i. p. and that each $X_n$ is Gauss distributed. Prove that $X_n \to X$ in the square mean.

**1.5.** Prove that the sample function of a Lévy process is continuous a. s. at every *fixed* time point.

Hint. Observe

$$\{ |X(t) - X(t-)| > 0 \} \subset \liminf_{n \to \infty} \{ |X(t) - X(t - 1/n)| > 0 \}.$$

**1.6.** If $X_1(t)$ and $X_2(t)$ are independent Lévy processes of Poisson type, then $X_1(t) + X_2(t)$ is also a Lévy process of Poisson type. Similarly for the Gauss type. Prove these facts.

**1.7.** Let $X(t)$ be a Lévy process of Gauss type such that $E(X(t)) = 0$ and $V(X(t)) = V(t)$. Prove that

$$S_n \equiv \sum_{k=1}^{n} \left( X\left(\frac{k}{n} t\right) - X\left(\frac{k-1}{n} t\right) \right)^2 \to V(t) \qquad \text{i. p.}$$

Hint. Compute the mean and the variance of $S_n$.

**1.8.** Let $X(t)$ be as in the previous problem. Assume that $V(t) > 0$ for all $t > 0$. Prove that almost all sample functions of this process are of unbounded variation on the time interval $[0, t]$ for every $t$ fixed.

Hint. By Problem 1.7 we can find a sequence of integers $p(1) < p(2) < p(3) < \cdots \to \infty$ such that $S_{p(n)} \to V(t)$ a. s. Observe the inequality

$$S_{p(n)} \leq \left( \sup_{\substack{|t_2 - t_1| \leq 1/p(n) \\ 0 \leq t_1 \leq t_2 \leq t}} |X(t_2) - X(t_1)| \right) \sum_{k=1}^{p(n)} \left| X\left(\frac{k}{p(n)} t\right) - X\left(\frac{k-1}{p(n)} t\right) \right|$$

and use the a. s. uniform continuity of the sample function on $[0, t]$.

**1.9.** Let $X(t,\omega)$ and $Y(t,\omega)$ be stochastic processes with almost all sample functions in $D$. Then the $\omega$-set (event) that

$$\text{“} X(t,\omega) = X(t-,\omega) \text{ or } Y(t,\omega) = Y(t-,\omega) \text{ for every } t \text{”}$$

is measurable. Prove this.

**1.10.** Let $X(t)$ be a Lévy process with the three components $m(t), V(t)$ and $n(\mathrm{d}s\mathrm{d}u)$. Set

$$S_\lambda(t) = \sum_{0<s\le t} |X(s) - X(s-)|^\lambda$$

and

$$\gamma_\lambda(t) = \int_{s=0}^{t} \int_{|u|\le 1} |u|^\lambda n(\mathrm{d}s\mathrm{d}u) .$$

for $\lambda > 0$. Prove the following.

(i) If $\gamma_\lambda(t) < \infty$, then $S_\lambda(t) < \infty$ a. s. and

$$E(\mathrm{e}^{-\alpha S_\lambda(t)}) = \exp\left\{ \int_{s=0}^{t} \int_{u\in R_0} (\mathrm{e}^{-\alpha|u|^\lambda} - 1)\, n(\mathrm{d}s\mathrm{d}u) \right\} .$$

(ii) If $\gamma_\lambda(t) = \infty$, then $S_\lambda(t) = \infty$ a. s.

(iii) If $\lambda \ge 2$, then $\gamma_\lambda(t) < \infty$ for all $t$.

Hint. Let $N(E)$ be the number of jumps appearing in the Lévy decomposition and $A_{m,k} = \{(s,u)\colon 0 < s \le t,\ k/m \le u < (k+1)/m\}$. Then

$$S_\lambda(t) = \int_{s=0}^{t} \int_{u\in R_0} |u|^\lambda N(\mathrm{d}s\mathrm{d}u) = \lim_{m\to\infty} \sum_{k=-\infty}^{\infty} |k/m|^\lambda N(A_{m,k}) .$$

**1.11.** Let $X(t)$ be a Lévy process whose sample path is increasing, $N(E)$ the number of jumps as above and $n(E) = E(N(E))$. Prove the following.

(i) $\quad X(t) = m(t) + \displaystyle\int_{s=0}^{t} \int_{u>0} u N(\mathrm{d}s\mathrm{d}u)$, where $m(t)$ is continuous and increasing and $\displaystyle\int_{s=0}^{t} \int_{u>0} \frac{u}{1+u} n(\mathrm{d}s\mathrm{d}u) < \infty$ for every $t$.

(ii) $\quad E(\mathrm{e}^{-pX(t)}) = \exp\left\{ -pm(t) - \displaystyle\int_{s=0}^{t} \int_{u>0} (1 - \mathrm{e}^{-pu}) n(\mathrm{d}s\mathrm{d}u) \right\}$ for $p > 0$.

**1.12.** Let $X(t)$ be a stable process with exponent $\alpha$ whose sample function is increasing and moves only with jumps a. s. Then prove that

(i) $\quad \alpha < 1$,

(ii) $\quad E(\mathrm{e}^{-pX(t)}) = \mathrm{e}^{-ctp^\alpha}$, $p > 0$, with some $c > 0$.

**1.13.** Let $\{B(t)\colon t \geq 0\}$ be a Brownian motion and let $\{X(t)\colon t \geq 0\}$ be the process in Problem 1.12. Suppose $\{X(t)\colon t \geq 0\}$ is independent of $\{B(t)\colon t \geq 0\}$. Then show that $Y(t) \equiv B(X(t))$ is a stable process with exponent $2\alpha$.[9]

Hint. First prove that $Y(t)$ is a Lévy process and then compute $E(\mathrm{e}^{\mathrm{i}zY(t)})$.

**1.14.** Let $X(t)$ be a stable process with exponent $\alpha$. Prove that

$$2 \neq \alpha < \lambda \implies P(S_\lambda(t) < \infty) = 1 \text{ for } t > 0,$$
$$2 \neq \alpha \geq \lambda \implies P(S_\lambda(t) = \infty) = 1 \text{ for } t > 0,$$

where $S_\lambda(t) = \sum_{0<s\leq t} |X(s) - X(s-)|^\lambda$.

## E.2 Chapter 2

**2.1.** Let $S = R^2 \cup \{\infty\}$ and set

$$p_t(x, \mathrm{d}y) = N_t(y - x)\mathrm{d}y \quad \text{for } t > 0 \quad (\mathrm{d}y = \mathrm{d}y_1\mathrm{d}y_2,\ x = (x_1, x_2)),$$
$$p_t(\infty, \{\infty\}) = 1, \qquad p_0(x, B) = \delta_x(B),$$

where

$$N_t(x) = \frac{1}{2\pi t}\mathrm{e}^{-|x|^2/(2t)} \qquad (|x|^2 = x_1^2 + x_2^2).$$

Prove that $p_t(x, \mathrm{d}y)$, $t \in [0, \infty)$, satisfy the conditions (T.0)–(T.5) in Section 2.1. (This $\{p_t(x, \mathrm{d}y)\}_t$ is called *two-dimensional Brownian transition probabilities.*)

**2.2.** Find a system of transition probabilities $\{q_t(x, B)\}$ on $S = R^2 \cup \{\infty\}$ which coincides with the two-dimensional Brownian transition probabilities for $t = 1$ but not for $t$ general.

Hint. Let $T_t$ be a rotation in $R^2$ by angle $2\pi t$ around the origin and set

$$q_t(x, \mathrm{d}y) = N_t(y - T_t x)\mathrm{d}y,$$
$$q_t(\infty, \{\infty\}) = 1, \qquad q_0(x, B) = \delta_x(B).$$

Verify (T.0)–(T.5) for the $\{q_t(x, \mathrm{d}y)\}$.

**2.3.** Let $S = [0, \infty]$ and set

$$p_t(a, B) = \int_B \frac{1}{\sqrt{2\pi t}} \left(\mathrm{e}^{-(b-a)^2/(2t)} + \mathrm{e}^{-(b+a)^2/(2t)}\right) \mathrm{d}b,$$
$$a \in [0, \infty),\ B \in \mathcal{B}[0, \infty),$$
$$p_t(a, \{\infty\}) = 0, \qquad a \in [0, \infty),$$
$$p_t(\infty, B) = 0, \qquad B \in \mathcal{B}[0, \infty),$$
$$p_t(\infty, \{\infty\}) = 1.$$

[9] The procedure to obtain $\{Y(t)\}$ from $\{B(t)\}$ and $\{X(t)\}$ is a special case of the transformation called *subordination.*

Prove that the family $\{p_t(a,B)\}$ satisfies all conditions (T.0)–(T.5) of transition probabilities in Section 2.1.

**2.4.** Prove that the generator $A$ of the transition semigroup given by the transition probabilities in Problem 2.3 is as follows.

$$\mathfrak{D}(A) = \Big\{u \in C[0,\infty]\colon u \in C^2 \text{ in } (0,\infty),\ u'(0+) = 0,\ \lim_{x\to\infty} u''(x) = 0,$$
$$\lim_{x\to 0} u''(x) \text{ exists and is finite}\Big\},$$

$$Au(x) = \begin{cases} (1/2)u''(x), & x \in (0,\infty) \\ \lim_{x\to 0}(1/2)u''(x), & x = 0 \\ 0, & x = \infty. \end{cases}$$

Hint. Compute $R_\alpha f$ and make the same argument as in the Example 2 in Section 2.2.

**2.5.** Prove that the probability law $P_a$, $a \in [0,\infty]$, of the path starting at $a$ of the Markov process determined by the transition probabilities in Problem 2.3 is obtained by

$$P_a(\Lambda) = Q_a(\omega\colon |\omega| \in \Lambda), \qquad \Lambda \in \mathcal{B},$$

where $|\omega|$ is the path defined by $|\omega|(t) \equiv |\omega(t)|$ and $Q_a$ is the probability law of the path of the Brownian motion starting at $a$. (Because of this fact, this Markov process is called *a Brownian motion with reflecting barrier at* 0.)

**2.6.** Prove that the family of all stopping times is closed under the following operations: $\sigma_1 \vee \sigma_2$, $\sigma_1 \wedge \sigma_2$, $\sigma_1 + \sigma_2$, and monotone limits.

**2.7.** Let $X(t)$ be a Brownian motion. Prove that if $a < b$ then

$$P_a\left(\max_{0\le s\le t} X_s \ge b\right) = 2P_a(X_t \ge b) = 2\int_b^\infty \frac{1}{\sqrt{2\pi t}} \mathrm{e}^{-(c-a)^2/(2t)} \mathrm{d}c\,.$$

Hint: Let $\sigma_b$ be the hitting time of $b$. Then

$$P_a(\sigma_b \le t) = P_a\left(\max_{0\le s\le t} X_s \ge b\right), \quad P_a(X_t \ge b) = P_a(X_t \ge b,\ \sigma_b \le t).$$

Use the example at the end of Section 2.7.

**2.8.** Prove that, for $a > 0$ and $B$ a Borel set in $[0,\infty)$,

$$\begin{aligned} P_a(X_t \in B,\ \sigma_0 > t) &= P_a(X_t \in B) - P_a(X_t \in -B) \\ &= \int_B \left(\frac{1}{\sqrt{2\pi t}}\mathrm{e}^{-(b-a)^2/(2t)} - \frac{1}{\sqrt{2\pi t}}\mathrm{e}^{-(b+a)^2/(2t)}\right) \mathrm{d}b\,, \end{aligned}$$

where $-B = \{-x\colon x \in B\}$.

Hint. Use the example at the end of Section 2.7.

**2.9.** Let $X(t)$ be a Brownian motion and let $\sigma_a$ be the hitting time of $(a, \infty)$. Prove that $\sigma_a$ is then a Lévy process, with $a$ regarded as time parameter, on $(\Omega, \mathcal{B}, P_0)$.

**2.10.** Prove that the Lévy process $\sigma_a$ in Problem 2.9 is a stable process with exponent $1/2$ with path increasing and moving only by jumps. Show[10] that for $a > 0$

$$\begin{aligned} E_0(\mathrm{e}^{-p\sigma_a}) &= \mathrm{e}^{-a\sqrt{2p}} \\ &= \exp\left(-\frac{a}{\sqrt{2\pi}}\int_0^\infty (1-\mathrm{e}^{-pu})u^{-3/2}\mathrm{d}u\right), \qquad p>0, \\ P_0(\sigma_a \in B) &= \frac{a}{\sqrt{2\pi}}\int_B \mathrm{e}^{-a^2/(2s)}s^{-3/2}\mathrm{d}s, \qquad B \in \mathcal{B}([0,\infty)). \end{aligned}$$

**2.11.** Let $X(t)$ be a $k$-dimensional Brownian motion, $k \le 2$. Use Theorem 2.19.6 to prove that

$$P_a(\omega\colon \{X_s(\omega)\colon s \ge t\} \text{ is dense in } R^k \text{ for all } t) = 1$$

for every $a$.

**2.12.** Let $X(t)$ be a $k$-dimensional Brownian motion, $k \ge 3$. Prove that

$$P_a(\omega\colon |X_t(\omega)| \to \infty \text{ as } t \to \infty) = 1$$

for every $a$.

[10] Added by the Editors.

# Appendix: Solutions of Exercises

## A.0 Chapter 0

**0.1.** Use the notation $\alpha\mu + \beta$ in Section 0.2.

(i) $\overline{\varphi(\xi)}$, $|\varphi(\xi)|^2$, $\varphi(\xi)^n$, and $e^{ib\xi}\varphi(a\xi)$ are the characteristic functions of $-\mu$, $(-\mu) * \mu$, $\mu^{n*}$, and $a\mu + b$, respectively.

(ii) Let $g(t) = \sum_{k=0}^{\infty} a_k t^k$ with $a_k$ satisfying $a_k \geq 0$ and $\sum_{k=0}^{\infty} a_k = 1$. Then $g(\varphi(\xi)) = \sum_{k=0}^{\infty} a_k \varphi(\xi)^k$, which is convergent uniformly in $\xi$ since $|\varphi(\xi)| \leq 1$. Hence $g(\varphi(\xi))$ is the characteristic function of $\nu = a_0\delta_0 + a_1\mu + a_2\mu^{2*} + \cdots$.

(iii) Note that $|\varphi(\xi)|^2 = \varphi_1(\xi)$ is the characteristic function of $(-\mu)*\mu = \nu$ and that $\varphi_1(\xi)$ is real. Then

$$
\begin{aligned}
\varphi_1(\xi) &= \int_{-\infty}^{\infty} \cos \xi x \, \nu(\mathrm{d}x), \\
1 - \varphi_1(2\xi) &= \int_{-\infty}^{\infty} (1 - \cos 2\xi x)\nu(\mathrm{d}x) = 2\int_{-\infty}^{\infty} \sin^2 \xi x \, \nu(\mathrm{d}x) \\
&= 2\int_{-\infty}^{\infty} (1 - \cos \xi x)(1 + \cos \xi x)\, \nu(\mathrm{d}x) \\
&\leq 4\int_{-\infty}^{\infty} (1 - \cos \xi x)\, \nu(\mathrm{d}x) = 4(1 - \varphi_1(\xi)).
\end{aligned}
$$

**0.2.** Use the grouping theorem (Theorem 0.1.4 (i)) and Theorem 0.1.1 (ii) to show the independence of $\mathcal{B}[X_1]$, $\mathcal{B}[X_2 + X_3]$, $\ldots$ .

**0.3.** A $\sigma$-algebra is clearly a multiplicative Dynkin class. Conversely, any multiplicative Dynkin class $\mathcal{C}$ is a $\sigma$-algebra, because
(1) $\Omega \in \mathcal{C}$,
(2) if $A \in \mathcal{C}$ then $\Omega - A = A^{\mathrm{C}} \in \mathcal{C}$,
(3) if $A_1, A_2, \ldots \in \mathcal{C}$, then

$$
\bigcup_{n=1}^{\infty} A_n = A_1 \cup (A_2 \cap A_1^{\mathrm{C}}) \cup (A_3 \cap A_1^{\mathrm{C}} \cap A_2^{\mathrm{C}}) \cup \cdots \in \mathcal{C}.
$$

---

These solutions were originally written by A. Grimvall. They were rewritten by the Editors.

Now $\mathcal{B}[\mathcal{C}]$ is a Dynkin class which contains $\mathcal{C}$. Because of minimality we have $\mathcal{D}[\mathcal{C}] \subset \mathcal{B}[\mathcal{C}]$. To prove $\mathcal{B}[\mathcal{C}] \subset \mathcal{D}[\mathcal{C}]$ it is enough to prove that $\mathcal{D}[\mathcal{C}]$ is multiplicative.
(1) $\Omega \in \mathcal{D}[\mathcal{C}]$ since $\Omega \in \mathcal{C}$.
(2) We want to show that if $A, B \in \mathcal{D}[\mathcal{C}]$ then $A \cap B \in \mathcal{D}[\mathcal{C}]$. Let $\mathcal{D}_A = \{B : B \in \mathcal{D}[\mathcal{C}],\ A \cap B \in \mathcal{D}[\mathcal{C}]\}$. If $A \in \mathcal{C}$ then $\mathcal{D}_A \supset \mathcal{C}$, $\mathcal{D}_A$ is a Dynkin class and therefore $\mathcal{D}_A = \mathcal{D}[\mathcal{C}]$. Now let $A$ be arbitrary in $\mathcal{D}[\mathcal{C}]$. The same method shows that $\mathcal{D}_A = \mathcal{D}[\mathcal{C}]$.

**0.4.** Let $Y = \limsup_{n\to\infty} (1/n)(X_1 + X_2 + \cdots + X_n)$. Since

$$Y = \limsup_{n\to\infty} (1/n)(X_k + X_{k+1} + \cdots + X_n) \qquad \text{for every fixed } k,$$

$Y$ is measurable with respect to $\bigvee_{n=k}^{\infty} \mathcal{B}[X_n]$ for every $k$. Hence $Y$ is measurable with respect to the tail $\sigma$-algebra $\bigwedge_{k=1}^{\infty} \bigvee_{n=k}^{\infty} \mathcal{B}[X_n]$, which is trivial according to Kolmogorov's 0-1 law. Let $c = \inf\{a \in [-\infty, \infty] : P(Y \le a) = 1\}$. Suppose that $c$ is finite. Then

$$P(Y \le c) = P(\textstyle\bigcap_n \{Y \le c + 1/n\}) = 1\,,$$
$$P(Y < c) \le \textstyle\sum_n P(Y \le c - 1/n) = 0\,.$$

Therefore $P(Y = c) = 1$. If $c = -\infty$ or $\infty$, a similar argument works.

**0.5.** The class $\mathcal{C}_X$ consisting of the full space $\Omega$ and all sets of the form $\{X \le a\}$ is a multiplicative class. The class $\mathcal{C}_Y$ similarly defined is also multiplicative. The classes $\mathcal{C}_X$ and $\mathcal{C}_Y$ generate $\mathcal{B}[X]$ and $\mathcal{B}[Y]$, respectively, and so we can apply Theorem 0.1.2.

**0.6.** Let $\mathcal{C}$ be the class which consists of all half-spaces i.e. all sets of the form $\{(x_1, \ldots, x_n) : \alpha_1 x_1 + \alpha_2 x_2 + \cdots + \alpha_n x_n \le k\}$ with non-zero vector $(\alpha_1, \ldots, \alpha_n)$ and $k \in R^1$. Let $\mu$ and $\nu$ be probability measures which are equal on $\mathcal{C}$. Let $\alpha_1, \alpha_2, \ldots, \alpha_n$ be fixed and let $k$ vary. It then follows from Lemma 0.1.1 that $\mu$ and $\nu$ are equal on all sets of the form

$$\{(x_1, x_2, \ldots, x_n) : \alpha x_1 + \cdots + \alpha_n x_n \in B\}, \qquad B \in \mathcal{B}[R^1].$$

Hence

$$\int \cdots \int e^{i(\alpha_1 x_1 + \cdots + \alpha_n x_n)} \mu(dx) = \int \cdots \int e^{i(\alpha_1 x_1 + \cdots + \alpha_n x_n)} \nu(dx)\,.$$

This means that $\mu$ and $\nu$ have a common characteristic function and so they are equal.

**0.7.** By the following three properties $\rho$ is a metric on $\mathfrak{D}$.
(a) $\rho(F, G) = \rho(G, F)$ because

$$F(x - \delta) - \delta \le G(x) \le F(x + \delta) + \delta \qquad \text{for all } x$$
$$\iff \quad G(x' - \delta) - \delta \le F(x') \le G(x' + \delta) + \delta \qquad \text{for all } x'.$$

(b) Clearly $\rho(F,G) \geq 0$. If $\rho(F,G) = 0$ then $F = G$, because $G(x) \leq F(x)$ by the right-continuity of $F$ and also $F(x) \leq G(x)$ by symmetry.

(c) Let $\rho(F,G) = a$, $\rho(G,H) = b$. We want to show that $\rho(F,H) \leq a + b$. For all $\varepsilon > 0$ and all $x$ we have

$$F(x - a - \varepsilon) - a - \varepsilon \leq G(x) \leq F(x + a + \varepsilon) + \varepsilon + a,$$
$$G(x - b - \varepsilon) - b - \varepsilon \leq H(x) \leq G(x + b + \varepsilon) + \varepsilon + b.$$

Therefore

$$F(x - a - b - 2\varepsilon) - 2\varepsilon - a - b \leq H(x) \leq F(x + a + b + 2\varepsilon) + 2\varepsilon + a + b,$$

that is, $\rho(F,H) \leq a + b + 2\varepsilon$.

Assume that $\rho(F_n, F) \to 0$ and let $x$ be a continuity point of $F$. Let us show that $F_n(x) \to F(x)$. Let $\varepsilon > 0$. Then there exists $\delta = \delta(\varepsilon) > 0$ with $\delta \leq \varepsilon$ such that $|F(y) - F(x)| < \varepsilon$ if $|y - x| \leq \delta$. Since $\rho(F_n, F) \to 0$ we have for $n$ large enough

$$F(x - \delta) - \delta \leq F_n(x) \leq F(x + \delta) + \delta$$

and hence

$$F(x) - 2\varepsilon \leq F_n(x) \leq F(x) + 2\varepsilon\,.$$

Conversely, suppose that $F_n(x) \to F(x)$ at all continuity points $x$ of $F$. Let us show that $\rho(F_n, F) \to 0$. Assume the contrary. Then there is $\varepsilon > 0$ such that $\rho(F_n, F) > \varepsilon$ for infinitely many $n$. Hence we can find a subsequence $\{F_{n'}\}$ of $\{F_n\}$ and a sequence $\{x_{n'}\}$ such that either $F_{n'}(x_{n'} - \varepsilon) - \varepsilon > F(x_{n'})$ or $F(x_{n'}) > \varepsilon + F_{n'}(x_{n'} + \varepsilon)$. We assume the former for infinitely many $n'$. Let $a$ and $b$ be two continuity points of $F$ such that $F(a) < \varepsilon$ and $F_n(b) > 1 - \varepsilon$. Then $F_n(a) < \varepsilon$ and $F_n(b) > 1 - \varepsilon$ for all $n$ large enough. It follows that $\{x_{n'}\}$ has no cluster point outside of $[a,b]$. Therefore we can assume that it converges to some point $x_0$. Let $c$ be a continuity point of $F$ in the interval $(x_0 - \varepsilon, x_0)$. Then

$$F_{n'}(c) - \varepsilon \geq F_{n'}(x_{n'} - \varepsilon) - \varepsilon > F(x_{n'}) > F(c) - \varepsilon/2$$

for infinitely many $n'$. This contradicts the assumption that $F_n(c) \to F(c)$.

**0.8.** (i) Let $A$ be a countable set and assume that $\sum_{\alpha \in A} V(X_\alpha) = K < \infty$. Let $F_n \uparrow A$, $F_n$ finite, $F_0 = \emptyset$. Let $Y_n = \sum^{\Delta}_{\alpha \in F_n \setminus F_{n-1}} X_\alpha$. Then $\{Y_n\}$ is independent and $V(\sum_{i=1}^n Y_i) = \sum_{\alpha \in F_n} V(X_\alpha) \to K$. Thus $\sum^{\Delta}_{\alpha \in F_n} X_\alpha$ converges a.s. and in $\mathcal{L}^2$-norm. Denote the limit by $S_F^{\Delta}$. Let $G_m \uparrow A$ be another sequence, $G_m$ finite, and denote the limit of $\sum^{\Delta}_{\alpha \in G_m} X_\alpha$ by $S_G^{\Delta}$. Then $S_G^{\Delta} = S_F^{\Delta}$ because, choosing $m = m(n)$ such that $G_m \supset F_n$, we have

$$P\left(\left|\sum\nolimits^{\Delta}_{\alpha \in G_m} X_\alpha - \sum\nolimits^{\Delta}_{\alpha \in F_n} X_\alpha\right| > \varepsilon\right) \to 0, \qquad n \to \infty$$

by Chesbyshev's inequality.

*Remark.* The assertion corresponding to Theorem 0.3.2 (ii) is not true in this $\mathcal{L}^2$-setting. Namely, if $\sum_{\alpha\in A} V(X_\alpha) = \infty$ and if $F_n \uparrow A$ with $F_n$ finite, $\sum_{\alpha\in F_n}(X_\alpha - c_\alpha)$ may be convergent a. s. for some choice of $c_\alpha$. For example, let $\{X_n\}_{n=1,2\ldots}$ be independent with

$$\varphi(\xi; X_n) = \exp\left(\int_n^{n+1} (e^{i\,\xi x} - 1)x^{-1-\alpha}dx\right), \qquad 0 < \alpha < 2$$

and let $S_n = \sum_{i=1}^n X_i$. By differentiation of $\varphi(\xi; S_n)$, we see that $V(S_n) = \int_1^{n+1} x^{1-\alpha}dx \to \infty$ as $n \to \infty$. Since $\varphi(\xi; S_n) \to \exp\left(\int_1^\infty (e^{i\,\xi x} - 1)x^{-1-\alpha}dx\right)$, which is a characteristic function $\varphi(\xi)$,

$$\delta(S_n) \to -\log\left(\frac{1}{\pi}\int_{R^1} |\varphi(\xi)|^2 \frac{d\xi}{1+\xi^2}\right) < \infty.$$

Thus $\{S_n - \gamma(S_n)\}$ is convergent a. s. by Theorem 0.3.2 (i).

(ii) Assume that $\sum_{\alpha\in A} V(X_\alpha) < \infty$ so that $\sideset{}{^\Delta}\sum_{\alpha\in A} X_\alpha$ can be defined. Let $F_n \uparrow A$, $F_n$ finite, and let $S_n^\Delta = \sideset{}{^\Delta}\sum_{\alpha\in F_n} X_\alpha$. Then, using (i) just proved and Theorem 0.3.2, we see that $\lim_{n\to\infty} \delta(S_n^\Delta) < \infty$ and so $\sideset{}{^\circ}\sum_{\alpha\in A} X_\alpha$ can be defined. Thus $S_n^\Delta \to \sideset{}{^\Delta}\sum_{\alpha\in A} X_\alpha$ a. s. and $S_n^\circ \to \sideset{}{^\circ}\sum_{\alpha\in A} X_\alpha$ a. s., where $S_n^\circ = \sideset{}{^\circ}\sum_{\alpha\in F_n} X_\alpha$. We know that $S_n^\circ = S_n^\Delta + c_n$ with some constant $c_n$. Hence $0 = \delta(S_n^\circ - S_n^\Delta) \to \delta(\sideset{}{^\circ}\sum_{\alpha\in A} X_\alpha - \sideset{}{^\Delta}\sum_{\alpha\in A} X_\alpha)$ by Theorem 0.2.6. Thus $\sideset{}{^\circ}\sum_{\alpha\in A} X_\alpha = \sideset{}{^\Delta}\sum_{\alpha\in A} X_\alpha + \text{const.}$

**0.9.** Proof of Theorem 0.3.4 (i). If $B = \bigcup_{j=1}^n B_j$ (disjoint finite union), then we can prove $S_B^\circ = \sideset{}{^\circ}\sum_{1\le j\le n} S_{B_j}^\circ$, approximating $B_j$ by finite subsets.

Next assume $B = \bigcup_{n=1}^\infty B_n$ (disjoint) and $\delta(B) < \infty$. Take finite subsets $F_{n,k}$ of $B_n$ such that $F_{n,k} \uparrow B_n$ as $k \to \infty$. Then $S_{F_{n,k}}^\circ \to S_{B_n}^\circ$ as $k \to \infty$. Let $\varepsilon > 0$ be given. Using Theorem 0.2.5, choose $k_1$ such that $\delta(S_{B_1}^\circ - S_{F_{1,k_1}}^\circ) < \varepsilon/2$, and $k_m > k_{m-1}$ such that $\delta(\sum_{j=1}^m (S_{B_j}^\circ - S_{F_{j,k_j}}^\circ)) < (1 - 2^{-m})\varepsilon$ for $m \ge 2$. Let $G_n = F_{1,k_n} \cup F_{2,k_n} \cup \cdots \cup F_{n,k_n}$. Then $G_n \uparrow B$ as $n \to \infty$, and so $S_{G_n}^\circ \to S_B^\circ$ a. s. as $n \to \infty$. We have

$$\begin{aligned}\delta(S_{G_n}^\circ - \sideset{}{^\circ}\sum_{j\le n} S_{B_j}^\circ) &= \delta(\sum_{j=1}^n (S_{B_j}^\circ - S_{F_j,k_n}^\circ)) \\ &\le \delta(\sum_{j=1}^n (S_{B_j}^\circ - S_{F_j,k_j}^\circ)) < \varepsilon\,.\end{aligned}$$

It is easily proved that $\delta(\sideset{}{^\circ}\sum_{j\le n} S_{B_j}^\circ) \le \delta(B)$. Therefore $\sideset{}{^\circ}\sum_{j\le n} S_{B_j}^\circ$ is convergent a. s. as $n \to \infty$. Passing to the limit we have $\delta(S_B^\circ - \sideset{}{^\circ}\sum_n S_{B_n}^\circ) \le \varepsilon$. Since $\varepsilon > 0$ is arbitrary, $\delta(S_B^\circ - \sideset{}{^\circ}\sum_n S_{B_n}^\circ) = 0$. Hence $S_B^\circ = \sideset{}{^\circ}\sum_n S_{B_n}^\circ + \text{const.}$ Taking central values, we see that the constant is zero.

Proof of Theorem 0.3.4 (ii). Assume that $B_n \uparrow B$ and $\delta(B) < \infty$. Let $C_1 = B_1$ and $C_n = B_n - B_{n-1}$ for $n \ge 2$. Then $B_n = \bigcup_{i=1}^n C_i$ (disjoint) and $B = \bigcup_{i=1}^\infty C_i$ (disjoint). Hence, by (i), $S_{B_n}^\circ = \sideset{}{^\circ}\sum_{i\le n} S_{C_i}^\circ \to \sideset{}{^\circ}\sum_i S_{C_i}^\circ = S_B^\circ$ a. s.

Before proving Theorem 0.3.4 (iii), we claim that if $B_n \downarrow \emptyset$ and $\delta(B_1) < \infty$, then $S_{B_n}^\circ \to 0$ a. s. We have $S_{B_1-B_n}^\circ \to S_{B_1}^\circ$ a. s. by (ii) since $B_1 - B_n \uparrow B_1$.

We have $S^\circ_{B_n} + S^\circ_{B_1 - B_n} + c_n = S^\circ_{B_1}$ with some constant $c_n$. Hence $S^\circ_{B_n} + c_n \to 0$ a. s. Therefore $c_n = \gamma(S^\circ_{B_n} + c_n) \to 0$. Hence $S^\circ_{B_n} \to 0$ a. s.

Finally let us prove Theorem 0.3.4 (iii). Let $B_n \downarrow B$ and $\delta(B_1) < \infty$. We have $S^\circ_{B_n - B} + S^\circ_B + c_n = S^\circ_{B_n}$ with some constant $c_n$. Since $B_n - B \downarrow \emptyset$, $S^\circ_{B_n - B} \to 0$ a. s. It follows that $S^\circ_{B_n} - c_n \to S^\circ_B$ a. s. Then $-c_n = \gamma(S^\circ_{B_n} - c_n) \to \gamma(S^\circ_B) = 0$. Hence $S^\circ_{B_n} \to S^\circ_B$ a. s.

The case with $\sum^\Delta$ in place of $\sum^\circ$ is similar.

**0.10.** (i) Let $\varphi_F(\xi) = \varphi(\xi; F)$ and $\varphi_{Y(t)}(\xi) = \varphi(\xi; Y(t))$. Then

$$\begin{aligned}
\varphi_{Y(t)}(\xi) &= E\left[\mathrm{e}^{\mathrm{i}\,\xi \sum_{i=1}^{N(t)} X_i}\right] = \sum_{k=0}^{\infty} E\left[\mathrm{e}^{\mathrm{i}\,\xi \sum_{i=1}^{k} X_i},\ N(t) = k\right] \\
&= \sum_{k=0}^{\infty} P(N(t) = k) E\left[\mathrm{e}^{\mathrm{i}\,\xi \sum_{i=1}^{k} X_i}\right] = \sum_{k=0}^{\infty} \frac{(\lambda t)^k}{k!} \mathrm{e}^{-\lambda t} \varphi_F(\xi)^k \\
&= \mathrm{e}^{-\lambda t (1 - \varphi_F(\xi))} \,.
\end{aligned}$$

(ii) Proof that, for $0 \le t_0 < t_1 < \cdots < t_n$, $\{Y(t_i) - Y(t_{i-1})\}_{i=1}^n$ is independent. We consider characteristic functions:

$$\begin{aligned}
&\varphi_{(Y(t_1)-Y(t_0), Y(t_2)-Y(t_1), \ldots, Y(t_n)-Y(t_{n-1}))}(\xi_1, \xi_2, \ldots, \xi_n) \\
&= E\left[\mathrm{e}^{\mathrm{i}\left(\xi_1 \sum_{j=N(t_0)+1}^{N(t_1)} X_j + \xi_2 \sum_{j=N(t_1)+1}^{N(t_2)} X_j + \cdots + \xi_n \sum_{j=N(t_{n-1})+1}^{N(t_n)} X_j\right)}\right] \\
&= \sum_{k_0, k_1, \ldots, k_n \ge 0} E\Bigg[\mathrm{e}^{\mathrm{i}\left(\xi_1 \sum_{j=k_0+1}^{k_0+k_1} X_j + \cdots + \xi_n \sum_{j=k_0+\cdots+k_{n-1}+1}^{k_0+\cdots+k_n} X_j\right)};\ N(t_0) = k_0, \\
&\qquad\qquad N(t_1) - N(t_0) = k_1, \ldots, N(t_n) - N(t_{n-1}) = k_n\Bigg] \\
&= \sum_{k_0, k_1 \cdots k_n \ge 0} P(N(t_0) = k_0) P(N(t_1) - N(t_0) = k_1) \cdots \\
&\qquad\qquad \cdots P(N(t_n) - N(t_{n-1}) = k_n) \prod_{j=1}^{n} \varphi_F(\xi_j)^{k_j} \\
&= \sum_{k_1, \ldots, k_n \ge 0} \prod_{j=1}^{n} \frac{(\lambda(t_j - t_{j-1}))^{k_j}}{k_j!} \mathrm{e}^{-\lambda(t_j - t_{j-1})} \prod_{j=1}^{n} \varphi_F(\xi_j)^{k_j} \\
&= \prod_{j=1}^{n} \sum_{k=0}^{\infty} \frac{(\lambda(t_j - t_{j-1}) \varphi_F(\xi_j))^k}{k!} \mathrm{e}^{-\lambda(t_j - t_{j-1})} \\
&= \prod_{j=1}^{n} \varphi_{Y(t_j) - Y(t_{j-1})}(\xi_j) \,.
\end{aligned}$$

This shows the independence.

**0.11.** Let $X_1, X_2, \ldots$ be independent and uniformly distributed on $[0,1]$ and let

$$Z_n(t) = \sqrt{n}(Y_n(t) - t) = \frac{nY_n(t) - nt}{\sqrt{n}} = \sum_{k=1}^{n} \frac{e_{[0,t]}(X_k) - t}{\sqrt{n}} ,$$

where $e_{[0,t]}$ is the indicator of $[0,t]$. Let $V_k(t) = e_{[0,t]}(X_k) - t$. Let $0 \le t_1 < t_2 < \cdots < t_n$. The random vectors $\mathbf{V}_k = (V_k(t_1), \ldots, .V_k(t_n))$, $k = 1, 2, \ldots$, are independent and identically distributed with mean vector 0 and, for $i \le j$,

$$\begin{aligned} E[V_k(t_i)V_k(t_j)] &= E[(e_{[0,t_i]}(X_k) - t_i)(e_{[0,t_j]}(X_k) - t_j)] \\ &= P(X_k \le t_i) - t_j P(X_k \le t_i) - t_i P(X_k \le t_j) + t_i t_j \\ &= t_i(1 - t_j) . \end{aligned}$$

The covariance matrix of $\mathbf{V}_k$ is

$$C = \begin{pmatrix} t_1(1-t_1) & t_1(1-t_2) & \ldots & t_1(1-t_n) \\ t_1(1-t_2) & t_2(1-t_2) & \ldots & t_2(1-t_n) \\ \vdots & \vdots & & \vdots \\ t_1(1-t_n) & t_2(1-t_n) & \ldots & t_n(1-t_n) \end{pmatrix} .$$

By the central limit theorem in $n$ dimensions,

$$(Z_n(t_1), Z_n(t_2), \ldots, Z_n(t_n)) = \textstyle\sum_{k=1}^{n} \mathbf{V}_k / \sqrt{n}$$

converges in law to a Gauss distribution with mean vector 0 and covariance matrix $C$. Let

$$\mathbf{B} = (B(t_1) - t_1 B(1), B(t_2) - t_2 B(1), \ldots, B(t_n) - t_n B(1)).$$

The mean vector of $\mathbf{B}$ is 0. For $i \le j$ we have

$$\begin{aligned} &E[(B(t_i) - t_i B(1))(B(t_j) - t_j B(1))] \\ &\quad = E\big[[-t_i(B(1) - B(t_j)) - t_i(B(t_j) - B(t_i)) + (1 - t_i)B(t_i)] \\ &\qquad\qquad [-t_j(B(1) - B(t_j)) + (1 - t_j)(B(t_j) - B(t_i)) + (1 - t_j)B(t_i)]\big] \\ &\quad = t_i t_j (1 - t_j) - t_i(1 - t_j)(t_j - t_i) + (1 - t_i)(1 - t_j)t_i \\ &\quad = t_i(1 - t_j) . \end{aligned}$$

That is, $\mathbf{B}$ has covariance matrix $C$. It is Gauss distributed, because a linear transformation maps $(B(t_1), B(t_2) - B(t_1), \ldots, B(t_n) - B(t_{n-1}))$ to $\mathbf{B}$.

**0.12.** Use

$$\frac{\varepsilon}{1+\varepsilon} P(|Y| > \varepsilon) \le E\left[\frac{|Y|}{1+|Y|}\right] \le \frac{\varepsilon}{1+\varepsilon} + P(|Y| > \varepsilon).$$

**0.13.** *Example 1.* Let $\{Z_n\}_n$ be independent identically distributed random variables with mean 0 and variance 1 and let

$$X_n = \frac{Z_1 + \cdots + Z_n}{\sqrt{2n \log\log n}}$$

for integers $n > \mathrm{e}$. Then, according to the law of the iterated logarithm,

$$\limsup_{n\to\infty} X_n = 1, \quad \liminf_{n\to\infty} X_n = -1 \quad \text{a. s.}$$

But, since $\sqrt{2\log\log n} X_n$ converges in law to the standard Gauss distribution by the central limit theorem,

$$P(|X_n| > \varepsilon) = P(\sqrt{2\log\log n}|X_n| > \varepsilon\sqrt{2\log\log n}) \to 0, \qquad n \to \infty.$$

*Example 2.* Consider the unit interval with Lebesgue measure as a probability space. Define $X_{2n-1}$ for $n = 1, 2, \ldots$, by $X_1 = e_{[0,1/2]}$, $X_3 = e_{[1/2,1]}$, $X_5 = e_{[0,1/3]}$, $X_7 = e_{[1/3,2/3]}$, $X_9 = e_{[2/3,1]}$, $X_{11} = e_{[0,1/4]}$, $\ldots$, and $X_{2n} = -X_{2n-1}$ for $n = 1, 2, \ldots$. Then $\{X_n\}$ has the desired properties.

**0.14.** (i) It follows from the definition of $d$ that $d(f_m, f) \to 0$ if and only if $\|f_m - f\|_n \to 0$ for all $n$.

(ii) Since $\frac{x}{1+x}$ is increasing for $x \geq 0$ we have

$$\frac{\|f+g\|_n}{1+\|f+g\|_n} \leq \frac{\|f\|_n + \|g\|_n}{1+\|f\|_n+\|g\|_n} \leq \frac{\|f\|_n}{1+\|f\|_n} + \frac{\|g\|_n}{1+\|g\|_n}.$$

Hence $d$ satisfies the triangle inequality.

Let us prove that $C$ is complete in the metric $d$. Let $\{f_p\}$ satisfy $d(f_q, f_p) \to 0$ as $p, q \to \infty$. Then $\|f_p - f_q\|_n \to 0$ for each $n$ and hence $\{f_p\}$ converges uniformly on $[-n, n]$ to a continuous function $f$. That is, $d(f_p, f) \to 0$.

To see the separability, note that any $f \in C$ is expressed as $f = g + \mathrm{i}\,h$ with real-valued continuous functions $g$, $h$. Given $\varepsilon > 0$, choose $n$ such that $\sum_{k>n} 2^{-k} < \varepsilon/2$ and apply Weierstrass's theorem on $[-n, n]$. Thus we can find real polynomials $\widetilde{g}$, $\widetilde{h}$ with rational coefficients such that $d(\widetilde{g}, g) < \varepsilon$ and $d(\widetilde{h}, h) < \varepsilon$. Since such polynomials are countable, this shows that $C$ is separable in the metric $d$.

(iii) *The "only if" part.* Let $M$ be conditionally compact.

Assume that (a) is not true. Then, for some $n$, $\sup_{f\in M} \|f\|_n = \infty$. Choose a sequence $f_p \in M$ such that $\|f_p\|_n \to \infty$ as $p \to \infty$. Then $\{f_p\}$ has no convergent subsequence, which is a contradiction.

Assume that (b) is not true. Then for some $n$ and some $\varepsilon > 0$ we can find a sequence $\{f_p\}$ in $M$ and pairs $(t_p, s_p)$ such that $|t_p|, |s_p| \leq n$, $|t_p - s_p| \to 0$ as $p \to \infty$, and $|f_p(t_p) - f_p(s_p)| > \varepsilon$ for all $p$. We can assume $t_p, s_p \to t$ with some $t$, and $d(f_p, f) \to 0$ with some $f \in C$ since $M$ is conditionally compact. Then

$$\begin{aligned}\varepsilon < |f_p(t_p) - f_p(s_p)| &\le |f_p(t_p) - f(t_p)| + |f_p(s_p) - f(s_p)| \\ &+ |f(t_p) - f(t)| + |f(s_p) - f(t)| \to 0\end{aligned}$$

as $p \to \infty$, which is a contradiction.

*The "if" part.* Assume that $M$ satisfies (a) and (b). Let $\{f_p\}$ be a sequence in $M$. We prove that $\{f_p\}$ has a convergent subsequence. Fix $n$ for the moment. Choose for each $m$ a $\delta_m$ such that

$$\sup\{|f(t) - f(s)|: |t|, |s| \le n,\ |t - s| < \delta_m,\ f \in M\} < 1/m$$

and let $-n = t_{m,0} < \cdots < t_{m,p(m)} = n$ be a subdivision of $[-n, n]$ with $|t_{m,i} - t_{m,i-1}| < \delta_m$, $i = 1, \ldots, p(m)$. Using (a), we can choose by the diagonal argument a subsequence $\{g_p\}$ of $\{f_p\}$ such that $\{g_p\}$ is convergent at all points $t_{m,i}$. For any $t \in [-n, -n]$ and $m$ we can find $t_{m,i}$ with $|t - t_{m,i}| < \delta_m$ and so

$$\begin{aligned}|g_p(t) - g_q(t)| &\le |g_p(t) - g_p(t_{m,i})| + |g_q(t) - g_q(t_{m,i})| + |g_p(t_{m,i}) - g_q(t_{m,i})| \\ &< 3/m\end{aligned}$$

for all $i$ if $p$ and $q$ are large enough. Thus $g_p(t)$ converges uniformly on $[-n, n]$. Again by the diagonal argument we get a subsequence which converges uniformly on all compact sets.

**0.15.** The mapping $\Phi$ is one-to-one by Lévy's inversion formula. If a sequence $\{\mu_n\}$ in $\mathfrak{P}$ converges to $\mu \in \mathfrak{P}$ in the Lévy metric $\rho$, then $\Phi\mu_n$ converges to $\Phi\mu$ uniformly on any finite interval, which is seen by essentially the same argument as the proof that $\rho(F_n, F) \to 0$ implies $\mu_n \to \mu$ (weak*) in the solution of Problem 0.43.

Conversely, assume that $\Phi\mu_n \to \Phi\mu$ uniformly on any finite interval (actually, we need only assume the pointwise convergence). Since $(\Phi\mu)(z)$ is continuous at $z = 0$ and $(\Phi\mu)(0) = 1$, for every $\varepsilon > 0$ there is $a > 0$ such that $(2a)^{-1}\int_{-a}^{a}(1 - (\Phi\mu)(z))\mathrm{d}z < \varepsilon$. Hence there is $n_0$ such that $(2a)^{-1}\int_{-a}^{a}(1 - (\Phi\mu_n)(z))\mathrm{d}z < \varepsilon$ for all $n \ge n_0$. Since

$$\begin{aligned}\frac{1}{2a}\int_{-a}^{a}(1 - (\Phi\mu_n)(z))\mathrm{d}z &= \frac{1}{2a}\int_{-a}^{a}\mathrm{d}z\int_{-\infty}^{\infty}(1 - \mathrm{e}^{\mathrm{i}zx})\mu_n(\mathrm{d}x) \\ &= \int_{-\infty}^{\infty}\frac{\mu_n(\mathrm{d}x)}{2a}\int_{-a}^{a}(1 - \mathrm{e}^{\mathrm{i}zx})\mathrm{d}z = \int_{-\infty}^{\infty}\left(1 - \frac{\sin ax}{ax}\right)\mu_n(\mathrm{d}x) \\ &\ge (1/2)\mu_n\{x: |x| \ge 2/a\},\end{aligned}$$

it follows that $\{\mu_n\}$ is tight (see Problem 0.44). Hence any subsequence $\{\mu_{n'}\}$ of $\{\mu_n\}$ contains a further subsequence $\{\mu_{n''}\}$ that converges to some $\nu$. Since $\Phi\mu_{n''} \to \Phi\nu$, we have $\Phi\nu = \Phi\mu$, that is, $\nu = \mu$. Therefore $\mu_n \to \mu$.

**0.16.** For any $\varphi \in \widetilde{\mathfrak{P}}$

$$|\varphi(z + h) - \varphi(z)| \le \sqrt{2|\varphi(h) - 1|},$$

because

$$
\begin{aligned}
|\varphi(z+h)-\varphi(z)| &= \left| \int_{-\infty}^{\infty} \mathrm{e}^{\mathrm{i}zx}(\mathrm{e}^{\mathrm{i}hx}-1)\mu(\mathrm{d}x) \right| \\
&\le \left( \int_{-\infty}^{\infty} |\mathrm{e}^{\mathrm{i}hx}-1|^2 \mu(\mathrm{d}x) \right)^{1/2} = \left( \int_{-\infty}^{\infty} 2(1-\cos hx)\mu(\mathrm{d}x) \right)^{1/2} \\
&= \sqrt{2(1-\operatorname{Re}\varphi(h))} \le \sqrt{2|1-\varphi(h)|}.
\end{aligned}
$$

If $M \subset \widetilde{\mathfrak{P}}$ is conditionally compact in the metric $d$, then by Problem 0.14 (iii) $M$ is equi-continuous on every bounded interval. Conversely, assume that $M$ is equi-continuous on some neighborhood of 0. Then $\sup_{\varphi \in M} |\varphi(h)-1| \to 0$ as $h \downarrow 0$. Thus, by the inequality above,

$$
\sup_{\varphi \in M,\, z \in R^1} |\varphi(z+h)-\varphi(z)| \to 0, \qquad h \downarrow 0,
$$

that is, $M$ is equi-continuous on the real line. Since $|\varphi(z)| \le 1$, $M$ is equi-bounded on the real line. Therefore $M$ is conditionally compact by Problem 0.14 (iii).

**0.17.** Let $\varphi(\xi) = \varphi(\xi;\mu)$. Assume that $\{\mu^{n*}\}_{n=1,2,\ldots}$ is conditionally compact. Then $\{\varphi(\xi)^n\}$ is conditionally compact in $\widetilde{\mathfrak{P}}$ by Problem 0.15. Then by Problem 0.16 $\{\varphi(\xi)^n\}$ is equi-continuous on any bounded interval. Hence there is a $\delta > 0$ such that $|1-\varphi(\xi)^n| < 1/2$ for all $n$ and all $|\xi| < \delta$. Fix $\xi$ for the moment and write $\varphi(\xi) = r\mathrm{e}^{\mathrm{i}\theta}$, $0 \le r \le 1$, $0 \le \theta < 2\pi$. Then $|1 - r^n \mathrm{e}^{\mathrm{i}n\theta}| < 1/2$ for all $n$. It follows that $r = 1$ and $\theta = 0$. Hence $\varphi(\xi) \equiv 1$ on $[-\delta, \delta]$. Since $|\varphi(\xi+h)-\varphi(\xi)| \le \sqrt{2|\varphi(h)-1|}$, we see that $\varphi(\xi) \equiv 1$ on $R^1$. Hence $\mu$ is the $\delta$-distribution at 0. The converse is obvious.

**0.18.** (i) Let $\mathfrak{M}$ be a family of Gauss distributions $N(m,v)$. If the parameters $m$, $v$ are bounded, then $\mathfrak{M}$ is conditionally compact, because for every sequence $N(m_n, v_n)$ in $\mathfrak{M}$ we can select a subsequence $N(m_{n'}, v_{n'})$ such that $m_{n'}$ and $v_{n'}$ tend to some $m$ and $v$ and so $N(m_{n'}, v_{n'}) \to N(m,v)$. Suppose that $\{m\colon N(m,v) \in \mathfrak{M}\}$ or $\{v\colon N(m,v) \in \mathfrak{M}\}$ is unbounded. Then there is a sequence $\mu_n = N(m_n, v_n) \in \mathfrak{M}$ such that $|m_n| \to \infty$ or $v_n \to \infty$. The sequence $\{\mu_n\}$ does not contain a convergent subsequence. Indeed, if $v_n \to \infty$ then $|\varphi(\xi;\mu_n)| = \mathrm{e}^{-v_n \xi^2/2} \to 0$ for $\xi \ne 0$. If $|m_n| \to \infty$ then $\{\varphi(\xi;\mu_n)\}$ is not equi-continuous in a neighborhood of 0, since $\varphi(\pi/m_n;\mu_n) < 0$ and $\varphi(0;\mu_n) = 1$.

(ii) A Cauchy distribution with parameters $c > 0$ and $m \in R^1$ has characteristic function $\mathrm{e}^{\mathrm{i}m\xi - c|\xi|}$ (see the solution of Problem 0.30). Hence the assertion may be proved in the same way as in (i).

(iii) Let $\mathfrak{M}$ be a family of Poisson distributions $p(\lambda)$. If the parameter $\lambda$ is bounded, then $\mathfrak{M}$ is conditionally compact as in (i). If $\lambda$ is unbounded then, choosing $\mu_n = p(\lambda_n)$ in $\mathfrak{M}$ with $\lambda_n \to \infty$, we see that $|\varphi(\xi, \mu_n)| = \mathrm{e}^{\lambda_n(\cos\xi - 1)} \to 0$ for $0 < \xi < 2\pi$ and hence $\{\mu_n\}$ does not contain a convergent subsequence.

**0.19.** It is enough to prove that (a) implies (c). Let $\mu_n$ be the distribution of $S_n$. Assume that $\mu_n \to \mu$. Then $\delta(\mu_n) \to \delta(\mu)$ and $\gamma(\mu_n) \to \gamma(\mu)$ by Theorem 0.2.3. It follows from Theorem 0.3.2 that $S_n - \gamma(S_n)$ is convergent a. s. Therefore $S_n$ is convergent a. s.

**0.20.** Given $\alpha > 0$, we can choose $\beta > 0$ such that $\alpha \int_{[0,\infty)} \mathrm{e}^{-\beta x}\mu(\mathrm{d}x) < 1$, since $\mu\{0\} = 0$. Then, by Fubini's theorem for nonnegative integrands,

$$\begin{aligned} I(\alpha,\beta) &\equiv \int_0^\infty \mathrm{e}^{-\beta t}\sum_{k=1}^\infty \alpha^k \mu^{k*}[0,t]\mathrm{d}t = \sum_{k=1}^\infty \alpha^k \int_0^\infty \mathrm{e}^{-\beta t}\mathrm{d}t \int_{[0,t]} \mu^{k*}(\mathrm{d}x) \\ &= \sum_{k=1}^\infty \alpha^k \int_0^\infty \mu^{k*}(\mathrm{d}x) \int_x^\infty \mathrm{e}^{-\beta t}\mathrm{d}t = \sum_{k=1}^\infty \alpha^k \frac{1}{\beta}\int_0^\infty \mathrm{e}^{-\beta x}\mu^{k*}(\mathrm{d}x) \\ &= \sum_{k=1}^\infty \frac{1}{\beta}\left(\alpha \int_0^\infty \mathrm{e}^{-\beta x}\mu(\mathrm{d}x)\right)^k. \end{aligned}$$

Therefore $I(\alpha,\beta)$ is finite. Hence $f(t) \equiv \sum_{k=1}^\infty \alpha^k\mu^{k*}[0,t] < \infty$ for almost all $t$. Since $f(t)$ is increasing, $f(t) < \infty$ for all $t > 0$.

**0.21.** Let $\mu$ be an infinitely divisible distribution with support $\subset [0,\infty)$. Let $\mu_n$ be such that $\mu = (\mu_n)^{n*}$. Let $X_{n,1},\ldots,X_{n,n}$ be independent and each $\mu_n$-distributed. Then $P(X_{n,1} \geq 0) = 1$, because otherwise

$$\mu(-\infty,0) = P(X_{n,1} + \cdots + X_{n,n} < 0) \geq P(X_{n,1} < 0)^n > 0.$$

Therefore $\mu_n$ has support in $[0,\infty)$. Let

$$L(\alpha) = \int_{[0,\infty)} \mathrm{e}^{-\alpha x}\mu(\mathrm{d}x), \quad L_n(\alpha) = \int_{[0,\infty)} \mathrm{e}^{-\alpha x}\mu_n(\mathrm{d}x)$$

for $\alpha > 0$. Then $L(\alpha) = L_n(\alpha)^n$ and

$$n(L_n(\alpha) - 1) = \frac{\mathrm{e}^{(1/n)\log L(\alpha)} - 1}{1/n} \to \log L(\alpha), \qquad n \to \infty.$$

Let

$$G_n(\mathrm{d}x) = \frac{nx}{1+x}\mu_n(\mathrm{d}x).$$

Then

$$\begin{aligned} \int_{(0,\infty)} (1-\mathrm{e}^{-\alpha x})\frac{1+x}{x}G_n(\mathrm{d}x) &= \int_{[0,\infty)} (1-\mathrm{e}^{-\alpha x})n\mu_n(\mathrm{d}x) \\ &= n(1 - L_n(\alpha)) \to -\log L(\alpha). \end{aligned}$$

Letting $\alpha = 1$, we see that $\int_0^\infty (1-\mathrm{e}^{-x})\frac{1+x}{x}G_n(\mathrm{d}x)$ is bounded. Since there exists a constant $c > 0$ such that $(1-\mathrm{e}^{-x})\frac{1+x}{x} \geq c$ for all $x \geq 0$, it follows that $G_n[0,\infty)$ is bounded. We can prove that $G_n(x,\infty)\to 0$ uniformly in $n$ as $x\to\infty$. Indeed, given $\varepsilon>0$ we can find $\delta>0$ such that $-\log L(\delta)<\varepsilon$. Then, for $n$ large enough, $\int_0^\infty (1-\mathrm{e}^{-\delta x})\frac{1+x}{x}G_n(\mathrm{d}x)<\varepsilon$. Since $(1-\mathrm{e}^{-\delta x})\frac{1+x}{x}\geq 1-\mathrm{e}^{-1}>1/2$ for $x\geq 1/\delta$, we have $G_n(1/\delta,\infty)<2\varepsilon$ for $n$ large enough. Hence there is $x$ such that $G_n(x,\infty)<2\varepsilon$ uniformly in $n$.

Now by the selection theorem we can find a subsequence $\{G_{p(n)}\}_n$ and a bounded measure $G$ such that

$$\int_{[0,\infty)} f(x)G_{p(n)}(\mathrm{d}x)\to\int_{[0,\infty)} f(x)G_n(\mathrm{d}x)$$

for all bounded continuous function $f$. It follows that

$$-\log L(\alpha)=\int_{[0,\infty)}(1-\mathrm{e}^{-\alpha x})\frac{1+x}{x}G(\mathrm{d}x)\,.$$

Now let $m=G(\{0\})$ and $n(\mathrm{d}x)=\frac{1+x}{x}G(\mathrm{d}x)$ on $(0,\infty)$. Then $m\geq 0$, $\int_0^\infty \frac{x}{1+x}n(\mathrm{d}x)<\infty$, and

$$L(\alpha)=\exp\left\{-\alpha m-\int_0^\infty (1-\mathrm{e}^{-\alpha x})\,n(\mathrm{d}x)\right\}\,.$$

Let $\{X_t\}$ be the homogeneous Lévy process whose value at $t=1$ has distribution $\mu$. Then $P(X_{1/n}\geq 0)=1$. Hence $P(X_r\geq 0)=1$ for all rational $r\geq 0$. Hence $P(X_t\geq 0)=1$ for all $t\geq 0$. It follows that $\{X_t\}$ has increasing paths. Denote by $\widetilde{m}$ and $\widetilde{n}$ the components of $\{X_t\}$ in Theorem 1.11.3 (ii). Applying then Theorem 1.11.1 we have

$$X_t=\widetilde{m}t+\iint\limits_{\substack{0<s\leq t\\ 0<u<\infty}} u\widetilde{N}(\mathrm{d}s\mathrm{d}u)\,,$$

where $\widetilde{m}\geq 0$, $\widetilde{N}$ is a Poisson random measure with mean measure $\mathrm{d}t\,\widetilde{n}(\mathrm{d}u)$, and $\int_{(0,\infty)}\frac{u}{1+u}\widetilde{n}(\mathrm{d}u)<\infty$. Then we can prove that

$$E(\mathrm{e}^{-\alpha X_t})=\exp\left\{t\left(-\widetilde{m}\alpha-\int_0^\infty (1-\mathrm{e}^{-\alpha x})\,\widetilde{n}(\mathrm{d}x)\right)\right\},\qquad \alpha>0.$$

Hence $\mu$ has Laplace transform

$$\exp\left\{-\widetilde{m}\alpha-\int_0^\infty (1-\mathrm{e}^{-\alpha x})\,\widetilde{n}(\mathrm{d}x)\right\}.$$

We can prove the uniqueness of the expression of the Laplace transform in this form. Thus $m=\widetilde{m}$ and $n=\widetilde{n}$. As $\widetilde{N}$ represents the number of jumps in time and size, $\widetilde{n}$ is the Lévy measure in Lévy's canonical form of the characteristic function of $\mu$.

Actually the latter half of the solution above gives the expression of $L(\alpha)$, so that the former half is unnecessary.

**0.22.** Notice that

$$\frac{2-\varphi(z)-\varphi(-z)}{z^2} = \int_{-\infty}^{\infty} \frac{2-\mathrm{e}^{\mathrm{i}xz}-\mathrm{e}^{-\mathrm{i}xz}}{z^2}\mu(\mathrm{d}x) = \int_{-\infty}^{\infty} \frac{2(1-\cos xz)}{z^2}\mu(\mathrm{d}x).$$

Since $1-\cos u \le u^2/2$ for all $u \in R^1$, $0 \le 2(1-\cos xz)/z^2 \le x^2$ for all $z \ne 0$ and $x \in R^1$. If $\int_{-\infty}^{\infty} x^2\mu(\mathrm{d}x) < \infty$, then by the dominated convergence theorem

$$\lim_{z\to 0} \frac{2-\varphi(z)-\varphi(-z)}{z^2} = \int_{-\infty}^{\infty} x^2\mu(\mathrm{d}x)\,.$$

If $\int_{-\infty}^{\infty} x^2\mu(\mathrm{d}x) = \infty$ then, by Fatou's lemma,

$$\liminf_{z\to 0} \frac{2-\varphi(z)-\varphi(-z)}{z^2} \ge \int_{-\infty}^{\infty} x^2\mu(\mathrm{d}x) = \infty\,.$$

**0.23.** (a) $\Longrightarrow$ (b). Since

$$\frac{\varphi(z+h)-\varphi(z)}{h} = \int_{-\infty}^{\infty} \frac{\mathrm{e}^{\mathrm{i}zx}(\mathrm{e}^{\mathrm{i}xh}-1)}{h}\mu(\mathrm{d}x)$$

and the integrand has modulus $\le |x|$, $\varphi(z)$ is differentiable and

$$\varphi'(z) = \int_{-\infty}^{\infty} \mathrm{i}x\mathrm{e}^{\mathrm{i}zx}\mu(\mathrm{d}x)$$

by the dominated convergence theorem. Now

$$\frac{\varphi'(z+h)-\varphi'(z)}{h} = \int_{-\infty}^{\infty} \frac{\mathrm{i}x\mathrm{e}^{\mathrm{i}zx}(\mathrm{e}^{\mathrm{i}hx}-1)}{h}\mu(\mathrm{d}x)$$

and the integrand has modulus $\le x^2$. Thus by the same theorem $\varphi'(z)$ is differentiable and

$$\varphi''(z) = -\int_{-\infty}^{\infty} x^2\mathrm{e}^{\mathrm{i}zx}\mu(\mathrm{d}x)\,.$$

The right-hand side is continuous in $z$ again by the same theorem.

(b) $\Longrightarrow$ (c). Trivial.

(c) $\Longrightarrow$ (a). We assume that $\varphi$ is differentiable on $R^1$ and $\varphi'(z)$ is differentiable at 0. By l'Hospital's rule

$$\lim_{z\to 0} \frac{2-\varphi(z)-\varphi(-z)}{z^2} = \lim_{z\to 0} \frac{\varphi'(-z)-\varphi'(z)}{2z}$$
$$= -\lim_{z\to 0} \frac{\varphi'(-z)-\varphi'(0)}{-2z} - \lim_{z\to 0} \frac{\varphi'(z)-\varphi'(0)}{2z} = -\varphi''(0)\,.$$

Now use Problem 0.22.

**0.24.** Let $X$ and $Y$ be independent random variables with distributions $\mu$ and $\nu$ respectively. Then $X+Y$ has the distribution $\mu * \nu$. We prove that $E[(X+Y)^2] < \infty$ if and only if $E[X^2] < \infty$ and $E[Y^2] < \infty$. It is easy to see the "if" part, since $(X+Y)^2 \leq 2(X^2+Y^2)$. Let us prove the "only if" part. Since $\int E[(X+y)^2]\nu(\mathrm{d}y) = \iint (x+y)^2 \mu(\mathrm{d}x)\nu(\mathrm{d}y) < \infty$, there is $y$ such that $E[(X+y)^2] < \infty$. Thus $E[X^2] < \infty$ since $X^2 \leq 2((X+y)^2 + y^2)$. Similarly $E[Y^2] < \infty$.

**0.25.** By using Problem 0.23 we can easily see that it is enough to prove that $f(z) = \int_{|u|\leq 1}(\mathrm{e}^{\mathrm{i}zu} - 1 - \mathrm{i}zu)n(\mathrm{d}u)$ is twice differentiable at $z = 0$. We have

$$\begin{aligned}\frac{f(z+h)-f(z)}{h} &= \int_{|u|\leq 1} \frac{\mathrm{e}^{\mathrm{i}zu}(\mathrm{e}^{\mathrm{i}hu}-1) - \mathrm{i}hu}{h} n(\mathrm{d}u) \\ &= \int_{|u|\leq 1} \frac{\mathrm{e}^{\mathrm{i}zu}(\mathrm{e}^{\mathrm{i}hu}-1-\mathrm{i}hu)}{h} n(\mathrm{d}u) + \int_{|u|\leq 1} \mathrm{i}u(\mathrm{e}^{\mathrm{i}zu}-1)\, n(\mathrm{d}u)\,.\end{aligned}$$

We use that

$$|\mathrm{e}^{\mathrm{i}\theta} - 1| \leq \theta, \quad |\mathrm{e}^{\mathrm{i}\theta} - 1 - \mathrm{i}\theta| \leq \theta^2/2 \qquad \text{for } \theta \in R^1.$$

Since $\int_{|u|\leq 1} u^2 n(\mathrm{d}u) < \infty$, the first integral tends to zero as $h \to 0$ and the second integral exists. Thus

$$f'(z) = \int_{|u|\leq 1} \mathrm{i}u(\mathrm{e}^{\mathrm{i}zu} - 1)\, n(\mathrm{d}u)$$

and we have

$$\frac{f'(h)-f'(0)}{h} = \int_{|u|\leq 1} \frac{\mathrm{i}u(\mathrm{e}^{\mathrm{i}hu}-1)}{h} n(\mathrm{d}u) \to -\int_{|u|\leq 1} u^2 n(\mathrm{d}u), \qquad h \to 0.$$

**0.26.** Let $f(z) = \int_{|u|>1}(\mathrm{e}^{\mathrm{i}zu} - 1)\, n(\mathrm{d}u)$ for $z \in R^1$.

*Sufficiency.* Assume that $\int_{|u|>1} u^2 n(\mathrm{d}u) < \infty$. By Problem 0.23 it is enough to prove that $f(z)$ is twice differentiable at 0. Notice that

$$\frac{f(z+h)-f(z)}{h} = \int_{|u|>1} \frac{\mathrm{e}^{\mathrm{i}zu}(\mathrm{e}^{\mathrm{i}hu}-1)}{h} n(\mathrm{d}u) \to \int_{|u|>1} \mathrm{i}u\mathrm{e}^{\mathrm{i}zu} n(\mathrm{d}u)$$

and that

$$\frac{f'(h)-f'(0)}{h} = \int_{|u|>1} \frac{\mathrm{i}u(\mathrm{e}^{\mathrm{i}hu}-1)}{h} n(\mathrm{d}u) \to -\int_{|u|>1} u^2 n(\mathrm{d}u).$$

*Necessity.* Assume that $\mu$ has finite second order moment. Then $\varphi(z)$ is twice differentiable (Problem 0.23). Since $\varphi(z) \neq 0$ and continuous and since $f(z)$ is continuous, we can for every $z_0$ find a branch of the logarithm such that

$\log \varphi(z) = f(z)$ in a neighborhood of $z_0$. Hence $f(z)$ is twice differentiable. Hence $\int_{|u|>1} e^{izu} n(du)$ is twice differentiable. It follows from Problem 0.23 that $\int_{|u|>1} u^2 n(du) < \infty$.

**0.27.** With some $m'$ we have $\mu = \mu_1 * \mu_2 * \mu_3$, where

$$\varphi(z;\mu_1) = \exp\{i m' z - vz^2/2\},$$

$$\varphi(z;\mu_2) = \exp\left\{\int_{|u|\le 1} (e^{izu} - 1 - izu)\, n(du)\right\},$$

$$\varphi(z;\mu_3) = \exp\left\{\int_{|u|>1} (e^{izu} - 1)\, n(du)\right\}.$$

Now apply the results of Problems 0.24, 0.25 and 0.26.

**0.28.** Notice that

$$e^{iuz} - 1 - \frac{izu}{1+u^2} = (e^{izu} - 1 - izu) + iz\frac{u^3}{1+u^2}.$$

**0.29.** It follows from the symmetry of $\mu$ that $\varphi(z) = \varphi(-z)$. In Lévy's canonical form it is written as

$$\exp\left\{imz - \frac{vz^2}{2} + \int_{-\infty}^{\infty}\left(e^{izu} - 1 - \frac{izu}{1+u^2}\right) n(du)\right\}$$
$$= \exp\left\{-imz - \frac{vz^2}{2} + \int_{-\infty}^{\infty}\left(e^{-izu} - 1 + \frac{izu}{1+u^2}\right) n(du)\right\}.$$

The integral in the right-hand side is $\int_{-\infty}^{\infty}\left(e^{izu} - 1 - \frac{izu}{1+u^2}\right) \check{n}(du)$. Therefore, because of the uniqueness of $m$, $v$, $n$, we have $m = 0$ and $n = \check{n}$. Thus

$$\varphi(z) = \exp\left\{-\frac{vz^2}{2} + \int_{0+}^{\infty} (\cos zu - 1)\, m(du)\right\}$$

where $m(du) = 2\,n(du)$.

**0.30.** (i) Negative binomial distribution $\mu$ with parameters $\lambda > 0$, $0 < \theta < 1$. First note that

$$(1-\theta)^{-\lambda} = \sum_{n=0}^{\infty} \frac{(-\lambda)(-\lambda-1)\cdots(-\lambda-n+1)}{n!}(-\theta)^n$$
$$= \sum_{n=0}^{\infty} \frac{\lambda(\lambda+1)\cdots(\lambda+n-1)}{n!}\theta^n = \sum_{n=0}^{\infty}\binom{\lambda+n-1}{n}\theta^n.$$

Thus $\sum_{n=0}^{\infty} \mu\{n\} = 1$ and $\mu$ has generating function

$$g(t) = (1-\theta)^{\lambda}(1-\theta t)^{-\lambda}.$$

The Laplace transform of $\mu$ is

$$L(\alpha) = g(\mathrm{e}^{-\alpha}) = (1-\theta)^{\lambda}(1-\theta\mathrm{e}^{-\alpha})^{-\lambda}.$$

Thus we can see that $\mu$ is infinitely divisible and

$$\begin{aligned}\log L(\alpha) &= \lambda\log(1-\theta) - \lambda\log(1-\theta\mathrm{e}^{-\alpha}) = -\lambda\sum_{k=1}^{\infty}\frac{\theta^k}{k} + \lambda\sum_{k=1}^{\infty}\frac{\theta^k\mathrm{e}^{-\alpha k}}{k} \\ &= \lambda\sum_{k=1}^{\infty}\frac{\theta^k(\mathrm{e}^{-\alpha k}-1)}{k} = \int_{0+}^{\infty}(\mathrm{e}^{-\alpha u}-1)n(\mathrm{d}u)\,,\end{aligned}$$

where $n$ is concentrated at the positive integers and $n(\{k\}) = \lambda\theta^k/k$. This gives the canonical form of $L(\alpha)$.

(ii) $\Gamma$-distribution $\mu(\mathrm{d}x) = \frac{1}{\Gamma(\lambda)}\mathrm{e}^{-x}x^{\lambda-1}\mathrm{d}x$ on $(0,\infty)$ $(\lambda > 0)$. The Laplace transform of $\mu$ is

$$L(\alpha) = \int_0^{\infty}\mathrm{e}^{-\alpha x}\frac{1}{\Gamma(\lambda)}\mathrm{e}^{-x}x^{\lambda-1}\mathrm{d}x = (1+\alpha)^{-\lambda}.$$

Hence $\mu$ is infinitely divisible. Now assume

$$\log L(\alpha) = -\lambda\log(1+\alpha) = -m\alpha - \int_0^{\infty}(1-\mathrm{e}^{-\alpha x})\,n(\mathrm{d}x)\,.$$

Differentiating (formally) with respect to $\alpha$ we get

$$\frac{-\lambda}{1+\alpha} = -m - \int_0^{\infty}\mathrm{e}^{-\alpha x}x\,n(\mathrm{d}x)\,.$$

Therefore we conjecture that $m = 0$ and $n(\mathrm{d}x) = \lambda\mathrm{e}^{-x}x^{-1}\mathrm{d}x$. It is now easily *proved* that

$$\begin{aligned}&\int_0^{\infty}(1-\mathrm{e}^{-\alpha u})\frac{\lambda\mathrm{e}^{-u}}{u}\mathrm{d}u = \lambda\int_0^{\infty}\mathrm{e}^{-u}\mathrm{d}u\int_0^{\alpha}\mathrm{e}^{-uv}\mathrm{d}v \\ &= \lambda\int_0^{\alpha}\mathrm{d}v\int_0^{\infty}\mathrm{e}^{-u(1+v)}\mathrm{d}u = \lambda\int_0^{\alpha}\frac{\mathrm{d}v}{1+v} = \lambda\log(1+\alpha) \\ &= -\log L(\alpha)\,.\end{aligned}$$

(iii) Cauchy distribution $\mu$ with parameters $c > 0$ and $m \in R^1$. Since

$$\int_0^{\infty}\mathrm{e}^{\mathrm{i}zx}\mathrm{e}^{-x}\mathrm{d}x = \frac{1}{1+\mathrm{i}z} \qquad \text{for } z \in R^1,$$

we have

$$\begin{aligned}\frac{1}{2}\int_{-\infty}^{\infty}\mathrm{e}^{\mathrm{i}zx}\mathrm{e}^{-|x|}\mathrm{d}x &= \frac{1}{2}\left(\int_0^{\infty}\mathrm{e}^{\mathrm{i}zx}\mathrm{e}^{-x}\mathrm{d}x + \int_0^{\infty}\mathrm{e}^{-\mathrm{i}zx}\mathrm{e}^{-x}\mathrm{d}x\right) \\ &= \frac{1}{2}\left(\frac{1}{1-\mathrm{i}z} + \frac{1}{1+\mathrm{i}z}\right) = \frac{1}{1+z^2}\,.\end{aligned}$$

In general, if a distribution $\nu$ has an integrable characteristic function $\varphi_\nu(z)$, then $\nu$ has density $\frac{1}{2\pi}\int_{-\infty}^{\infty} \mathrm{e}^{-\mathrm{i}xz}\varphi_\nu(z)\mathrm{d}z$ (Fourier inversion). Thus

$$\frac{1}{2\pi}\int_{-\infty}^{\infty} \mathrm{e}^{-\mathrm{i}xz}\frac{\mathrm{d}z}{1+z^2} = \frac{1}{2}\mathrm{e}^{-|x|}.$$

Changing the roles of $x$ and $z$, we get

$$\frac{1}{\pi}\int_{-\infty}^{\infty} \mathrm{e}^{\mathrm{i}zx}\frac{\mathrm{d}x}{1+x^2} = \mathrm{e}^{-|z|}.$$

It follows that the Cauchy distribution $\mu$ has characteristic function

$$\varphi(z) = \int_{-\infty}^{\infty} \mathrm{e}^{\mathrm{i}zx}\mu(\mathrm{d}x) = \frac{c}{\pi}\int_{-\infty}^{\infty} \mathrm{e}^{\mathrm{i}zx}\frac{\mathrm{d}x}{(x-m)^2+c^2} = \mathrm{e}^{imz-c|z|}.$$

This shows that $\mu$ is infinitely divisible. For $z > 0$

$$\begin{aligned}\int_{0+}^{\infty}(\cos zu - 1)\frac{du}{u^2} &= z\int_{0+}^{\infty}\frac{\cos u - 1}{u^2}du \\ &= z\left[-\frac{1}{u}(\cos u - 1)\right]_{0+}^{\infty} - z\int_{0+}^{\infty}\frac{\sin u}{u}du = -\frac{z\pi}{2}.\end{aligned}$$

Hence

$$\int_{-\infty}^{\infty}(\cos zu - 1)\frac{c}{\pi u^2}du = -c|z| \qquad \text{for } z \in R^1.$$

Now we have

$$\int_{-\infty}^{\infty}\left(\mathrm{e}^{\mathrm{i}zu} - 1 - \frac{\mathrm{i}zu}{1+u^2}\right)\frac{c}{\pi u^2}du = -c|z|,$$

since the integral of the imaginary part is zero. Hence $\varphi(z)$ has Lévy formula with $m$ being the parameter $m$ of the Cauchy distribution and $v = 0$, $n(du) = \dfrac{c}{\pi u^2}du$.

**0.31.** Let $0 < \alpha < 2$. Then $\displaystyle\int_{0+}^{\infty}\frac{1-\cos u}{u^{\alpha+1}}du$ is finite and positive. Let $1/c$ denote this value. Then

$$|z|^\alpha = c\int_{0+}^{\infty}\frac{1-\cos zu}{u^{\alpha+1}}du \qquad \text{for } z \in R^1.$$

Therefore

$$|z|^\alpha = -\frac{c}{2}\int_{-\infty}^{\infty}\left(\mathrm{e}^{\mathrm{i}zu} - 1 - \frac{\mathrm{i}zu}{1+u^2}\right)\frac{1}{|u|^{\alpha+1}}du,$$

since the integral of the imaginary part is zero. Recalling Lévy's formula, we see that $\mathrm{e}^{-|z|^\alpha}$ is the characteristic function of an infinitely divisible distribution.

For $\alpha = 2$, $e^{-|z|^2}$ is the characteristic function of a Gauss distribution.

**0.32.** It is enough to prove that

$$\int_B X \mathrm{d}P = \int_B E(X \mid \mathcal{D}) \mathrm{d}P \tag{1}$$

for all $B \in \mathcal{D} \vee \mathcal{E}$. Let $D \in \mathcal{D}$ and $E \in \mathcal{E}$. Then, using the independence of $\mathcal{D} \vee \mathcal{C}$ and $\mathcal{E}$ and the $\mathcal{C}$-measurability of $X$, we have

$$\begin{aligned} E(X, D \cap E) &= E(X, D)\, P(E) = E[E(X \mid \mathcal{D}), D]\, P(E) \\ &= E[E(X \mid \mathcal{D}), D \cap E]\,. \end{aligned}$$

Now note that the class of all $B \in \mathcal{D} \vee \mathcal{E}$ satisfying (1) is a Dynkin class containing the multiplicative class $\mathcal{G} = \{D \cap E \colon D \in \mathcal{D},\, E \in \mathcal{E}\}$ and that $\mathcal{D} \vee \mathcal{E}$ is the $\sigma$-algebra generated by $\mathcal{G}$.

**0.33.** We know that $S_n = \sum_{i=1}^n (X_i - E(X_i))$ is a martingale $\mathcal{B}[X_1, \ldots, X_n]$. Since $|x|^p$ is a convex function for $p \geq 1$, $\{|S_n|^p\}_n$ is a submartingale (Theorem 0.7.1). Therefore $f(n) = E(|S_n|^p)$ increases with $n$.

**0.34.** Since $X_S$ is measurable ($\mathcal{B}_S$), it is enough to show that

$$\int_B X_S \mathrm{d}P = \int_B X_m \mathrm{d}P \qquad \text{for all } B \in \mathcal{B}_S\,.$$

We have

$$\begin{aligned} E(X_S, B) &= \sum_{i=1}^m E(X_S, B \cap \{S = i\}) = \sum_{i=1}^m E(X_i, B \cap \{S = i\}) \\ &= \sum_{i=1}^m E(X_m, B \cap \{S = i\}) = E(X_m, B), \end{aligned}$$

since $B \cap \{S = i\} \in \mathcal{B}_i$ and $E(X_m \mid \mathcal{B}_i) = X_i$ for $i \leq m$.

**0.35.** *Existence.* Define $A_1 \equiv 0$,

$$A_n = \sum_{k=2}^n E(X_k - X_{k-1} \mid \mathcal{B}_{k-1}) \qquad \text{for } n \geq 2,$$

and then $M_n = X_n - A_n$. Then (b) is trivial; (c) follows from the submartingale property as

$$A_n - A_{n-1} = E(X_n - X_{n-1} \mid \mathcal{B}_{n-1}) \geq 0.$$

Let us show (a): $M_n$ is measurable ($\mathcal{B}_n$) and

$$\begin{aligned} E(M_n \mid \mathcal{B}_{n-1}) &= E[X_n - E(X_n - X_{n-1} \mid \mathcal{B}_{n-1}) - A_{n-1} \mid \mathcal{B}_{n-1}] \\ &= E(X_n \mid \mathcal{B}_{n-1}) - E(X_n - X_{n-1} \mid \mathcal{B}_{n-1}) - A_{n-1} \\ &= X_{n-1} - A_{n-1} = M_{n-1}\,. \end{aligned}$$

*Uniqueness.* Assume that we have two decompositions

$$X_n = M_n + A_n = M'_n + A'_n$$

satisfying (a)–(c). Then $M_1 = M'_1$ since $A_1 = A'_1 \equiv 0$. If $M_{n-1} = M'_{n-1}$, then $E(M_n - M'_n \mid \mathcal{B}_{n-1}) = M_{n-1} - M'_{n-1} = 0$ by (a) and thus $M_n = M'_n$. Hence $M_n = M'_n$ for all $n$ and so $A_n = A'_n$ for all $n$.

**0.36.** Let $X_n = M_n + A_n$ be Doob's decomposition. Then we have $E(X_T \mid \mathcal{B}_S) = E(M_T \mid \mathcal{B}_S) + E(A_T \mid \mathcal{B}_S)$ and $E(A_T \mid \mathcal{B}_S) \geq 0$. We have $\mathcal{B}_S \subset \mathcal{B}_T$ since $S \leq T$. Thus, by Problem 0.34,

$$E(M_T \mid \mathcal{B}_S) = E[E(M_m \mid \mathcal{B}_T) \mid \mathcal{B}_S] = E(M_m \mid \mathcal{B}_S) = M_S\ .$$

(For an alternative solution use Theorem 0.7.4.)

**0.37.** Let $S = \inf\{n\colon X_n > c\}$. Following the idea in the hint and using Problem 0.36, we get

$$EX_m^+ \geq EX_{S\wedge m}^+ \geq E(X_S^+,\, S \leq m) \geq cP(S \leq m) = cP(\sup_{n\leq m} X_n > c).$$

Let $m \uparrow \infty$ to complete the proof.

**0.38.** (i) Assume that, for some $p > 1$, $E(|X_\lambda|^p) \leq K$ for every $\lambda \in \Lambda$. Then

$$\sup_\lambda E(|X_\lambda|,\, |X_\lambda| > a) \leq \sup_\lambda \frac{E(|X_\lambda|^p)}{a^{p-1}} \leq \frac{K}{a^{p-1}} \to 0$$

as $a \to \infty$.

To consider the case $p = 1$, assume that $P(X_n = n) = 1/n$ and $P(X_n = 0) = 1 - 1/n$ for $n = 1, 2, \ldots$. Then $E(|X_n|) = 1$ for all $n$ but $\sup_n E(|X_n|,\, |X_n| > a) = 1$ for all $a > 0$.

(ii) Assume that $E(|X_n - X|) \to 0$ and $X_n, X \in L^1$. We have $\int_A |X| \mathrm{d}P \to 0$ as $P(A) \to 0$, because

$$\begin{aligned}\int_A |X|\mathrm{d}P &= \int_{A\cap\{|X|>b\}} |X|\mathrm{d}P + \int_{A\cap\{|X|\leq b\}} |X|\mathrm{d}P \\ &\leq \int_{|X|>b} |X|\mathrm{d}P + bP(A)\ .\end{aligned}$$

Now we note that

$$E(|X_n|,\, |X_n| > a) \leq E(|X_n - X|) + E(|X|,\, |X_n| > a)\ .$$

Let $\varepsilon > 0$ be given. Since $P(|X_n| > a) \leq (1/a)E(|X_n|)$ and since $\sup E(|X_n|) < \infty$, we can find $a_0$ such that $E(|X|,\, |X_n| > a) < \varepsilon/2$ for all $n$ and all $a \geq a_0$. We can also find $N_0$ such that $E(|X_n - X|) < \varepsilon/2$ for all $n \geq N_0$. Hence $E(|X_n|,\, |X_n| > a) < \varepsilon$ for all $n \geq N_0$ if $a \geq a_0$. Since, for fixed $n$,

$E(|X_n|, |X_n| > a) \to 0$ as $a \to \infty$, we get $\sup_n E(|X_n|, |X_n| > a) < \varepsilon$ for $a$ big enough. That is, $\{X_n\}$ is uniformly integrable. Since

$$P(|X_n - X| > \varepsilon) \le (1/\varepsilon)E(|X_n - X|) \to 0 \qquad \text{as } n \to \infty\,,$$

$X_n \to X$ i. p.

Conversely, assume that $\{X_n\}$ is uniformly integrable and that $X_n \to X$ i. p. Then $\sup_n E(|X_n|) < \infty$. We have $X \in L^1$ since, choosing a subsequence $X_{n'} \to X$ a. s., we get

$$E(|X|) \le \liminf E(|X_{n'}|) < \infty$$

by Fatou's lemma. For any $c > 0$ we have

$$E(|X_n - X|) \le I_{n,c} + E(|X_n|, |X_n| > c) + E(|X|, |X| > c)$$

with $I_{n,c} = E(|X_n e_{\{|X_n| \le c\}} - X e_{\{|X| \le c\}}|)$. As $c \to \infty$, $E(|X|, |X| > c) \to 0$ and $E(|X_n|, |X_n| > c) \to 0$ uniformly in $n$. Fix $c$ such that $P(|X| = c) = 0$. For any subsequence $\{X_{n'}\}$ of $\{X_n\}$, we can choose a subsequence $\{X_{n''}\}$ of $\{X_{n'}\}$ such that $X_{n''} \to X$ a. s. Hence $X_{n''} e_{\{X_{n''} \le c\}} \to X e_{\{|X| \le c\}}$ a. s. and it follows that $I_{n'',c} \to 0$, by the dominated convergence theorem. Thus we see that $I_{n,c} \to 0$ as $n \to \infty$. Therefore $E(|X_n - X|) \to 0$ as $n \to \infty$.

**0.39.** Let $X_n$, $n = 1, 2, \ldots$, be a martingale $\{\mathcal{B}_n\}$ and suppose that $\{E|X_n|\}$ is bounded.

(i) Given $a < b$, notice that

$$P\left(\liminf_{n\to\infty} X_n < a < b < \limsup_{n\to\infty} X_n\right) = 0\,.$$

Indeed, if this probability is not zero, then $P(U_{ab}(X_n) = \infty) > 0$ and hence $E[U_{ab}(X_n)] = \infty$, which contradicts the inequality

$$E[U_{ab}(X_n)] \le \frac{1}{b-a} \sup_n E[(X_n - a)^+] \le \frac{1}{b-a}\left(\sup_n E(|X_n|) + |a|\right) < \infty\,.$$

Now we see that

$$P\left(\bigcup_{a,b:\ \text{rational},\, a<b} \left\{\liminf_{n\to\infty} X_n < a < b < \limsup_{n\to\infty} X_n\right\}\right) = 0\,.$$

Therefore

$$P\left(\liminf_{n\to\infty} X_n < \limsup_{n\to\infty} X_n\right) = 0\,,$$

that is, there is $X_\infty$ such that $X_n \to X_\infty$ a. s. Since Fatou's lemma shows that

$$E(|X_\infty|) \le \liminf_{n\to\infty} E(|X_n|) < \infty\,,$$

$X_\infty$ is in $L^1$ and is finite a. s.

(ii) We assume that $\{X_n\}$ is uniformly integrable. By Problem 0.38, $X_n \to X_\infty$ in $L^1$-norm. In order to show that $\{X_n\}_{n=1,2,\ldots,\infty}$ is a martingale, we need only prove that

$$E(X_\infty, B) = E(X_n, B) \qquad \text{for every } B \in \mathcal{B}_n .$$

Notice that $|E(X_\infty, B) - E(X_m, B)| \le E(|X_\infty - X_m|, B) \le E(|X_\infty - X_m|) \to 0$ as $m \to \infty$ and that $E(X_m, B) = E(X_n, B)$ for all $m \ge n$.

**0.40.** Let $\{B(t), t \in [0,\infty)\}$ be a Brownian motion. Let $\mathcal{B}_t = \mathcal{B}[B(s)\colon 0 \le s \le t]$. To show that $\{aB(t)+b\}_t$ is a martingale $\{\mathcal{B}_t\}$, it is enough to prove that for $0 \le s < t$

$$E(aB(t)+b, C) = E(aB(s)+b, C) \quad \text{i.e.} \quad E(B(t)-B(s), C) = 0$$

for all $C \in \mathcal{B}_s$. Since $B(t)-B(s)$ is independent of $\mathcal{B}_s$,

$$E(B(t)-B(s), C) = E(B(t)-B(s))P(C) = 0 .$$

**0.41.** Let $B_i(t)$, $i = 1,2$, be independent Brownian motions and $W_t = B_1(t) + \mathrm{i}B_2(t)$. We set $\mathcal{B}_t = \mathcal{B}[W_s\colon 0 \le s \le t]$. We want to show that ${W_t}^n$ is a martingale $\{\mathcal{B}_t\}$ for any integer $n \ge 1$. It is enough to show that for $s < t$ and $C \in \mathcal{B}_s$

$$E({W_t}^n - {W_s}^n, C) = 0 .$$

Since $W_t = (W_t - W_s) + W_s$, it is enough to prove that

$$E[(W_t - W_s)^p {W_s}^q, C] = 0$$

for all positive integers $p$ and nonnegative integers $q$. Since $W_t - W_s$ is independent of $\mathcal{B}_s$, the left-hand side is

$$= E[(W_t - W_s)^p]\, E[{W_s}^q, C] .$$

Now notice that

$$E[(W_t - W_s)^p] = \int_{-\infty}^{\infty}\int_{-\infty}^{\infty} (x+\mathrm{i}y)^p \frac{1}{2\pi(t-s)} \mathrm{e}^{-(x^2+y^2)/(2(t-s))} \mathrm{d}x\mathrm{d}y$$
$$= \int_0^\infty \mathrm{d}r \int_0^{2\pi} r^p(\cos p\theta + \mathrm{i}\sin p\theta) \frac{r}{2\pi(t-s)} \mathrm{e}^{-r^2/(2(t-s))} \mathrm{d}\theta = 0 .$$

(ii) Let $u(x,y)$ be a polynomial in $x$ and $y$ with real coefficients such that $\dfrac{\partial^2 u}{\partial x^2} + \dfrac{\partial^2 u}{\partial y^2} = 0$. Then there is a unique (up to a constant term) function $v(x,y)$ such that the pair $u, v$ satisfies the Cauchy–Riemann equations. This function $v$ is a polynomial with real coefficients. The complex-valued function $f$ defined by

$$f(x+\mathrm{i}y) = u(x,y) + \mathrm{i}v(x,y)$$

is analytic in the whole complex plane. Since there are constants $n$ and $c$ such that

$$|f(x+\mathrm{i}y)| \le c(1+|x+\mathrm{i}y|^n)\,,$$

$f$ is a polynomial with complex coefficients. Therefore it follows from (i) that $f(W_t)$ is a martingale. Taking the real part we find that $u(B_1(t), B_2(t))$ is a martingale.

**0.42.** Let $\mu_k$, $k = 1,2,3,\ldots$, be infinitely divisible distributions and let $\mu_k \to \mu$ as $k \to \infty$. We want to prove that $\mu$ is infinitely divisible. There exist distributions $\mu_{k,n}$ such that

$$\mu_k = (\mu_{k,n})^{n*}, \qquad n = 1,2,3,\ldots\,.$$

Fix $n$. By Theorem 0.3.5 $\{\mu_{k,n} - \gamma(\mu_{k,n})\}_k$ is conditionally compact. Thus there is a subsequence $\{\mu_{k',n}\}$ such that

$$\mu_{k',n} - \gamma(_{k',n}) \to \nu_n$$

for some $\nu_n$. It follows that

$$(\mu_{k',n} - \gamma(\mu_{k',n}))^{n*} = (\mu_{k',n})^{n*} - n\gamma(\mu_{k',n}) \to {\nu_n}^{n*}.$$

Since $(\mu_{k',n})^{n*} = \mu_{k'} \to \mu$, we see that $n\gamma(\mu_{k',n}) \to c_n$ for some $c_n$. Hence $\mu_{k',n} \to \nu_n + c_n/n$ and so $(\mu_{k',n})^{n*} \to (\nu_n + c_n/n)^{n*}$. Thus $\mu = (\nu_n + c_n/n)^{n*}$.

**0.43.** The weak* topology of $\mathfrak{P}$ is given by the system of neighborhoods of a point $\mu_0 \in \mathfrak{P}$ consisting of

$$U_{f_1,\ldots,f_n,\varepsilon}(\mu_0) = \{\mu \in \mathfrak{P} \colon |\textstyle\int f_i(x)\mu(\mathrm{d}x) - \int f_i(x)\mu_0(\mathrm{d}x)| < \varepsilon,\ i = 1,\ldots,n\},$$
$$f_1,\ldots,f_n \in C(R^1),\ \varepsilon > 0.$$

Thus a sequence $\{\mu_n\}_{n=1,2\ldots}$ in $\mathfrak{P}$ converges to $\mu \in \mathfrak{P}$ in the weak* topology if and only if

$$\int_{R^1} f(x)\mu_n(\mathrm{d}x) \to \int_{R^1} f(x)\mu(\mathrm{d}x)$$

for all $f \in C(R^1)$. Denote the distribution functions of $\mu_n$ and $\mu$ by $F_n$ and $F$. Let us show that

$$\mu_n \to \mu \quad (\text{weak*}) \quad \Longleftrightarrow \quad \rho(F_n, F) \to 0,$$

where $\rho$ is the Lévy metric.

*Proof of* ($\Longrightarrow$). By Problem 0.7 it is enough to show that $F_n(x_0) \to F(x_0)$ for each continuity point $x_0$ of $F$. Let $x_1 < x_0 < x_2$. Choose $f_1$ and $f_2$ in $C(R^1)$ satisfying $0 \le f_i \le 1$ for $i = 1,2$, $f_1 = 1$ on $(-\infty, x_1]$ and 0 on $[x_0, \infty)$, and $f_2 = 1$ on $(-\infty, x_0]$ and 0 on $[x_2, \infty)$. Then

$$\liminf_{n\to\infty} F_n(x_0) \geq \lim_{n\to\infty} \int f_1(x)\mu_n(\mathrm{d}x) = \int f_1(x)\mu(\mathrm{d}x) \geq F(x_1)\,,$$

$$\limsup_{n\to\infty} F_n(x_0) \leq \lim_{n\to\infty} \int f_2(x)\mu_n(\mathrm{d}x) = \int f_2(x)\mu(\mathrm{d}x) \leq F(x_2)\,,$$

and so $\lim_{n\to\infty} F_n(x_0) = F(x_0)$ if $x_0$ is a continuity point of $F$.

*Proof of* ($\Longleftarrow$). Let $f \in C(R^1)$. Let $0 < \varepsilon < 1/2$. Choose continuity points $x_0 < x_1 < \cdots < x_N$ of $F$ such that $F(x_0) < \varepsilon$, $1 - F(x_N) < \varepsilon$, and $|f(x) - f(x_i)| < \varepsilon$ for all $x \in [x_{i-1}, x_i]$ and $i = 1, \ldots, N$. Then choose $n_0$ so large that $|F(x_i) - F_n(x_i)| < \varepsilon/N$ for all $n \geq n_0$ and $i = 0, \ldots, N$. Then

$$\left|\int_{(-\infty,x_0]} f(x)(\mu - \mu_n)(\mathrm{d}x)\right| \leq \|f\| \left(2\varepsilon + \frac{\varepsilon}{N}\right) \leq 3\|f\|\varepsilon\,,$$

$$\left|\int_{(x_N,\infty)} f(x)(\mu - \mu_n)(\mathrm{d}x)\right| \leq \|f\| \left(2\varepsilon + \frac{\varepsilon}{N}\right) \leq 3\|f\|\varepsilon\,,$$

$$\begin{aligned}
\left|\int_{(x_0,x_N]} f(x)(\mu - \mu_n)(\mathrm{d}x)\right| &\leq \sum_{i=1}^{\infty} \left|\int_{(x_{i-1},x_i]} f(x)(\mu - \mu_n)(\mathrm{d}x)\right| \\
&\leq \sum_{i=1}^{\infty} [\, |f(x_i)((F(x_i) - F(x_{i-1})) - (F_n(x_i) - F_n(x_{i-1})))| \\
&\qquad + \varepsilon(F(x_i) - F(x_{i-1})) + \varepsilon(F_n(x_i) - F_n(x_{i-1}))\,] \\
&\leq 2\|f\|\varepsilon + 2\varepsilon
\end{aligned}$$

for $n \geq n_0$. This shows that $\lim_{n\to\infty} \int f(x)\mu_n(\mathrm{d}x) = \int f(x)\mu(\mathrm{d}x)$.

Next let us show that the metric $\rho$ is complete. Assume that $\{f_n\}$ in $\mathfrak{D}$ is a Cauchy sequence with respect to $\rho$, that is, $\rho(F_n, F_m) \to 0$ as $n, m \to \infty$. Since $0 \leq F_n(x) \leq 1$, we can select by the diagonal argument a subsequence $\{F_{n_k}\}$ for which $F_{n_k}(r)$ tends to some $F^0(r)$ for all rational $r$. Define $F(x) = \inf_{r>x} F^0(r)$ for all real $x$. Then $F(x)$ is increasing and right-continuous. For $r' < x < r$ we have

$$F^0(r') \leq \liminf_{k\to\infty} F_{n_k}(x) \leq \limsup_{k\to\infty} F_{n_k}(x) \leq F^0(r)$$

and so the same inequalities hold with $F(x-)$ and $F(x)$ in place of $F^0(r')$ and $F^0(r)$. Hence, if $x$ is a continuity point of $F$, then $F_{n_k}(x) \to F(x)$. We have $\lim_{x\to-\infty} F(x) = 0$. Indeed, for any $\varepsilon > 0$ there is $N$ such that $\rho(F_n, F_m) < \varepsilon$ for all $n, m \geq N$. Choose $x_0$ satisfying $F_N(x_0) < \varepsilon$. If $r < x_0 - \varepsilon$ and $n_k \geq N$, then $F_{n_k}(r) \leq F_{n_k}(x_0 - \varepsilon) \leq F_N(x_0) + \varepsilon < 2\varepsilon$ and so $F(x) < 2\varepsilon$ for $x < x_0 - \varepsilon$. Similarly we have $\lim_{x\to\infty} F(x) = 1$. Hence $F \in \mathfrak{D}$ and $\rho(F, F_{n_k}) \to 0$. Since

$$\rho(F, F_n) \leq \rho(F, F_{n_k}) + \rho(F_{n_k}, F_n)\,,$$

we see that $\rho(F, F_n) \to 0$.

**0.44.** Suppose that $\mathfrak{M}$ is tight. As in the last part of the solution of Problem 0.43, for any sequence $\{\mu_n\}$ in $\mathfrak{M}$ we can find a subsequence $\{\mu_{n_k}\}$ of $\{\mu_n\}$ and a right-continuous increasing function $F(x)$ on $R^1$ such that the distribution function $F_{n_k}(x)$ of $\mu_{n_k}$ tends to $F(x)$ on the set of continuity points of $F$. Then the tightness ensures that $F(x)$ is the distribution function of a probability measure $\mu$. Thus $\mu_{n_k} \to \mu$.

Conversely, suppose that $\mathfrak{M}$ is not tight. That is, there exist $\varepsilon > 0$ and a sequence $\{\mu_n\}$ in $\mathfrak{M}$ satisfying $\mu_n(-n, n) < 1 - \varepsilon$. If there are a subsequence $\{\mu_{n_k}\}$ of $\{\mu_n\}$ and $\mu \in \mathfrak{P}$ such that $\mu_{n_k} \to \mu$, then for each $x$ with $\mu\{x\} = \mu\{-x\} = 0$

$$1 - \varepsilon > \mu_{n_k}(-n_k, n_k) \geq \mu_{n_k}(-x, x) \qquad \text{for large } n_k$$

and $\mu_{n_k}(-x, x) \to \mu(-x, x)$, which implies $\mu(R^1) \leq 1 - \varepsilon$, a contradiction. Therefore $\mathfrak{M}$ is not conditionally compact.

**0.45.** For $\mathcal{B}_1, \mathcal{B}_2 \in \mathfrak{B}$ we define $\mathcal{B}_1 \prec \mathcal{B}_2$ by $\mathcal{B}_1 \subset \mathcal{B}_2$. Then $\mathfrak{B}$ is a complete lattice with respect to this relation because, for any set $\Lambda$ of indices, the $\sigma$-algebra $\bigcap_{\lambda\in\Lambda} \mathcal{B}_\lambda$ is the greatest lower bound $\bigwedge_{\lambda\in\Lambda} \mathcal{B}_\lambda$ and the $\sigma$-algebra generated by $\bigcup_{\lambda\in\Lambda} \mathcal{B}_\lambda$ is the least upper bound $\bigvee_{\lambda\in\Lambda} \mathcal{B}_\lambda$ .

## A.1 Chapter 1

**1.1.** Let $X_{1,n}, \ldots, X_{m,n}$ be independent for each $n$. Then

$$E(e^{i(\xi_1 X_{1,n} + \cdots + \xi_m X_{m,n})}) = E(e^{i\xi_1 X_{1,n}}) \cdots E(e^{i\xi_m X_{m,n}}) .$$

Since $X_{k,n} \to X_k$ i.p. as $n \to \infty$, we get

$$E(e^{i(\xi_1 X_1 + \cdots + \xi_m X_m)}) = E(e^{i\xi_1 X_1}) \cdots E(e^{i\xi_m X_m}) ,$$

and so $X_1, \ldots, X_m$ are independent.

**1.2.** Let $\mu_n$ be Cauchy distributions with parameters $c_n, m_n$, that is, $\varphi(\xi; \mu_n) = e^{im_n\xi - c_n|\xi|}$ (Problem 0.30 (iii)). Let $\mu_n \to \mu$ as $n \to \infty$. Then $\varphi(\xi; \mu_n) \to \varphi(\xi; \mu)$. Since $|\varphi(\xi; \mu_n)| = e^{-c_n|\xi|} \to |\varphi(\xi; \mu)|$, $\{c_n\}$ tends to some $c \in [0, \infty]$. We have $c < \infty$, since otherwise $\varphi(\xi; \mu) = 0$ for all $\xi \neq 0$. Next we can see that $\{m_n\}$ is bounded. Indeed, if there is a subsequence $\{m_{n'}\}$ with $|m_{n'}| \to \infty$, then

$$\int_0^z e^{im_{n'}\xi} d\xi \to \int_0^z \varphi(\xi; \mu) e^{c|\xi|} d\xi$$

while

$$\left| \int_0^z e^{im_{n'}\xi} d\xi \right| = \left| \frac{e^{im_{n'}z} - 1}{i\, m_{n'}} \right| \leq \frac{2}{|m_{n'}|} \to 0,$$

which is a contradiction. Now we can see that $\{m_n\}$ is convergent to some $m$. If not, we can find $a_1 \neq a_2$ and subsequences $m_{n'} \to a_1$, $m_{n''} \to a_2$, from which it follows that

$$\varphi(\xi;\mu) = \mathrm{e}^{\mathrm{i}a_1\xi - c|\xi|} = \mathrm{e}^{\mathrm{i}a_2\xi - c|\xi|}$$

and $\mathrm{e}^{\mathrm{i}\xi(a_1-a_2)} \equiv 1$, a contradiction. Hence $\mu$ is Cauchy with parameters $c, m$. Here a Cauchy distribution with parameters $0, m$ means the $\delta$-distribution at $m$.

The proof for Poisson distributions is similar and easier.

**1.3.** We may assume that $X(s,\omega)$ is a $D$-function in $s$ for all $\omega$. Hence, for each $\omega$, $X(s,\omega)$ is bounded on $[0,t]$ and measurable in $s$ and so $f(s)X(s,\omega)$ is integrable on $[0,t]$. Let $0 = s_{n,0} < s_{n,1} < \cdots < s_{n,k_n} = t$ be a sequence of nonrandom subdivisions such that $\sup_i |s_{n,i} - s_{n,i-1}| \to 0$. Define $X_n(s,\omega) = X(s_{n,i},\omega)$ for $s_{n,i-1} < s \leq s_{n,i}$ . Then $\int_0^t f(s)X_n(s,\omega)\mathrm{d}s \to \int_0^t f(s)X(s,\omega)\mathrm{d}s$ for each $\omega$ and

$$\int_0^t f(s)X_n(s,\omega)\mathrm{d}s = \sum_{i=1}^{k_n} X(s_{n,i},\omega)\int_{s_{n,i-1}}^{s_{n,i}} f(s)\mathrm{d}s\,,$$

which is measurable in $\omega$.

**1.4.** Assume that $X_n$ is $N(m_n, v_n)$-distributed, and that $X_n \to X$ i. p. as $n \to \infty$. By the same method as in the solution of Problem 1.2 we can prove that $X$ is $N(m,v)$-distributed for some $m \in R^1$ and $v \geq 0$. Moreover $m_n \to m$ and $v_n \to v$.

In general a family of random variables $\{X_\lambda \colon \lambda \in \Lambda\}$ is uniformly integrable if and only if $\sup_\lambda E(|X_\lambda|) < \infty$ and $\sup_\lambda E(|X_\lambda|,\, B) \to 0$ as $P(B) \to 0$ (see the solution of Problem 0.38 (ii)).

Since

$$E[(X - X_n)^2] = E[((X-m) - (X_n - m_n))^2] - (m - m_n)^2,$$

we can and do assume that $m = m_n = 0$ in the proof that $E[(X - X_n)^2] \to 0$. Since

$$\begin{aligned} E(X_n^2,\, |X_n| \geq a) &= \int_{|x|\geq a} \frac{x^2}{\sqrt{2\pi v_n}}\,\mathrm{e}^{-x^2/(2v_n)}\mathrm{d}x \\ &= \int_{|y|\geq a/\sqrt{v_n}} \frac{v_n y^2}{\sqrt{2\pi}}\,\mathrm{e}^{-y^2/2}\mathrm{d}y \leq \int_{|y|\geq a/\sqrt{K}} \frac{K y^2}{\sqrt{2\pi}}\,\mathrm{e}^{-y^2/2}\mathrm{d}y \end{aligned}$$

with $K = \sup_n v_n < \infty$, we see that $\{X_n^2\}$ is uniformly integrable. Therefore

$$\sup_n E[(X - X_n)^2,\, B] \leq 2E(X^2,\, B) + 2\sup_n E(X_n^2,\, B) \to 0$$

as $P(B) \to 0$. That is, $\{(X - X_n)^2\}$ is uniformly integrable. Since $(X - X_n)^2 \to 0$ i. p., we have $E[(X - X_n)^2] \to 0$ as $n \to \infty$ by Problem 0.38.

**1.5.** Let $\varepsilon > 0$. Since

$$\{ |X_t - X_{t-}| > \varepsilon \} \subset \liminf_{n\to\infty} \{ |X_t - X_{t-1/n}| > \varepsilon \} ,$$

we get

$$P(|X_t - X_{t-}| > \varepsilon) \le \liminf_{n\to\infty} P(|X_t - X_{t-1/n}| > \varepsilon)$$

by Fatou's lemma applied to indicators. Now use the continuity in probability to get $P(|X_t - X_{t-}| > \varepsilon) = 0$.

**1.6.** Let $Y(t) = X_1(t) + X_2(t)$. By Theorem 1.1.3, $Y(t)$ is an additive process. Since $Y(t)$ is obviously continuous in probability and almost all sample functions are $D$-functions, $Y(t)$ is a Lévy process. The increment $Y(t) - Y(s) = X_1(t) - X_1(s) + X_2(t) - X_1(s)$ is Poisson distributed since $X_1(t) - X_1(s)$ and $X_2(t) - X_1(s)$ are independent and Poisson distributed. Thus $Y(t)$ is a Lévy process of Poisson type. Similarly for the Gauss type.

**1.7.** For fixed $t$ and $n$, denote $t_k = (k/n)t$ for $k = 0, \ldots, n$. Since $X(t_k) - X(t_{k-1})$ is Gauss distributed with mean 0 and variance $V(t_k) - V(t_{k-1})$, we have

$$E[(X(t_k) - X(t_{k-1}))^2] = V(t_k) - V(t_{k-1}) ,$$
$$E[(X(t_k) - X(t_{k-1}))^4] = 3[V(t_k) - V(t_{k-1})]^2 .$$

Hence $E(S_n) = V(t)$ and

$$\begin{aligned}
E(S_n^2) &= \sum_{k,j=1}^{n} E[(X(t_k) - X(t_{k-1}))^2 (X(t_j) - X(t_{j-1}))^2] \\
&= \sum_k E[(X(t_k) - X(t_{k-1}))^4] \\
&\quad + \sum_{k\neq j} E[(X(t_k) - X(t_{k-1}))^2]\, E[(X(t_j) - X(t_{j-1}))^2] \\
&= 3\sum_k [V(t_k) - V(t_{k-1})]^2 + \sum_{k\neq j} [V(t_k) - V(t_{k-1})]\, [V(t_j) - V(t_{j-1})] \\
&= \left( \sum_k [V(t_k) - V(t_{k-1})] \right)^2 + 2\sum_k [V(t_k) - V(t_{k-1})]^2.
\end{aligned}$$

It follows that

$$\begin{aligned}
\operatorname{Var}(S_n) = E(S_n^2) - [E(S_n)]^2 &= 2\sum_k [V(t_k) - V(t_{k-1})]^2 \\
&\le 2 \max_k |V(t_k) - V(t_{k-1})|\, V(t),
\end{aligned}$$

which tends to 0 as $n \to \infty$ since $V(s)$ is uniformly continuous on $[0, t]$. Hence by Chebyshev's inequality

$$P(|S_n - V(t)| > \varepsilon) \le \operatorname{Var}(S_n)/\varepsilon^2 \to 0 \qquad \text{as } n \to \infty.$$

**1.8.** There is a subsequence $\{S_{n_j}\}$ of $\{S_n\}$ such that $S_{n_j} \to V(t) > 0$ a. s. Let

$$K_n(\omega) = \sup\{|X(t_2) - X(t_1)|\colon |t_2 - t_1| \le 1/n,\ 0 \le t_1 \le t_2 \le t\}.$$

Then

$$S_n \le K_n \sum_{k=1}^{n} \left|X\left(\tfrac{k}{n}t\right) - X\left(\tfrac{k-1}{n}t\right)\right|.$$

Since, by the uniform continuity of the sample function, $K_n(\omega) \to 0$ a. s. as $n \to \infty$, we get

$$\sum_{k=1}^{n_j} \left|X\left(\tfrac{k}{n_j}t\right) - X\left(\tfrac{k-1}{n_j}t\right)\right| \to \infty \qquad \text{a. s.}$$

Therefore almost all sample functions are of unbounded variation on $[0, t]$.

**1.9.** Since the given probability space $(\Omega, \mathcal{B}, P)$ is assumed to be complete, we can assume that *all* sample functions are in $D$. Let

$$B = \{\omega\colon X_t(\omega) \text{ and } Y_t(\omega) \text{ have a common jump}\}\,.$$

We want to show that $B \in \mathcal{B}$. Let

$$B_n^k = \{|X_t - X_{t-}| > 1/n \text{ and } |Y_t - Y_{t-}| > 1/n \text{ for some } t \le k\},$$

$$A_n^k = \bigcap_{q=1}^{\infty} \bigcup_{\substack{r,s \in Q \cap [0,k] \\ |r-s| < 1/q}} \{\omega\colon |X_r - X_s| > 1/n \text{ and } |Y_r - Y_s| > 1/n\}.$$

It follows immediately that $B_n^k \subset A_n^k$ and that $A_n^k \in \mathcal{B}$. If $\omega \in A_n^k$, then we can find sequences $\{r_j\}$ and $\{s_j\}$ such that $r_j$ and $s_j$ both tend to some $c \in [0, k]$ and $|X_{r_j} - X_{s_j}| > 1/n$, $|Y_{r_j} - Y_{s_j}| > 1/n$. Hence, if $\omega \in A_n^k$ then $X_t(\omega)$ and $Y_t(\omega)$ have a common jump of size $\ge 1/n$ at $t = c$. Thus $A_n^k \subset B_{n+1}^k$. Therefore $B^k \equiv \bigcup_n B_n^k = \bigcup_n A_n^k \in \mathcal{B}$. Hence $B = \bigcup_k B^k \in \mathcal{B}$.

**1.10.** Let $A_{m,k}$ be as in the hint and $B_\varepsilon = \{(s, u)\colon 0 < s \le t \text{ and } |u| > \varepsilon\}$ for $\varepsilon > 0$. Let

$$S_{\lambda,\varepsilon}(t) = \sum_{\substack{0 < s \le t \\ |X_s - X_{s-}| > \varepsilon}} |X_s - X_{s-}|^\lambda,$$

a finite sum for each $\omega$. Then

$$S_{\lambda,\varepsilon}(t) = \lim_{m\to\infty} \sum_{k=-\infty}^{\infty} |k/m|^\lambda N(A_{m,k} \cap B_\varepsilon)\,.$$

Hence, for $\alpha > 0$,

$$
\begin{aligned}
E\left[e^{-\alpha S_{\lambda,\varepsilon}(t)}\right] &= \lim_{m\to\infty} E\left[\exp\left(-\alpha \sum_k |k/m|^\lambda N(A_{m,k}\cap B_\varepsilon)\right)\right] \\
&= \lim_{m\to\infty} \prod_k E\left[e^{-\alpha |k/m|^\lambda N(A_{m,k}\cap B_\varepsilon)}\right] \\
&= \lim_{m\to\infty} \prod_k \exp\left(-n(A_{m,k}\cap B_\varepsilon)(1-e^{-\alpha|k/m|^\lambda})\right) \\
&= \exp\left\{\int_{s=0}^t \int_{|u|>\varepsilon} \left(e^{-\alpha|u|^\lambda}-1\right) n(dsdu)\right\}.
\end{aligned}
$$

Here we used the fact that if $Y$ is a Poisson $(\lambda)$ distributed random variable, then $E(e^{-\alpha Y}) = \exp(-\lambda(1-e^{-\alpha}))$. Letting $\varepsilon \downarrow 0$, we get

$$
E\left[e^{-\alpha S_\lambda(t)}\right] = \exp\left\{-\int_{s=0}^t \int_{u\in R_0} \left(1-e^{-\alpha|u|^\lambda}\right) n(dsdu)\right\}
$$

for $\alpha > 0$, where the integral and $S_\lambda(t)$ are possibly infinite. Notice that

$$
P(S_\lambda(t) < \infty) = \lim_{\alpha\downarrow 0} E\left[e^{-\alpha S_\lambda(t)}\right].
$$

(i) Suppose that $\gamma_\lambda(t) < \infty$. Then

$$
\lim_{\alpha\downarrow 0} \int_{s=0}^t \int_{u\in R_0} \left(1-e^{-\alpha|u|^\lambda}\right) n(dsdu) = 0
$$

since $1-e^{-|x|} \le 1 \wedge |x|$. Hence $P(S_\lambda(t) < \infty) = 1$.

(ii) Suppose that $\gamma_\lambda(t) = \infty$. Then

$$
\int_{s=0}^t \int_{u\in R_0} \left(1-e^{-\alpha|u|^\lambda}\right) n(dsdu) = \infty \qquad \text{for } \alpha > 0,
$$

since $1-e^{-|x|} \ge |x|/2$ for small $|x|$. Hence $P(S_\lambda(t) < \infty) = 0$.

(iii) Notice that $\int_{s=0}^t \int_{|u|\le 1} |u|^2 n(dsdu) < \infty$.

**1.11.** (i) The proof is included in Section 1.11.

(ii) The idea of the proof is exactly the same as in the solution of Problem 1.10.

**1.12.** The process $X(t)$ is a homogeneous Lévy process with the Lévy measure $n$ described in Section 1.12. Since the sample function is increasing, $n$ is a measure on $(0,\infty)$ satisfying $\int_0^\infty \frac{u}{1+u} n(du) < \infty$ as in Section 1.11. Therefore $n(du) = \gamma_+ u^{-\alpha-1}du$ with $0 < \alpha < 1$ and $\gamma_+ > 0$. We have

$$X_t = mt + \iint_{\substack{0<s\le t\\ u\in R_0}} uN(\mathrm{d}s\mathrm{d}u),$$

$$E(\mathrm{e}^{-pX_t}) = \exp\left[t\left(-mp - \gamma_+ \int_0^\infty (1-\mathrm{e}^{-pu})u^{-\alpha-1}\mathrm{d}u\right)\right]$$

with $m \ge 0$. Since $X_t$ moves only by jumps, we have $m = 0$ and thus

$$E(\mathrm{e}^{-pX_t}) = \mathrm{e}^{-ctp^\alpha}$$

with $c = \gamma_+ \int_0^\infty (1-\mathrm{e}^{-u})u^{-\alpha-1}\mathrm{d}u$.

**1.13.** We will use Theorem 0.6.5 to prove that the process $Y_t \equiv B(X_t)$ has independent increments. That theorem is valid even if $X$ and $Y$ there are not real-valued; we can use it with $X$ and $Y$ replaced by $\mathbf{X} = \mathbf{B}(\omega) = \{B(\cdot,\omega)\}$ and $\mathbf{Y} = (X_s, X_t)$. Let $0 \le s < t$ and $\psi(\mathbf{B}(\omega), t) = B(t,\omega)$. Then $Y_t(\omega) = B(X_t(\omega),\omega) = \psi(\mathbf{B}(\omega), X_t(\omega))$ and

$$\begin{aligned}
E\left(\mathrm{e}^{\mathrm{i}zY_s+\mathrm{i}u(Y_t-Y_s)}\right) &= E\left[\mathrm{e}^{\mathrm{i}zB(X_s)}\mathrm{e}^{\mathrm{i}u(B(X_t)-B(X_s))}\right]\\
&= E\left[E\left(\mathrm{e}^{\mathrm{i}z\psi(\mathbf{B}(\omega),X_s(\omega))}\mathrm{e}^{\mathrm{i}u(\psi(\mathbf{B}(\omega),X_t(\omega))-\psi(\mathbf{B}(\omega),X_s(\omega)))}\,\Big|\,(X_s,X_t)\right)\right]\\
&= E\left[E\left(\mathrm{e}^{\mathrm{i}z\psi(\mathbf{B}(\omega),\xi)}\mathrm{e}^{\mathrm{i}u(\psi(\mathbf{B}(\omega),\eta)-\psi(\mathbf{B}(\omega),\xi))}\right)_{\xi=X_s,\,\eta=X_s}\right]\\
&= E\left[E\left(\mathrm{e}^{\mathrm{i}zB_\xi}\mathrm{e}^{\mathrm{i}u(B_\eta-B_\xi)}\right)_{\xi=X_s,\,\eta=X_s}\right]\\
&= E\left[E\left(\mathrm{e}^{\mathrm{i}zB_\xi}\right)E\left(\mathrm{e}^{\mathrm{i}u(B_\eta-B_\xi)}\right)_{\xi=X_s,\,\eta=X_s}\right] \quad \text{(since } \{X_t\} \text{ is increasing)}\\
&= E\left[\left(\mathrm{e}^{-\xi z^2/2}\mathrm{e}^{-(\eta-\xi)u^2/2}\right)_{\xi=X_s,\,\eta=X_s}\right]\\
&= E\left(\mathrm{e}^{-X_s z^2/2}\mathrm{e}^{-(X_t-X_s)u^2/2}\right)\\
&= E\left(\mathrm{e}^{-z^2X_s/2}\right)E\left(\mathrm{e}^{-u^2(X_t-X_s)/2}\right)
\end{aligned}$$

As special cases we get

$$E\left(\mathrm{e}^{\mathrm{i}zY_s}\right) = E\left(\mathrm{e}^{-z^2X_s/2}\right), \quad E\left(\mathrm{e}^{\mathrm{i}u(Y_t-Y_s)}\right) = E\left(\mathrm{e}^{-u^2(X_t-X_s)/2}\right).$$

Hence

$$E\left(\mathrm{e}^{\mathrm{i}zY_s+\mathrm{i}u(Y_t-Y_s)}\right) = E\left(\mathrm{e}^{\mathrm{i}zY_s}\right)E\left(\mathrm{e}^{\mathrm{i}u(Y_t-Y_s)}\right),$$

that is, $Y_s$ and $Y_t - Y_s$ are independent. By a similar argument we get, for $0 \le t_0 < t_1 < \cdots < t_n$, that $Y_{t_0}, Y_{t_1} - Y_{t_0}, \ldots, Y_{t_n} - Y_{t_{n-1}}$ are independent. Hence $\{Y_t\}$ has independent increments. Since

$$E\left(\mathrm{e}^{-u^2(X_t-X_s)/2}\right) = \mathrm{e}^{-c(t-s)2^{-\alpha}|u|^{2\alpha}},$$

we see that

$$E\left(e^{iu(Y_t-Y_s)}\right) = E\left(e^{iuY_{t-s}}\right)$$

and that $\{Y_t\}$ is continuous in probability. Obviously sample functions of $\{Y_t\}$ are $D$-functions. It follows that $\{Y_t\}$ is a homogeneous Lévy process with

$$E\left(e^{izY_t}\right) = e^{-2^{-\alpha}ct|z|^{2\alpha}}.$$

Now we see that, for any $a > 0$, $\{Y_{at}\}$ and $\{a^{1/(2\alpha)}Y_t\}$ are identical in law. Therefore $\{Y_t\}$ is stable with exponent $2\alpha$.

**1.14.** Use the form of the Lévy measures in Section 1.12 and the result of Problem 1.10.

## A.2 Chapter 2

**2.1.** (T.0), (T.1) and (T.2) are trivially satisfied.

(T.3) Recall that $E = R^2 \cup \{\infty\}$ is the one-point compactification of $R^2$. If $t = 0$, (T.3) is trivial. Let $t > 0$ and let $f(x)$ be a continuous function on $E$. We have

$$\int_E p_t(x, dy)f(y) = \int_{R^2} \frac{1}{2\pi t} e^{-|y-x|^2/(2t)} f(y)dy = \int_{R^2} \frac{1}{2\pi t} e^{-|y|^2/(2t)} f(x+y)dy$$

for $x \in R^2$, and

$$\int_E p_t(\infty, dy)f(y) = f(\infty).$$

It follows from the bounded convergence theorem that

$$\int_E p_t(x, dy)f(y) \to \int_E p_t(x^0, dy)f(y)$$

as $x \to x^0 \subset E$. When $x^0 = \infty$, recall that $f(x+y) \to f(\infty)$ as $x \to \infty$.

(T.4) If $x = \infty$, this is trivial. Let $x \in R^2$. It is enough to assume that $U(x)$ is an $\varepsilon$-neighborhood of $x$ in the Euclidean metric. Then

$$\begin{aligned} p_t(x, U(x)) &= \int_{|y-x|<\varepsilon} \frac{1}{2\pi t} e^{-|y-x|^2/(2t)} dy = \int_{|y|<\varepsilon} \frac{1}{2\pi t} e^{-|y|^2/(2t)} dy \\ &= \int_{|z|<\varepsilon/\sqrt{t}} \frac{1}{2\pi} e^{-|z|^2/2} dz \qquad (y/\sqrt{t} = z), \end{aligned}$$

which tends to 1 as $t \downarrow 0$.

(T.5) Let $x \in R^2$ and $t, s > 0$, as the other cases are trivial. The Gauss distribution $N(m, v)$ on $R^1$ has the property that $N(m, v) * N(m', v') = N(m + m', v + v')$. Thus (when $m = m' = 0$),

(1) $\displaystyle\int \frac{1}{\sqrt{2\pi v}} \mathrm{e}^{-\eta^2/(2v)} \frac{1}{\sqrt{2\pi v'}} \mathrm{e}^{-(\xi-\eta)^2/(2v')} \mathrm{d}\eta = \frac{1}{\sqrt{2\pi(v+v')}} \mathrm{e}^{-\xi^2/(2(v+v'))}$.

In order to show that

$$\int_{R^2} N_t(y-x)\mathrm{d}y \int_B N_s(z-y)\mathrm{d}z = \int_B N_{t+s}(z-x)\mathrm{d}z$$

for $B \in \mathcal{B}(R^2)$, it is enough to show

$$\int_{R^2} N_t(y-x)N_s(z-y)\mathrm{d}y = N_{t+s}(z-x)\ .$$

This follows from

$$\int_{R^2} N_t(y)N_s(z-y)\mathrm{d}y = N_{t+s}(z)\ . \tag{2}$$

But (2) is written as

$$\int_{-\infty}^{\infty}\int_{-\infty}^{\infty} \frac{1}{2\pi t} \mathrm{e}^{-(y_1^2+y_2^2)/(2t)} \frac{1}{2\pi s} \mathrm{e}^{-((z_1-y_1)^2+(z_2-y_2)^2)/(2s)} \mathrm{d}y_1 \mathrm{d}y_2$$
$$= \frac{1}{2\pi(t+s)} \mathrm{e}^{-(z_1^2+z_2^2)/(2(t+s))},$$

which is a consequence of (1).

**2.2.** We use $q_t(x, \mathrm{d}y)$ in the hint. Since $T_1 x = x$ for all $x \in R^2$, we have $q_1(x,\mathrm{d}y) = p_1(x,\mathrm{d}y)$, where $p_t(x,\mathrm{d}y)$ is defined in Problem 2.1. Let us show that $q_t(x,\mathrm{d}y)$ satisfies (T.0)–(T.5). Among these (T.0), (T.1) and (T.2) are trivial.

(T.3) Note that

$$\int_E q_t(x,\mathrm{d}y)\, f(y) = \int_{R^2} p_t(T_t x, \mathrm{d}y)\, f(y)$$

for $x \in R^2$. Since $T_t x \to T_{x^0}$ as $x \to x^0$, (T.3) for $q_t(x,\mathrm{d}y)$ is reduced to (T.3) for $p_t(x,\mathrm{d}y)$.

(T.4) Let $x \in R^2$ and $\varepsilon > 0$. Choose $t$ so small that $|x - T_t x| < \varepsilon/2$. Then

$$\int_{|y-x|<\varepsilon} q_t(x,\mathrm{d}y) = \int_{|y-x|<\varepsilon} N_t(y - T_t x)\mathrm{d}y = \int_{|y+T_t x - x|<\varepsilon} N_t(y)\mathrm{d}y$$
$$\geq \int_{|y|<\varepsilon/2} N_t(y)\mathrm{d}y \to 1 \qquad \text{as } t \downarrow 0.$$

(T.5) Let $x \in R^2$. Then

$$\begin{aligned}\int_E q_t(x, \mathrm{d}y) q_s(y, B) &= \int_{R^2} N_t(y - T_t x)\mathrm{d}y \int_{B\cap R^2} N_s(z - T_s y)\mathrm{d}z \\ &= \int_{R^2} N_t(T_s^{-1}y' - T_t x)\mathrm{d}y' \int_{B\cap R^2} N_s(z - y')\mathrm{d}z \\ &= \int_{R^2} N_t(y' - T_s T_t x)\mathrm{d}y' \int_{B\cap R^2} N_s(z - y')\mathrm{d}z \\ &= \int_{B\cap R^2} N_{t+s}(z - T_{t+s}x)\mathrm{d}z = q_{t+s}(B).\end{aligned}$$

We have used the fact that $N_t(\cdot)$ is $T_t$-invariant and $T_s T_t = T_{t+s}$.

**2.3.** Let $N_t(x) = \frac{1}{\sqrt{2\pi t}} \mathrm{e}^{-x^2/(2t)}$ for $t > 0$ and $x \in R^1$. Then $N_t(-x) = N_t(x)$ and $\int_{-\infty}^{\infty} N_t(x)\mathrm{d}x = 1$. We have

$$p_t(x, B) = \int_{B\cap[0,\infty)} (N_t(y - x) + N_t(y + x))\mathrm{d}y$$

for $t > 0$, $x \in [0, \infty)$, $B \in \mathcal{B}([0, \infty])$. (T.0) and (T.2) are evident.

(T.1) Clearly $p_t(x, \cdot\,)$ is a measure. The total mass for $t > 0$ and $x \in [0, \infty)$ is

$$\begin{aligned}&= \int_0^{\infty} (N_t(y - x) + N_t(y + x))\mathrm{d}y = \int_0^{\infty} N_t(y - x)\mathrm{d}y + \int_{-\infty}^{0} N_t(-y + x)\mathrm{d}y \\ &= \int_0^{\infty} N_t(y - x)\mathrm{d}y + \int_{-\infty}^{0} N_t(y - x)\mathrm{d}y = 1.\end{aligned}$$

(T.3) Let $f \in C[0, \infty]$. For $x \in [0, \infty)$ we have

$$\begin{aligned}\int_S p_t(x, \mathrm{d}y) f(y) &= \int_0^{\infty} (N_t(y - x) + N_t(y + x)) f(y)\mathrm{d}y \\ &= \int_{-x}^{\infty} N_t(y) f(y + x)\mathrm{d}y + \int_x^{\infty} N_t(y) f(y - x)\mathrm{d}y.\end{aligned}$$

Hence, if $x_0 \in [0, \infty)$ then $\int_S p_t(x, \mathrm{d}y) f(y) \to \int_S p_t(x_0, \mathrm{d}y) f(y)$ as $x \to x_0$; if $x \to \infty$ then $\int_S p_t(x, \mathrm{d}y) f(y) \to \int_{-\infty}^{\infty} N_t(y) f(\infty)\mathrm{d}y = f(\infty)$.

(T.4) Let $\varepsilon > 0$. We have

$$\begin{aligned}&p_t(\infty, (1/\varepsilon, \infty]) = 1 \qquad \text{for } t \geq 0, \\ &p_t(0, [0, \varepsilon)) = 2\int_0^{\varepsilon} N_t(y)\mathrm{d}y \to 1 \qquad \text{as } t \downarrow 0.\end{aligned}$$

If $x \in (0, \infty)$ and $\varepsilon < x$, then

$$\begin{aligned}p_t(x, (x - \varepsilon, x + \varepsilon)) &= \int_{x-\varepsilon}^{x+\varepsilon} (N_t(y - x) + N_t(y + x))\mathrm{d}y \\ &= \int_{-\varepsilon}^{\varepsilon} N_t(y)\mathrm{d}y + \int_{2x-\varepsilon}^{2x+\varepsilon} N_t(y)\mathrm{d}y \to 1 \qquad \text{as } t \downarrow 0.\end{aligned}$$

(T.5) It is enough to show that

$$\int_0^\infty (N_t(y-x)+N_t(y+x))(N_s(z-y)+N_s(z+y))\mathrm{d}y \\ = N_{t+s}(z-x)+N_{t+s}(z+x) \qquad \text{for } x,z\in[0,\infty). \tag{1}$$

The left-hand side equals $I_1+I_2+I_3+I_4$, where

$$I_1=\int_0^\infty N_t(y-x)N_s(z-y)\mathrm{d}y,\quad I_2=\int_0^\infty N_t(y-x)N_s(z+y)\mathrm{d}y,$$
$$I_3=\int_0^\infty N_t(y+x)N_s(z-y)\mathrm{d}y,\quad I_4=\int_0^\infty N_t(y+x)N_s(z+y)\mathrm{d}y.$$

Since

$$I_2=\int_0^\infty N_t(y-x)N_s(-z-y)\mathrm{d}y,$$
$$I_3=\int_{-\infty}^0 N_t(-y+x)N_s(z+y)\mathrm{d}y=\int_{-\infty}^0 N_t(y-x)N_s(-z-y)\mathrm{d}y,$$
$$I_4=\int_{-\infty}^0 N_t(-y+x)N_s(z-y)\mathrm{d}y=\int_{-\infty}^0 N_t(y-x)N_s(z-y)\mathrm{d}y,$$

we obtain

$$I_1+I_4=\int_{-\infty}^\infty N_t(y-x)N_s(z-y)\mathrm{d}y=N_{t+s}(z-x),$$
$$I_2+I_3=\int_{-\infty}^\infty N_t(y-x)N_s(-z-y)\mathrm{d}y=N_{t+s}(-z-x)=N_{t+s}(z+x).$$

This shows (1).

**2.4.** Let $f\in C[0,\infty]$. Then

$$R_\alpha f(\infty)=f(\infty)/\alpha$$

and

$$R_\alpha f(a)=\int_0^\infty\int_0^\infty \mathrm{e}^{-\alpha t}\frac{1}{\sqrt{2\pi t}}\left(\mathrm{e}^{-(y-a)^2/(2t)}+\mathrm{e}^{-(y+a)^2/(2t)}\right)f(y)\mathrm{d}y\mathrm{d}t$$
$$=\frac{1}{\sqrt{2\alpha}}\int_0^\infty(\mathrm{e}^{-\sqrt{2\alpha}|y-a|-\sqrt{2\alpha}|y+a|})f(y)\mathrm{d}y \qquad \text{for } a\in[0,\infty),$$

using the evaluation of the integral in Example 1, Section 2.2. Let

$$\mathfrak{D}=\Big\{u\in C[0,\infty]\colon u\in C^2 \text{ in } (0,\infty),\ u'(0+)=0,\ \lim_{x\to\infty}u''(x)=0, \\ \lim_{x\to 0}u''(x) \text{ exists and is finite}\Big\}.$$

1. First we prove that if $u = R_\alpha f$, $f \in C[0,\infty]$, then $u \in \mathfrak{D}$. For $a \in [0,\infty)$

$$u(a) = \frac{1}{\sqrt{2\alpha}} \left[ \int_0^a e^{-\sqrt{2\alpha}(a-y)} f(y) dy + \int_a^\infty e^{-\sqrt{2\alpha}(y-a)} f(y) dy \right. \\ \left. + \int_0^\infty e^{-\sqrt{2\alpha}(y+a)} f(y) dy \right]$$

and so

$$u'(a) = -\int_0^a e^{-\sqrt{2\alpha}(a-y)} f(y) dy + \int_a^\infty e^{-\sqrt{2\alpha}(y-a)} f(y) dy \\ - \int_0^\infty e^{-\sqrt{2\alpha}(y+a)} f(y) dy\,,$$

$$u''(a) = -2f(a) + \sqrt{2\alpha} \left[ \int_0^a e^{-\sqrt{2\alpha}(a-y)} f(y) dy + \int_a^\infty e^{-\sqrt{2\alpha}(y-a)} f(y) dy \right. \\ \left. + \int_0^\infty e^{-\sqrt{2\alpha}(y+a)} f(y) dy \right].$$

Thus $u \in C^2$ in $(0,\infty)$, $u'(0+) = 0$, and

$$\alpha u - (1/2)u'' = f \qquad \text{in } (0,\infty). \tag{1}$$

It follows that $\lim_{a\to 0} u''(a)$ exists and is finite ($u, f \in C[0,\infty]$) and that

$$\lim_{a\to\infty} u''(a) = \lim_{a\to\infty} 2(\alpha u(a) - f(a)) = 2(\alpha u(\infty) - f(\infty)) = 0\,.$$

2. Now we prove that if $u \in \mathfrak{D}$, then $u = R_\alpha f$ for some $f \in C[0,\infty]$. This will verify $\mathfrak{D} = \mathfrak{D}(A)$. Define $f$ by

$$f(x) = \begin{cases} \alpha u(0) - (1/2)\lim_{x\to 0} u''(x), & x = 0 \\ \alpha u(x) - (1/2)u''(x), & x \in (0,\infty) \\ \alpha u(\infty), & x = \infty\,. \end{cases}$$

Then $f \in C[0,\infty]$ and $v = R_\alpha f$ satisfies $\alpha v - (1/2)v'' = f$ in $(0,\infty)$ and $v'(0+) = 0$. Let $w = u - v$. Then $\alpha w - (1/2)w'' = 0$ in $(0,\infty)$. Hence

$$w(x) = Ae^{-\sqrt{2\alpha x}} + Be^{\sqrt{2\alpha x}}, \qquad x \in (0,\infty)\,.$$

Since $w \in C[0,\infty]$, $B = 0$. Since $0 = w'(0+) = -\sqrt{2\alpha}A$, $A = 0$. Thus $u \equiv v$ in $(0,\infty)$. Because of continuity $u \equiv v$ in $[0,\infty]$.

3. If $u \in \mathfrak{D}(A)$, then $(\alpha - A)u = f$ with $f = R_\alpha^{-1}u$ and (1) shows that

$$Au(x) = \begin{cases} (1/2)u''(x), & x \in (0,\infty) \\ \lim_{x\to 0}(1/2)u''(x), & x = 0 \\ 0, & x = \infty. \end{cases}$$

**2.5.** Let $p_t(x, \mathrm{d}y)$ be the transition probabilities in Problem 2.4 and $q_t(x, \mathrm{d}y)$ the Brownian transition probabilities. Let $a \in [0, \infty)$. If $B \in \mathcal{B}([0, \infty))$ and $t > 0$, then

$$p_t(a, B) = q_t(a, B \cup (-B))$$

since $p_t(a, B) = q_t(a, B) + q_t(-a, B)$ and $q_t(-a, B) = q_t(a, -B)$. Hence, for any bounded, Borel measurable function $f$ on $[0, \infty)$

$$\int_{[0,\infty)} p_t(a, \mathrm{d}x) f(x) = \int_{[0,\infty)} q_t(a, \mathrm{d}x) f(x) + \int_{(-\infty,0]} q_t(a, \mathrm{d}x) f(-x).$$

If $B_1, B_2 \in \mathcal{B}([0, \infty))$ and $0 < t_1 < t_2$, then

$$\begin{aligned}
\iint_{x_1 \in B_1,\, x_2 \in B_2} & p_{t_1}(a, \mathrm{d}x_1) p_{t_2 - t_1}(x_1, \mathrm{d}x_2) = \int_{B_1} p_{t_1}(a, \mathrm{d}x_1) p_{t_2 - t_1}(x_1, B_2) \\
&= \int_{B_1} p_{t_1}(a, \mathrm{d}x_1) q_{t_2 - t_1}(x_1, B_2 \cup (-B_2)) \\
&= \int_{B_1} q_{t_1}(a, \mathrm{d}x_1) q_{t_2 - t_1}(x_1, B_2 \cup (-B_2)) \\
&\quad + \int_{-B_1} q_{t_1}(a, \mathrm{d}x_1) q_{t_2 - t_1}(-x_1, B_2 \cup (-B_2)) \\
&= \int_{B_1 \cup (-B_1)} q_{t_1}(a, \mathrm{d}x_1) q_{t_2 - t_1}(x_1, B_2 \cup (-B_2)) ,
\end{aligned}$$

since $q_{t_2 - t_1}(-x_1, B_2 \cup (-B_2)) = q_{t_2 - t_1}(x_1, B_2 \cup (-B_2))$. Repeating this procedure, we get the following: if $B_1, \ldots, B_n \in \mathcal{B}([0, \infty))$ and $0 < t_1 < \cdots < t_n$, then

$$\begin{aligned}
\int_{x_1 \in B_1} \cdots \int_{x_n \in B_n} & p_{t_1}(a, \mathrm{d}x_1) p_{t_2 - t_1}(x_1, \mathrm{d}x_2) \cdots p_{t_n - t_{n-1}}(x_{n-1}, \mathrm{d}x_n) \\
&= \int_{x_1 \in B_1 \cup (-B_1)} \cdots \int_{x_n \in B_n \cup (-B_n)} q_{t_1}(a, \mathrm{d}x_1) q_{t_2 - t_1}(x_1, \mathrm{d}x_2) \\
&\qquad \cdots q_{t_n - t_{n-1}}(x_{n-1}, \mathrm{d}x_n) .
\end{aligned}$$

This means that $P_a(\Lambda) = Q_a(\omega\colon |\omega| \in \Lambda)$ for $\Lambda = \{\omega\colon \omega(t_1) \in B_1, \ldots, \omega(t_n) \in B_n\}$. Now this identity is generalized to all $\Lambda \in \mathcal{B}$ by Lemma 0.1.1.

**2.6.** If $\sigma_1$ and $\sigma_2$ are stopping times (with respect to $\{\overline{\mathcal{B}}_t\}$), then $\sigma_1 \vee \sigma_2$, $\sigma_1 \wedge \sigma_2$ and $\sigma_1 + \sigma_2$ are stopping times because

$$\begin{aligned}
&\{\sigma_1 \vee \sigma_2 \le t\} = \{\sigma_1 \le t\} \cap \{\sigma_2 \le t\} \in \overline{\mathcal{B}}_t , \\
&\{\sigma_1 \wedge \sigma_2 \le t\} = \{\sigma_1 \le t\} \cup \{\sigma_2 \le t\} \in \overline{\mathcal{B}}_t , \\
&\{\sigma_1 + \sigma_2 < t\} = \textstyle\bigcup_{r \in Q \cap (0,t)} (\{\sigma_1 < r] \cap \{\sigma_2 < t - r\}) \in \overline{\mathcal{B}}_t,
\end{aligned}$$

where $Q$ is the set of rational numbers. If $\{\sigma_n\}$ is a sequence of stopping times increasing (or decreasing) to $\sigma$, then $\sigma$ is a stopping time because

$$\{\sigma \le t\} = \textstyle\bigcap_n \{\sigma_n \le t\} \in \overline{\mathcal{B}}_t \quad (\text{or } \{\sigma < t\} = \bigcup_n \{\sigma_n < t\} \in \overline{\mathcal{B}}_t).$$

**2.7.** Let $a < b$. Let $\sigma_b$ be the hitting time of $b$. Then

$$P_a(X_t \ge b) = P_a(X_t \ge b,\ \sigma_b \le t)\,.$$

Since $P_a(\sigma_b = t) \le P_a(X_t = b) = 0$, we get

$$P_a(X_t \ge b,\ \sigma_b \le t) = E_a\left[p_{t-\sigma_b}(X_{\sigma_b}, [b,\infty)),\ \sigma_b \le t\right]$$

from the example at the end of Section 2.7. But

$$p_{t-\sigma_b}(X_{\sigma_b}, [b,\infty)) = p_{t-\sigma_b}(b, [b,\infty)) = 1/2 \qquad \text{for } t > \sigma_b\,.$$

Therefore

$$P_a(X_t \ge t) = (1/2)P_a(\sigma_b \le t) = (1/2)P_a\left(\max_{0\le s\le t} X_s \ge b\right).$$

**2.8.** Let $a > 0$ and $B \in \mathcal{B}([0,\infty))$. Then

$$\begin{aligned}
P_a(X_t \in B,\ \sigma_0 > t) &= P_a(X_t \in B) - P_a(X_t \in B,\ \sigma_0 \le t)\\
&= P_a(X_t \in B) - \int_0^t p_{t-s}(0,B)P_a(\sigma_0 \in ds) \quad \text{(Example in Section 2.7)}\\
&= P_a(X_t \in B) - \int_0^t p_{t-s}(0,-B)P_a(\sigma_0 \in ds)\\
&= P_a(X_t \in B) - P_a(X_t \in -B,\ \sigma_0 \le t)\\
&= P_a(X_t \in B) - P_a(X_t \in -B)\,.
\end{aligned}$$

The last equality holds because the sample path is continuous. Then note that $P_a(X_t \in -B) = P_{-a}(X_t \in B)$.

**2.9.** Let us see that

$$P_0(\sigma_0 = 0) = 1\,,$$
$$P_0(\sigma_a < \infty) = 1 \qquad \text{for } a \in [0,\infty).$$

Indeed, using Problem 2.7, we have for $\varepsilon > 0$

$$\begin{aligned}
P_0(\sigma_0 < \varepsilon) &= P_0\left(\max_{0\le s\le\varepsilon} X_s > 0\right) = \lim_{h\downarrow 0} P_0\left(\max_{0\le s\le\varepsilon} X_s \ge h\right)\\
&= \lim_{h\downarrow 0} 2P_0(X_\varepsilon \ge h) = 1\,,
\end{aligned}$$

$$\begin{aligned}
P_0(\sigma_a < \infty) &= \lim_{k\to\infty} P_0(\sigma_a < k) = \lim_{k\to\infty} P_0\left(\max_{0\le s\le k} X_s > a\right)\\
&= \lim_{k\to\infty} 2P_0(X_k > a) = \lim_{k\to\infty} 2P_0(X_1 > k^{-1/2}a) = 1\,.
\end{aligned}$$

Consider $\omega$ such that $X_t(\omega)$ is continuous in $t$, $X_0(\omega) = 0$, $\sigma_0(\omega) = 0$, and $\sigma_a(\omega) < \infty$ for all $a > 0$. Since $\sigma_a(\omega) = \inf\{t > 0\colon X_t(\omega) > a\}$ by definition, $X_{\sigma_a(\omega)}(\omega) = a$, $\sigma_a(\omega)$ is strictly increasing in $a$, and

$$\sigma_{a+h}(\omega) = \sigma_a(\omega) + \sigma_{a+h}(\theta_{\sigma_a}\omega) \qquad \text{for } a \ge 0,\ h \ge 0.$$

Moreover $\sigma_a(\omega)$ is right continuous in $a$, because for each $a$ there is a sequence $s_n \downarrow 0$ such that $X_{\sigma_a+s_n} > a$, which implies $\sigma_{a+h} < \sigma_a + s_n$ for small $h > 0$.

For $B_1, B_2 \in \mathcal{B}([0,\infty))$ and $h > 0$, we have

$$(1) \quad P_0(\sigma_a \in B_1,\ \sigma_{a+h} - \sigma_a \in B_2) = P_0(\sigma_a \in B_1)\, P_0(\sigma_{a+h} - \sigma_a \in B_2)\,.$$

Indeed,

$$\begin{aligned} P_0(\sigma_a \in B_1,\ \sigma_{a+h} - \sigma_a \in B_2) &= P_0(\sigma_a \in B_1,\ \sigma_{a+h}(\theta_{\sigma_a}\omega) \in B_2) \\ &= E_0\left[P_{X(\sigma_a)}(\sigma_{a+h} \in B_2),\ \sigma_a \in B_1\right] \quad \text{(by strong Markov property)} \\ &= E_0\left[P_a(\sigma_{a+h} \in B_2),\ \sigma_a \in B_1\right] = P_0(\sigma_a \in B_1)\, P_a(\sigma_{a+h} \in B_2) \end{aligned}$$

and $P_0(\sigma_{a+h} - \sigma_a \in B_2) = P_a(\sigma_{a+h} \in B_2)$ as a special case. This shows (1) and also

$$(2) \qquad P_0(\sigma_{a+h} - \sigma_a \in B_2) = P_0(\sigma_h \in B_2)\,.$$

By (1), $\sigma_a$ and $\sigma_{a+h} - \sigma_a$ are independent. Similarly we can prove that $\{\sigma_a\}_{a\in[0,\infty)}$ has independent increments. The process $\{\sigma_a\}$ is continuous in probability, because for $a < b$

$$\begin{aligned} P_0(\sigma_b - \sigma_a > \varepsilon) &= P_0(\sigma_{b-a} > \varepsilon) \qquad \text{(by (2))} \\ &= 1 - 2P_0(X_\varepsilon \geq b - a) \qquad \text{(by Problem 2.7)} \\ &\to 0 \qquad \text{as } b - a \to 0. \end{aligned}$$

Therefore $\{\sigma_a\}$ under $P_0$ is a Lévy process.

**2.10.** By Problem 2.9 and by (2) in its solution, $\{\sigma_a\}$ under $P_0$ is a homogeneous Lévy process. For each $c > 0$ the two processes $\{X_{ct}\}$ and $\{c^{1/2}X_t\}$ under $P_0$ are identical in law; thus

$$\begin{aligned} P_0(\sigma_{ca} \in B) &= P_0(\inf\{t\colon X_t > ca\} \in B) = P_0(\inf\{t\colon X_{c^{-2}t} > a\} \in B) \\ &= P_0(c^2 \inf\{t\colon X_t > a\} \in B) = P_0(c^2\sigma_a \in B) \end{aligned}$$

for $B \in \mathcal{B}([0,\infty))$. It follows that $\{\sigma_{ca}\}$ and $\{c^2\sigma_a\}$ under $P_0$ are homogeneous Lévy processes identical in law. Hence $\{\sigma_a\}$ under $P_0$ is a stable process with exponent $1/2$ (this fact is also a consequence of the formula of $P_0(\sigma_a \leq t)$ below). We have

$$\begin{aligned} P_0(\sigma_a \leq t) &= \int_a^\infty \frac{2}{\sqrt{2\pi t}} \mathrm{e}^{-u^2/(2t)} \mathrm{d}u \qquad \text{(Problem 2.7)} \\ &= \frac{a}{\sqrt{2\pi}} \int_0^t \mathrm{e}^{-a^2/(2s)} s^{-3/2} \mathrm{d}s \quad \text{(change of variables } u = a\sqrt{t/s}\text{)}. \end{aligned}$$

Denoting $F_a(t) = P_0(\sigma_a \le t)$, we have for $p > 0$

$$
\begin{aligned}
E_0(\mathrm{e}^{-p\sigma_a}) &= \int_0^\infty \mathrm{e}^{-pt}\mathrm{d}F_a(t) = \int_0^\infty p\,\mathrm{e}^{-pt}F_a(t)\mathrm{d}t \\
&= \int_0^\infty p\,\mathrm{e}^{-pt}\mathrm{d}t \int_a^\infty \frac{2}{\sqrt{2\pi t}}\mathrm{e}^{-u^2/(2t)}\mathrm{d}u \\
&= \int_a^\infty p\,\mathrm{d}u \int_0^\infty \frac{2}{\sqrt{2\pi t}}\mathrm{e}^{-pt}\mathrm{e}^{-u^2/(2t)}\mathrm{d}t \\
&= \sqrt{2p}\int_a^\infty \mathrm{e}^{-\sqrt{2p}u}\mathrm{d}u \qquad \text{(Example 1, Section 2.2)} \\
&= \mathrm{e}^{-a\sqrt{2p}}.
\end{aligned}
$$

The formula

$$
E_0(\mathrm{e}^{-p\sigma_a}) = \exp\left(-\frac{a}{\sqrt{2\pi}}\int_0^\infty (1-\mathrm{e}^{-pu})u^{-3/2}\mathrm{d}u\right)
$$

is obtained from the following:

$$
\begin{aligned}
\int_0^\infty (1-\mathrm{e}^{-pu})u^{-3/2}\mathrm{d}u &= \int_0^\infty u^{-3/2}\mathrm{d}u \int_0^u p\,\mathrm{e}^{-pv}\mathrm{d}v \\
&= p\int_0^\infty \mathrm{e}^{-pv}\mathrm{d}v \int_v^\infty u^{-3/2}\mathrm{d}u = 2p\int_0^\infty \mathrm{e}^{-pv}v^{1/2}\mathrm{d}v \\
&= 2\sqrt{p}\,\Gamma(1/2) = 2\sqrt{\pi p}\,.
\end{aligned}
$$

This expression shows that $m(a) = 0$ in the decomposition of $\sigma_a$ in Problem 1.11 (i), that is, the process $\sigma_a$ moves only by jumps. This follows also from the fact that $\{\sigma_{ca}\}$ is identical in law with $\{c^2\sigma_a\}$, not with $\{c^2\sigma_a + ba\}$, $b \ne 0$.

**2.11.** Let $k = 1$ or 2. The hitting time $\sigma_U$ of the ball $U = U(b,r)$ is finite a. s. $(P_a)$ for every $a \in R^k$ by Theorem 2.19.6. Thus for any $t > 0$

$$
P_a(\sigma_U(\theta_t\omega) < \infty) = E_a[P_{X_t}(\sigma_U < \infty)] = 1.
$$

Let $U_1, U_2, \ldots$ be an enumeration of the family of balls $U(b,r)$ with $b \in Q$ $(k=1)$ or $b \in Q \times Q$ $(k=2)$ and with $r \in Q \cap (0,\infty)$, where $Q$ is the set of rationals. Let

$$
\Omega_0 = \{\omega\colon \sigma_{U_n}(\theta_t\omega) < \infty \text{ for all } n = 1,2,\ldots \text{ and } t \in Q\cap[0,\infty)\}.
$$

Then $P_a(\Omega_0) = 1$ for all $a \in R^k$. If $\omega \in \Omega_0$, then $\{X_s(\omega)\colon s \ge t\}$ is dense in $R^k$ for all $t$.

**2.12.** Fix $a \in R^k$. Let $\sigma_n$ and $\tau_n$ be the hitting time of and the exit time from the ball $U(a,n)$. We have $P_a(\tau_n < \infty) = 1$ by Theorem 2.19.1 and $P_a(\tau_n \to \infty) = 1$ by the continuity of paths. Notice that

$$
\begin{aligned}
P_a(|X_t(\omega)| &\text{ does not tend to } \infty \text{ as } t \to \infty) \\
&= P_a(\text{there is } n \text{ such that } \sigma_n(\theta_t\omega) < \infty \text{ for all } t) \\
&\leq \sum_{n=1}^{\infty} P_a(\sigma_n(\theta_t\omega) < \infty \text{ for all } t)
\end{aligned}
$$

and that

$$
\begin{aligned}
P_a(\sigma_n(\theta_t\omega) < \infty \text{ for all } t) &= P_a(\sigma_n(\theta_{\tau_m}\omega) < \infty \text{ for all } m) \\
&= \lim_{m\to\infty} P_a(\sigma_n(\theta_{\tau_m}\omega) < \infty)
\end{aligned}
$$

Now let $k \geq 3$. Then, by Theorem 2.19.6, $P_a(\sigma_n(\theta_{\tau_m}\omega) < \infty) = (n/m)^{k-2} \to 0$ as $m \to \infty$. Therefore $P_a(|X_t| \to \infty) = 1$.

# Index